Microbial Functional Foods and Nutraceuticals

Microbial Functional Foods and Nutraceuticals

Edited by

Vijai Kumar Gupta
Tallinn University of Technology, Tallinn, Estonia

Helen Treichel
Universidade Federal da Fronteira Sul (UFFS), Brazil

Volha (Olga) Shapaval
Norwegian University of Life Sciences, Aas, Norway

Luiz Antonio de Oliveira
Instituto Nacional de Pesquisas da Amazônia in Manaus, Brazil

Maria G. Tuohy
National University of Ireland Galway, Galway, Ireland

Registered Offices
John Wiley & Sons, Inc., 111 River Street, Hoboken, NJ 07030, USA
John Wiley & Sons Ltd, The Atrium, Southern Gate, Chichester, West Sussex, PO19 8SQ, UK

Editorial Office
The Atrium, Southern Gate, Chichester, West Sussex, PO19 8SQ, UK

For details of our global editorial offices, customer services, and more information about Wiley products visit us at www.wiley.com.

Wiley also publishes its books in a variety of electronic formats and by print-on-demand. Some content that appears in standard print versions of this book may not be available in other formats.

Library of Congress Cataloging-in-Publication Data

Names: Gupta, Vijai Kumar, editor. | Treichel, Helen, editor. | Shapaval, Volha, editor. | Oliveira, Luiz Antonio de, editor. | Tuohy, Maria G., editor.
Title: Microbial functional foods and nutraceuticals / edited by Vijai K. Gupta, Helen Treichel, Volha Shapaval, Luiz Antonio de Oliveira, Maria Tuohy.
Description: Hoboken, NJ : John Wiley & Sons, 2017. | Includes index. |
Identifiers: LCCN 2017027918 (print) | LCCN 2017040991 (ebook) |
    ISBN 9781119048978 (pdf) | ISBN 9781119048992 (epub) | ISBN 9781119049012 (cloth)
Subjects: LCSH: Functional foods. | Microorganisms–Health aspects.
Classification: LCC QP144.F85 (ebook) | LCC QP144.F85 M537 2017 (print) |
    DDC 572/.429–dc23
LC record available at https://lccn.loc.gov/2017027918

Cover Design: Wiley
Cover Image: (Background) Courtesy of Vijai Kumar Gupta and Maria Tuohy; (Image insets) © 3Dalia/Shutterstock; © Alexander Raths/Shutterstock; © deliormanli/iStockphoto

Set in 10/12pt Warnock by SPi Global, Pondicherry, India
Printed and bound in Malaysia by Vivar Printing Sdn Bhd

10 9 8 7 6 5 4 3 2 1

Contents

List of Contributors

Nissreen Abu-Ghannam
School of Food Science and
Environmental Health
Dublin Institute of Technology
Dublin
Ireland

Faisal Alsenani
Algae Biotechnology Laboratory
School of Agriculture and Food
Sciences
University of Queensland
Brisbane
Queensland
Australia

Cheviri N. Ambarish
Department of Biosciences
Mangalore University
Mangalagangotri
Mangalore
Karnataka
India

Sarika Amdekar
Department of Microbiology
Barkatullah University
Habibganj
Bhopal
Madhya Pradesh
India

Marimuthu Anandharaj
Agricultural Biotechnology
Research Center
Academia Sinica
Taipei
Taiwan

Ozlem Ateş
Department of Genetics and
Bioengineering
Nisantasi University
Istanbul
Turkey

Michelle Fleet
School of Science and
Engineering
Teesside University
Middlesbrough
UK

Charu Gupta
Amity Institute for Herbal
Research and Studies
Amity University UP
Noida
India

Vijai Kumar Gupta
Tallinn University of Technology
Tallinn
Estonia

Md Nazmul Islam
Algae Biotechnology Laboratory
School of Agriculture and Food
Sciences
University of Queensland
Brisbane
Queensland
Australia

Namera C. Karun
Department of Biosciences
Mangalore University
Mangalagangotri
Mangalore
Karnataka
India

Mayuri Khare
Department of Microbiology
Barkatullah University
Habibganj
Bhopal
Madhya Pradesh
India

Essam Kotb
Research Laboratory of
Bacteriology
Department of Microbiology
Faculty of Science
Zagazig University
Zagazig
Egypt

Avnish Kumar
Department of Biotechnology
Dr. B. R. Ambedkar University
Agra
India

Simpal Kumari
Department of Biotechnology
Mushroom Training and Research
Centre

Faculty of Science
Veer Bahadur Singh Purvanchal
University
Jaunpur (UP)
India

Shashank Mishra
Department of Bio-Engineering
Birla Institute of Technology
Mesra Ranchi Jharkhand
India;
Biotech Park
Lucknow
Uttar Pradesh
India

Ram Naraian
Department of Biotechnology
Mushroom Training and Research
Centre
Faculty of Science
Veer Bahadur Singh Purvanchal
University
Jaunpur (UP)
India

Chandra Nayak
Department of Studies in
Biotechnology
Mangalore University
Mangalagangotri
Mangalore
Karnataka
India

Ebru Toksoy Oner
IBSB
Department of Bioengineering
Marmara University
Istanbul
Turkey

Zhongli Pan
Department of Biological and
Agricultural Engineering

University of California
Davis
California
USA;
Healthy Processed Foods
Research Unit
USDA-ARS-WRRC
Albany
USA

Neha Panjiar
Department of Bio-Engineering
Birla Institute of Technology
Mesra Ranchi Jharkhand
India

Dhan Prakash
Amity Institute for Herbal
Research and Studies
Amity University UP
Noida
India

**Vinay Basavegowda
Raghavendra**
Department of Biological and
Agricultural Engineering
University of California
Davis
California
USA

Pattanathu K.S.M. Rahman
School of Science and
Engineering
Teesside University
Middlesbrough
UK

Rizwana Parveen Rani
Gandhigram Rural Institute
Department of Biology
Tamil Nadu
India

Peer M. Schenk
Algae Biotechnology Laboratory
School of Agriculture and Food
Sciences
University of Queensland
Brisbane
Queensland
Australia

Emer Shannon
School of Food Science and
Environmental Health
Dublin Institute of Technology
Dublin
Ireland

Vivek Kumar Shrivastava
Department of Microbiology
College of Life Sciences
Cancer Hospital and Research
Institute
Gwalior
Madhya Pradesh
India

Vinod Singh
Department of Microbiology
Barkatullah University
Habibganj
Bhopal
Madhya Pradesh
India

Kandikere R. Sridhar
Department of Biosciences
Mangalore University
Mangalagangotri
Mangalore
Karnataka
India

Manas R. Swain
Department of Biotechnology
Indian Institute of Technology
Madras
Chennai
India

Chandrasekar Venkitasamy
Department of Biological and
Agricultural Engineering
University of California
Davis
California
USA

Priyanka Verma
Department of Microbiology
Akal College of Basic Science
Eternal University
Baru Sahib
India

Ajar Nath Yadav
Department of Biotechnology
Akal College of Agriculture
Eternal University
Baru Sahib
India

1

Microalgae as a Sustainable Source of Nutraceuticals

Md Nazmul Islam, Faisal Alsenani, and Peer M. Schenk*

Algae Biotechnology Laboratory, School of Agriculture and Food Sciences, University of Queensland, Brisbane, Queensland, Australia

*Corresponding author e-mail: p.schenk@uq.edu.au

Introduction

Nutraceutical is a broad term which describes any food product with increased health benefits and which exceeds the usual health benefits of normal foods (Borowitzka 2013). Many bioactive constituents of food have been commercialized in the form of pharmaceutical products (pills, capsules, solutions, gels, liquors, powders, granules, etc.) which contribute to enhanced human health. However, these products cannot be categorized solely as "food" or "pharmaceutical" and a new hybrid term between nutrients and pharmaceuticals, "nutraceuticals," has been introduced (Palthur et al. 2010).

A generally accepted term for nutraceuticals is "food supplements." Another closely related term is "functional foods," defined as "products derived from natural sources which can also be fortified, whose consumption is likely to benefit human health" (Burja et al. 2008). However, it is a widely held view that there appears to be a boundary between nutraceuticals and functional foods. For example, when a bioactive compound is added in a food formulation, i.e., 200 mg of carotenoids dissolved in 1 L of juice, this may result in a new potential functional food, whereas the same amount of carotenoids encapsulated in a tablet or capsule is considered a nutraceutical (Espin et al. 2007). Additionally, nutraceuticals can either be whole food products (e.g., *Spirulina* in tablet form) or dietary supplements where the nutraceutical compound(s) may be concentrated to provide the claimed health benefits (e.g., astaxanthin extracted from *Haematococcus* microalgae is available in the market).

Microbial Functional Foods and Nutraceuticals, First Edition. Edited by Vijai Kumar Gupta, Helen Treichel, Volha (Olga) Shapaval, Luiz Antonio de Oliveira, and Maria G. Tuohy.
© 2018 John Wiley & Sons Ltd. Published 2018 by John Wiley & Sons Ltd.

Therefore, the emphasis on searching for nutraceuticals that contribute to improved human health has increased worldwide. Microalgae have become a popular target in the research community and biotechnology industry based on findings that many microalgal strains are very good sources of various nutraceuticals, such as vitamins, carotenoids, polyunsaturated fatty acids (PUFAs), phytosterols, etc. (Hudek et al. 2014). Moreover, the use of microalgal biomass has attracted attention because they grow fast regardless of the land's suitability for farming. In principle, microalgae cultivation can be carried out independent of freshwater supply and does not compete with arable land or biodiverse landscapes. In fact, many microalgae with health benefits are marine or brackish water algae (Lim et al. 2012). Microalgae are therefore considered an ideal source for the sustainable production of physiologically active compounds (Abdelaziz et al. 2013; Hudek et al. 2014). Many of them have a surprising capability of enduring adverse environmental conditions by means of their secondary metabolites, and some of these conditions lead to high accumulation of these compounds (e.g., *Dunaliella salina* produces high levels of β-carotene under highly saline conditions; Borowitzka 2013).

Prior studies have noted the health benefits of algal nutraceuticals which include improved immunity, neurological development, increased health of different organs including bones, teeth, intestine, etc. Algal nutraceuticals were also found to be effective in fighting obesity and cholesterol, and decrease blood pressure and maintain optimum heart condition. Several studies also documented some antiviral and anticancer properties (Venugopal 2008). In this chapter, we will provide a brief overview of the different nutraceutical compounds currently reported to be available in microalgae.

Microalgae-Derived Nutraceuticals

Pigments

Stengel and his team (2011) pointed out a few of the major categories of microalgal pigments which are closed tetrapyrroles such as chlorophylls a and b (chlorins), porphyrins (chlorophyll c), open chain tetrapyrroles (phycobilin pigments), and carotenoids (polyisoprenoids; carotenes and xanthophylls). Among them, the most targeted pigment groups are carotenoids and phycobilins which are already widely used by industry (Stengel et al. 2011).

Generally, carotenoids are powerful antioxidants and provide photo-protection to cells. Most of them have a 40-carbon polyene chain as their molecular backbone (del Campo et al. 2007; Guedes et al. 2011a). A recent study by our group showed that when induced by external stimuli, various microalgae can produce significant amounts of carotenoids. Among a few hundred Australian microalgal strains, 12 rapidly growing strains were screened for carotenoid profiles and *D. salina, Tetraselmis suecica, Isochrysis galbana*, and *Pavlova salina* were found to be good sources of various carotenoids at 4.68–6.88 mg/g dry weight (DW) even without external stimuli (Ahmed et al. 2014).

A considerable amount of work has been done worldwide to screen microalgae for carotenoid production. For example, lutein is dominant in *Muriellopsis* sp., *Scenedesmus almeriensis*, and *Chlorella* sp. (Blanco et al. 2007; Borowitzka 2013; del Campo et al. 2001; Fernandez-Sevilla et al. 2010); astaxanthin, canthaxanthin and lutein are abundant in *Chlorella zofingiensis*; canthaxanthin in *Scenedesmus komareckii*; aplanospores in *D. salina*; echinenone in *Botryococcus braunii*; and fucoxanthin in *Phaeodactylum tricornutum* (Table 1.1). However, apart from β-carotene and astaxanthin, large-scale production for these microalgae-synthesized carotenoids is still in consideration due to the high costs of downstream processing for extraction and purification (Borowitzka 2013). Furthermore, the cyanobacterium *Synechocooocccus* and the microalga *Nannochloropsis gaditana* present a good source of β-carotene, zeaxanthin, violaxanthin, vaucheriaxanthin, and chlorophyll a (Macías-Sánchez et al. 2007).

Under stress condition, two well-known species of microalgae, *D. salina* and *Haematococcus pluvialis*, accumulate considerable amounts of β-carotene (up to 14% DW under extreme conditions) and astaxanthin (2–3% DW), respectively (Ibanez and Cifuentes 2013). Moreover, *D. salina* is the leading producer of carotenoids among all types of food sources and therefore, this salt-tolerant microalga is currently the most popular species for the algae biotechnology industry (see Table 1.1). On a special note, β-carotene is converted *in vivo* into provitamin A (retinol) which is an essential part of multivitamin preparations (Guedes et al. 2011a; Spolaore et al. 2006).

Astaxanthin, one of the other highly sought after carotenoids, is currently commercially produced from the freshwater microalga *H. pluvialis* by several companies. This carotenoid has very strong antioxidant and anti-inflammatory properties. Thus, it has a role in protecting against macular and protein degradation, Parkinson's disease, reduced vision, cancer, rheumatoid arthritis, etc. (Hudek et al. 2014; Nakajima et al. 2008). Microalgal astaxanthin, although typically synthetically produced, is also commercially used in aquaculture, especially for salmon and trout. This is mainly because astaxanthin included in aquaculture feed enhances muscle color of salmonids. Moreover, when small amounts of microalgal biomass obtained from the genera *Chlorella, Scenedesmus* or *Spirulina*

Table 1.1 Selected microalgae-derived bioactive compounds: their sources and function.

Bioactive compounds		Microalga source	Function	References
Pigments	Lutein	*Muriellopsis* sp. *Scenedesmus almeriensis* *Chlorella* sp.	Prevention of cancer, protection from macular degeneration and cognitive impairment	Guedes et al. 2011a,c Borowitzka 2013
	Astaxanthin	*Chlorella zofigiensis* *Dunaliella salina* *Haematococcus pluvialis*	Antioxidant and anti-inflammatory properties, effective against cancer, protein degradation, Parkinson's disease, reduced vision, rheumatoid arthritis	Hudek et al. 2014, Ibanez and Cifuentes 2013, Borowitzka 2013
	β-carotene	*Nannochloropsis gaditana* *Dunaliella salina* *Haematococcus pluvialis*	Prevention of breast cancer and macular degeneration	Macías-Sánchez et al. 2007, Stengel et al. 2011, Guedes et al. 2011a
PUFAs	EPA	*Phaeodactylum tricornutum* *Nannochloropsis* sp. *Monodus subterraneus*	Neural development, prevention of cardiovascular disease, lessen and protect against COPD, asthma, rheumatoid arthritis, atherosclerosis, Crohn's disease and cystic fibrosis.	Stengel et al. 2011, Janssen and Kiliaan 2014,
	DHA	*Crypthecodinium cohnii* *Pavlova salina* *Isochrysis galbana*	Facilitate child and infant development	Guedes et al. 2011b, Borowitzka 2013
	Linolenic acid	*Dunaliella salina* *Spirulina platensis*		

Proteins	Phycobiliproteins, hormone-like bioactive peptides, 2–20, and essential amino acids, etc.	*Haematococcus pluvialis Chlamydomonas reinhardtii Chlorella* sp. *Spirulina* sp. *Scenedesmus* sp. *Dunaliella* sp. *Oscillatoria* sp. *Chlamydomonas* sp.	Anticancer, immunomodulating, hepatoprotective, anti-inflammatory, antioxidant, antimicrobial properties, protection of DNA Production of recombinant proteins, e.g., for type 1 diabetes detection, etc.	Himaya et al. 2012, Stengel et al. 2011, Dewapriya and Kim 2014, Gong et al. 2011, Mayfield et al. 2007, Agyei and Danquah 2011, Sheih et al. 2009
Vitamins	Cobalamin (vitamin B12)	*Dunaliella tertiolecta Chlorella* sp. *Spirulina* sp.	Antioxidant activities, role in body function, immunity, digestive system, etc.	Fabregas and Herrero 1990, Watanabe et al. 2002,Durmaz et al. 2007
	Tocopherols (vitamin E)	*Dunaliella tertiolecta Porphyridium cruentum*		
	Thiamin (vitamin B1)	*Tetraselmis suecica*		
	Nicotinic acid (vitamin B3)	*Tetraselmis suecica*		
	Ascorbic acid (vitamin C)	*Tetraselmis suecica*		
	Biotin (vitamin B7)	*Chlorella* sp.		

(Continued)

Table 1.1 (Continued)

Bioactive compounds		Microalga source	Function	References
Polysaccharides	Sulfated polysaccharides Insoluble fiber and others	*Porphyridium* sp. *Spirulina platensis* *Phaeodactylum* sp. *Chlorella stigmatophora* *Gyrodinium impudicum*	Antiviral, antioxidant, antitumor, anti-inflammatory, antihyperlipidemia and anticoagulant activities	de Jesus Raposo et al. 2013, Huheihel et al. 2002, Yim et al. 2005
Phenolic compounds	Heptadecane, tetradecane, etc.	*Nitzschia laevis Nostoc ellipsosporum Nostoc piscinale Chlorella protothecoides Synechococcus* sp. *Chlorella vulgaris Anabaena cylindrica Tolypothrix tenuis Chlorella pyrenoidosa Crypthecodinium cohnii Chlamydomonas nivalis*	Anticancer, antidiabetic action, lessen the risks of cardiovascular and neurodegenerative diseases, protection from biotic and abiotic stress	Stengel et al. 2011, Kim et al. 2009, Li et al. 2007
Phytosterols	Brassicasterol, sitosterol, stigmasterol, etc.	*Dunaliella tertiolecta Dunaliella salina*	Cholesterol-lowering and neuromodulatory activities, fights nervous system anomalies, autoimmune encephalomyelitis, amyotrophic lateral sclerosis	Francavilla et al. 2012

COPD, chronic obstructive pulmonary disease; DHA, docosahexaenoic acid; EPA, eicosapentaenoic acid; PUFA, polyunsaturated fatty acid.

were added to cattle feed, significant improvements in the animals' immune system were noticed (Guerin et al. 2003; Plaza et al. 2009; Pulz and Gross 2004). Microalgal carotenoids are also considered to be effective against cancer. For example, lutein, zeaxanthin and β-carotene were found to act against premenopausal breast cancer, whereas cryptoxanthin and α-carotene showed effectiveness against cervical cancer (see Table 1.1). It has also been recorded that lycopene has a role in preventing and treating prostate and stomach cancer (Stengel et al. 2011).

Several strains of *Chlorella* are reportedly good sources of lutein, α-carotene, β-carotene, zeaxanthin, violaxanthin, antheraxanthin, zeaxanthin, and astaxanthin. Evidence suggests that carotenoid extracts of *C. ellipsoidea* and *C. vulgaris* showed anticancer properties by inducing apoptosis in colon cancer cells (Cha et al. 2008). Additionally, protection from macular degeneration and cognitive impairment was seen in transgenic mice when they were fed lutein- and β-carotene-rich *Chlorella* extracts (Guedes et al. 2011a).

Polyunsaturated Fatty Acids

Microalgae play a critical role in the nutraceutical industry as they are considered as good sources of fatty acids with health benefits, especially the long-chain polyunsaturated fatty acids (PUFAs). PUFAs are defined as the fatty acids containing more than 18 carbon atoms and more than one double bond in their structures. Examples are α-linolenic acid (ALA; C18:3 (ω-3)), eicosapentaenoic acid (EPA; C20:5 (ω-3)), arachidonic acid (AA; C20:4 (ω-6)), docosapentaenoic acid (DPA; C22:5), and docosahexaenoic acid (DHA; C22:6 (ω-3)) (Ryckebosch et al. 2012). Health benefits have been demonstrated for the ω-3 fatty acids ALA, EPA, and DHA.

Interestingly, humans and most higher organisms cannot independently produce long PUFAs so these fatty acids need to be provided by diet. They are an indispensable part of a healthy diet as they have a salient role in neural development and neurodegeneration (Janssen and Kiliaan 2014). Moreover, ω-3 PUFA therapy has been a research focus for its role in prevention of cardiovascular diseases (Lavie et al. 2009).

Other beneficial roles of PUFAs are free radical scavenging, anti-inflammation, antimicrobial, antiviral, and anticancer activities. They can also lessen the extent of chronic obstructive pulmonary disease (COPD), asthma, rheumatoid arthritis, atherosclerosis, etc. Beside these, they were also found to alleviate symptoms of cystic fibrosis and Crohn's disease (Stengel et al. 2011). Likewise, DHA (ω-3) and arachidonic acid (ω-6) are two striking component of tissues in the brain, nervous system, and eye. Hence, they play very important roles in neuronal development and child growth. A few studies have demonstrated that children display improved behaviour when their diet is supplemented with PUFAs (Janssen and Kiliaan 2014).

Several microalgal strains have been recorded as outstanding sources of various PUFAs (see Table 1.1). *Crypthecodinium cohnii* can produce DHA-rich oil (up to 39% of total fatty acids) which is extensively used in infant formula (Borowitzka 2013). *Isochrysis galbana* and *Pavlova salina* also produce considerable amounts of DHA (Guedes et al. 2011b). *Phaeodactylum tricornutum* is composed of around 30–45% PUFA where 20–40% of total fatty acid is EPA (Fajardo et al. 2007). Other useful EPA-producing microalgae include *Nannochloropsis* sp. (up to 40% of total fatty acids) and *Monodus subterraneus* in which EPA is around 24% of its total fatty acid content (Borowitzka 2013; Cohen 1999; Sharma and Schenk 2015).

Furthermore, PUFAs of *Dunaliella salina* such as palmitic, linolenic, and oleic acids may contribute up to 85% of its fatty acid content (Herrero et al. 2006) whereas *Spirulina platensis* (also called *Arthrospira platensis*) has been reported as a good source of γ-linolenic acid, palmitoleic lauric acid, and DHA (see Table 1.1) (Tokuşoglu and Ünal 2003).

Proteins

Several genera of microalgae such as *Chlorella, Spirulina, Scenedesmus, Dunaliella, Micractinium, Oscillatoria, Chlamydomonas,* and *Euglena* contain proteins in very high amounts which can constitute more than 50% of their DW (Becker 2007; see Table 1.1). This makes microalgae an interesting protein-rich feedstock for biotech and feed industries. Even the defatted, algal biomass left over after oil extraction can be sustainably used for bioactive proteins (Dewapriya and Kim 2014).

Microalgae have a high content of essential amino acids such as lysine, leucine, isoleucine, and valine. These four amino acids account for 35% of the essential amino acids in human muscle protein. Thus, microalgae can be used as a dietary supplement to meet the protein requirement of both humans and animals (Dewapriya and Kim 2014).

Ju and his co-workers (2012) conducted a series of trials where shrimps were fed with defatted *H. pluvialis* as a supplement. They found that the microalgae-supplemented shrimps were healthier compared to the control group fed a commercial diet. A few cyanobacteria, such as *S. platensis* and *Porphyridium* sp., were reported to have hepatoprotective, anti-inflammatory, immune-modulating, anticancer, and antioxidant properties (see Table 1.1). It has been concluded that the protein cluster, phycobiliprotein, exerts such action with the help of its algal pigments (Stengel et al. 2011).

As well as microalgal bioactive peptides, protein fragments 2–20 amino acids long can act like a hormone after being activated by hydrolysis (Himaya et al. 2012). These peptides were found to have antioxidant, anticancer, antihypertensive, immunomodulatory, antimicrobial, and cholesterol-lowering effects (Agyei and Danquah 2011). For example, *C. vulgaris*-derived peptides have protective effects on DNA and preventive effects on cell peroxidation and thus can prevent diseases like cancer and cardiac disorder (see Table 1.1) (Sheih et al. 2009). Therefore, microalgae are treated especially to generate bioactive peptides. For example, *P. lutheri* biomass was successfully fermented to generate small peptides with high antioxidant activity (Ryu et al. 2012).

Another prospective area is the development of therapeutic recombinant proteins from microalgae. This is suitable for the biotechnology industry compared to other available protein expression systems as it offers cost-effective and relatively easy procedures. It does not involve expensive bacteria and yeast-based bioreactors. Moreover, microalgae bioreactors for recombinant protein allow post-transcriptional and post-translational modifications of biosynthesized proteins (Dewapriya and Kim 2014).

Focus has been placed on *Chlamydomonas reinhardtii* for the production of recombinant therapeutic proteins because of its exceptional precise protein folding and assembling (Mayfield et al. 2007). Furthermore, different recombinant proteins such as HSV8-lsc and HSV8-scFv (a human IgA antiherpes monoclonal antibody), and human glutamic acid decarboxylase 65 (hGAD65), a key autoantigen for detecting type 1 diabetes, were developed from *C. reinhardtii* (Gong et al. 2011).

Vitamins

Some microalgae contain both water- and lipid-soluble vitamins in higher amounts than conventional vitamin-rich foods and thus ingestion of these microalgae can meet the requirement of certain vitamins in humans and animals (Fabregas and Herrero 1990).

In the early 1990s, a seminal study in this area by Fabregas and Herrero (1990) involved screening of microalgae for their vitamin contents and the authors concluded that the concentration of four vitamins, β-carotene (provitamin A), tocopherol (vitamin E), thiamin (vitamin B1) and folic acid, was higher in many microalgae than in conventional foods. For example, *Tetraselmis suecica* was found to be a rich source of thiamin (vitamin B1), pyridoxine (vitamin B6), nicotinic acid (vitamin B3), pantothenic acid (vitamin B5), and ascorbic acid (vitamin C), whereas *Dunaliella tertiolecta* was a good source for β-carotene (provitamin A), riboflavin (vitamin B2), cobalamin (vitamin B12), and tocopherol (vitamin E). Moreover, biotin (vitamin B7) was found in high concentrations in *Chlorella* (see Table 1.1) (Fabregas and Herrero 1990). Another investigation showed that 9–18% of *Chlorella* strains contain high amounts of vitamin B12 (Shim et al. 2008). *Spirulina* is also considered a good source of vitamin B12; however, *Chlorella* maintains a better bioavailability for vitamin B12 (Watanabe et al. 2002). Vitamin E (tocopherol) was found to be present in significant amounts in the red microalga *Porphyridium cruentum*, in which α-tocopherol and γ-tocopherol contents were 55.2 μg/g DW and 51.3 μg/g DW respectively (Durmaz et al. 2007).

Minerals

Minerals are naturally occurring substances which can either be assimilated as compounds or can remain in elemental form. These include nitrogen, phosphorus, sodium, chlorine, potassium, calcium, sulfur, fluorine, zinc, iron, cobalt, copper, magnesium, iodine, molybdenum, manganese, and selenium (Marschner 1995).

Though microalgae can assimilate minerals and trace elements, only a small amount of research work on the mineral content of microalgae has been carried out. Tokuşoglu and Ünal (2003) and Fabregas and his team (1986) are the two main groups who have conducted detailed studies on minerals in microalgae. They reported significant quantities of zinc, phosphorus, sodium, iron, potassium, manganese, magnesium, and calcium in *D. tertiolecta*, *Isochrysis galbana*, *Tetraselmis suecica*, *C. vulgaris*, *S. platensis*, and *Chlorella stigmatophora*.

The recommended daily intake for calcium (Ca) is 1000 mg/day for adults, 800 mg/day for children (>8 years) and 270 mg/day for infants. Therefore, a daily intake of approximately 92 g of *I. galbana* will meet Ca requirements for an adult (Table 1.2) (Tokuşoglu and Ünal 2003). Similarly, consumption of 142 g of *S. platensis* or 37 g of *C. vulgaris* per day fulfills the suggested potassium (K) intake for adults (RDA 2000 mg/day for adults, 1600 mg/day for children, and 700 mg/day for infants) (Tokuşoglu and Ünal 2003). It is also evident that 40 g of *C. vulgaris* microalgae achieved the dietary intake requirements of phosphorus (P): 700 mg/day for adults, 500 mg/day for children (>8 years), and 275 mg/day for infants (7–12 months) (Tokuşoglu and Ünal 2003).

Likewise, daily ingestion of around 60 g and 47 g of *I. galbana* satisfies the RDA for magnesium (Mg) for an adult male and female respectively (RDA 420 mg and 320 mg), while only about 11 g of these microalgae provide the Mg demand for infants (75 mg/day) (Tokuşoglu and Ünal 2003). Daily intake of around 4.5 g of *I. galbana*, 10.5 g of *S. platensis*, and 4 g of *C. vulgaris*, respectively, will provide the advised daily requirement of iron (Fe) for children, adults, and infants (Tokuşoglu and Ünal 2003). In addition, these microalgae have been proved to contain adequate amount of selenium (Se), manganese (Mn), and zinc (Zn). It has been shown that pregnant and lactating women can obtain their prescribed quantity of Zn from only 8 g daily intake of *T. suecica* (Tokuşoglu and Ünal 2003).

Table 1.2 Examples of microalgae quantities required to meet the Recommended Daily Amount (RDA) of minerals. Source: Adapted from Tokuşoglu and Ünal (2003).

Minerals	Microalgae source	Required daily amount of microalgae to fulfill the adult demand	Recommended daily dose of minerals (RDA)		
			Adult	Child	Infant
Calcium (Ca)	*Isochrysis galbana*	92 g	1000 mg	800 mg	270 mg
Potassium (K)	*Spirulina platensis*	142 g	2000 mg	1600 mg	700 mg
Phosphorus (P)	*Chlorella vulgaris*	40 g	700 mg	500 mg	275 mg
Magnesium (Mg)	*Isochrysis galbana*	60 g (male), 47 g (female)	420 mg	320 mg	75 mg

Polysaccharides

Polysaccharides from some microalgae have been reported to have anti-viral, antioxidant, antitumor, anti-inflammatory, antihyperlipidemia, and anticoagulant activities (de Jesus Raposo et al. 2013).

For example, sulfated polysaccharides of *Phorphiridium* sp. reportedly display antiviral activity against herpes simplex virus (type 1 and 2), both *in vitro* and *in vivo*. This polysaccharide is composed of xylose, glucose, mannose, pentose, galactose, and methylated galactose (Huheihel et al. 2002). *S. platensis* produces calcium-spirulan which is an intracellular polysaccharide. This also works as an antiviral agent by inhibiting viral replication, by preventing virus from penetrating into host cells. Moreover, polysaccharides released from *S. platensis* also show activity against *Vaccinia* virus and an *Ectromelia* virus (de Jesus Raposo et al. 2013).

Polysaccharides from *Porphyridium* protect against oxidative damage caused by linoleic acid and ferrous sulfate ($FeSO_4$). Similarly, *Porphyridium-*, *Phaeodactylum-*, and *C. stigmatophora*-derived polysaccharides have been reported to show anti-inflammatory and immunomodulatory activities (see Table 1.1) (de Jesus Raposo et al. 2013). One of the significant roles of microalgal polysaccharides is the ability to prevent some cancer cell growth. *Gyrodinium impudicum* released homopolysaccharides that restrained tumor cell growth, both *in vitro* and *in vivo* (Yim et al. 2005). Likewise, the *Spirulina*-derived polysaccharide calcium-spirulan was found to be effective against pulmonary cancer growth (de Jesus Raposo et al. 2013).

Phenolic and Volatile Compounds

Phenolic compounds are secondary metabolites which can protect microalgae against biotic and abiotic stresses, such as grazing, bacterial infection, UV radiation, or metal contamination (Stengel et al. 2011). Many naturally occurring phytochemicals are composed of phenolic compounds and a great deal of previous research has concluded that some can exert anticancer, antidiabetic action. It has also been claimed that they can lessen the risks of cardiovascular and neurodegenerative diseases (Kim et al. 2009).

Previous studies showed that *Nitzschia laevis*, *Nostoc ellipsosporum*, *Nostoc piscinale*, *Chlorella protothecoides*, *Synechococcus* sp., *C. vulgaris*, *Anabaena cylindrica*, *Tolypothrix tenuis*, *Chlorella pyrenoidosa*, *Crypthecodinium cohnii*, and *Chlamydomonas nivalis* have total phenolic contents that are similar to or higher than in conventional fruit and vegetable food sources, such as oriental plum, apples, and strawberries (see Table 1.1) (Li et al. 2007). It has also been found that phenolic-rich *Spirulina maxima* extract protects liver from CCl4-induced lipid peroxidation *in vitro* (El-Baky et al. 2009). *Spirulina* also possesses antimicrobial activity due to the presence of phenolic compounds, such as heptadecane and tetradecane (Ozdemir et al. 2004). Additionally, β-cyclocitral, α- and β-ionone, neophytadiene, and phytol are volatile compounds which have antimicrobial properties, that are common in *D. salina* (Herrero et al. 2006).

Sterols

Scientists emphasize the importance of phytosterols which can only be taken up exclusively through diet. Based on their cholesterol-lowering property, phytosterols have been added to different commercial food items such as margarine, yogurts, and milk. Additionally, sterols have a role in nervous system anomalies such as autoimmune encephalomyelitis, amyotrophic lateral sclerosis, etc. (Francavilla et al. 2012). Similarly, animal feed rich in phytosterols can prevent or interrupt the onset of Alzheimer's disease in animal models (Francavilla et al. 2012).

In a recent study, a notable amount of phytosterols, such as brassicasterol, sitosterol, stigmasterol, etc., were found in two microalgal strains, *D. tertiolecta* and *D. salina*. These sterol compounds were found to have *in vivo* neuromodulatory activity (Francavilla et al. 2012).

Microalgae in the Pharmaceutical and Food Industries

We have already discussed the production of recombinant protein for the development of antibodies, immunotoxins, and therapeutics from *C. reinhardtii*. Another crucial application of microalgae is the production of oral vaccines (Guarnieri and Pienkos 2015). Different observational studies suggest that plant-derived vaccines have added advantages over traditional vaccines. Plant vaccines are usually less costly, even 1000 times lower compared to the cost of conventional vaccine production (Specht and Mayfield 2014). Using microalgae as a vaccine source makes the vaccine less prone to mammalian pathogenic contamination. Moreover, microalgae-derived vaccines have higher durability, as antigens are protected by cell walls from proteolytic degradation (Guarnieri and Pienkos 2015). The first reported vaccines derived from microalgae were VP1 (structural protein of foot and mouth disease) and the beta subunit of chlorella toxin (CTB) (Sun et al. 2003). One of the recent approaches to vaccine production has been the design of chimeric proteins having a mucosal adjuvant (CtxB) and a malaria transmission-blocking vaccine candidate (Pfs25) (Guarnieri and Pienkos 2015). In 2013, an interesting investigation into algal vaccines by Tran and his co-workers (2013) resulted in the production of anticancer immunotoxin where the antigen was designed to reduce the effect of B-cell lymphoma.

In the meantime, in the pharmaceutical industry, microalgae have also been widely used as additives in cosmetic products. *Chlorella*, *Spirulina*, *Nannochloropsis*, and *D. salina* are used to produce antiaging and anti-irritant preparations. *Chlorella* extracts were found to be stimulating collagen synthesis, whereas *Spirulina*-synthesized proteins were reported to remove the signs of skin aging (Spolaore et al. 2006).

Generally, microalgae have been predominantly used as feed in aquaculture. However, recent advances in our knowledge of the nutritional profile of hundreds of microalgal strains have allowed them to take a major place in the human food industry as well. Currently, *Spirulina* and *Chlorella* are approved as food by the EU and FDA and are available in the market for human consumption. They are also sold in the form of beverages, biscuits, etc. (Spolaore et al. 2006). For example, Synergy is an Australian company which commercializes *Spirulina* in the product name of Spirulina Premium®. Similarly, in the USA, Cyanotech uses the product name Spirulina Pacifica® to sell *Spirulina* as food. *Haematococcus pluvialis*, *Nannochloropsis* sp., and *Scenedesmus* sp. have also been developed as food supplements, but are not yet approved in all countries. Clearly, more research is needed for the identification of safe microalgae for food processing (Bourdichon et al. 2012).

Conclusion and Future Prospects

Despite having high ecological value and outstanding nutritional characteristics, microalgae are still one of the most unplumbed groups of organisms. It is estimated that only around 5% of the entire group of microalgae are maintained in formal collections. Moreover, approximately 5–10% have been scrutinized for chemical content, and only a few are currently used by the biotechnology industry (Guedes et al. 2011b). Therefore, there is a huge need for further research to identify and commercialize uninvestigated strains of microalgae and to identify novel metabolites with various health benefits from this large group of unexplored micro-organisms.

One of the major drawbacks in microalgae production systems is the high processing cost. Additional improvements are necessary to design economic and efficient microalgae production systems. Thus, one urgent issue is to develop new sets of bioreactors or open ponds from low-priced materials to ensure economic feasibility and sustainability. Harvesting of microalgae, although currently mostly carried out by centrifugation (*Chlorella*) or filtration (*Spirulina*) for edible microalgae, can be made much more cost-effective, for example by using induced settling techniques. Furthermore, most of the current microalgae extraction techniques deal with a single product, thereby affecting other products that may provide value. Advancement and integration of extraction procedures is required to fix this issue (Vanthoor-Koopmans et al. 2014). Consequently, economic and environmentally friendly materials and processes can facilitate sustainable production of nutraceuticals from microalgae which will undoubtedly continue to create interest in the biotech industry.

Another major gap in microalgae research is the mystery of their genomes. Whole-genome sequencing is missing for most of the microalgal species and even apparently related species may only share a very low degree of homology, with most work focusing on *C. reinhardtii* (Guarnieri and Pienkos 2015). Completion of genome sequencing for more microalgae will provide a much better understanding of the genes responsible for their bioactive compound production. As a result, this may help to induce the precise production of target compounds of microalgae by genetic manipulation or selective breeding.

At the same time, very little is known about phenolic and volatile compounds from microalgae which have many potential health benefits for humans. Extensive information is required to know more about their structure. Moreover, improvement of analytical techniques is needed for their precise identification, quantification, and isolation. The high presence of carotenoids in microalgae suggests that the genes responsible for carotenoid synthesis pathways should be better investigated in order to develop molecular tools to produce more carotenoids.

A pressing problem is the global fish harvest which fails to fulfill the need of PUFAs for most people because of globally diminishing fish populations. Therefore, microalgae, the primary producers of PUFAs, are an obvious option to meet the global PUFA demand. Microalgae have already been used as an alternative source for PUFAs in the industry but unfortunately, the production of PUFAs from autotrophic microalgae is currently still expensive (Ryckebosch et al. 2012). Thus, much attention should be focused on this sector.

To conclude, it can be said that recent advancement in the research of microalgal nutraceuticals proves that we are right on the edge of a breakthrough in the biotechnology industry. The versatility and the huge potential of microalgae could make a significant difference in the energy, pharmaceutical, cosmetics, and food industries in the forthcoming years.

References

Abdelaziz, A. E., Leite, G. B. and Hallenbeck, P. C. (2013) Addressing the challenges for sustainable production of algal biofuels: I. Algal strains and nutrient supply. *Environ. Technol.*, 34(13–16): 1783–1805.

Agyei, D. and Danquah, M. K. (2011) Industrial-scale manufacturing of pharmaceutical-grade bioactive peptides. *Biotechnol. Adv.*, 29(3): 272–277.

Ahmed, F., Fanning, F., Netzel, M.,Turner, W., Li, Y. and Schenk, P. M. (2014) Profiling of carotenoids and antioxidant capacity of microalgae from subtropical coastal and brackish waters. *Food Chem.*, 165: 300–306.

Becker, E. W. (2007) Micro-algae as a source of protein. *Biotechnol. Adv.*, 25(2): 207–210.

Blanco, A. M., Moreno, J., Del Campo, J. A., Rivas, J. and Guerrero, M. G. (2007) Outdoor cultivation of lutein-rich cells of Muriellopsis sp. in open ponds. *Appl. Microbiol. Biotechnol.*, 73(6): 1259–1266.

Borowitzka, M. (2013) High-value products from microalgae – their development and commercialisation. *J. Appl. Phycol.*, 25(3): 743–756.

Bourdichon, F., Berger, B., Casaregola, S., Farrokh, C., Frisvad, J. and Gerds, M. (2012) Safety demonstration of microbial food cultures (MFC) in fermented food products. *Bull. Int. Dairy Fed.*, 455: 1–62.

Burja, A. M. and Radianingtyas, H. (2008) Nutraceuticals and functional foods from marine microbes: an introduction to a diverse group of natural products isolated from marine macroalgae, microalgae, bacteria, fungi, and cyanobacteria. In: Barrow, C. J. and Shahidi, F. (eds) *Marine Nutraceuticals and Functional Foods*. Boca Raton: CRC Press, pp. 367–403.

Cha, K. H., Koo, S. Y. and Lee, D. U. (2008) Antiproliferative effects of carotenoids extracted from Chlorella ellipsoidea and *Chlorella vulgaris* on human colon cancer cells. *J. Agric. Food Chem.*, 56(22): 10521–10526.

Cohen, Z. (1999) *Production of Polyunsaturated Fatty Acids by the Microalga Porphyridium cruentum.* London: Taylor and Francis.

De Jesus Raposo, M. F., de Morais, R. M. S. C. and de Morais, A. M. M. B. (2013) Bioactivity and applications of sulphated polysaccharides from marine microalgae. *Marine Drugs*, 11(1): 233–252.

Del Campo, J. A., Rodriguez, H., Moreno, J., Vargas, M. A., Rivas, J. and Guerrero, M. G. (2001) Lutein production by *Muriellopsis* sp. in an outdoor tubular photobioreactor. *J. Biotechnol.*, 85(3): 289–295.

Del Campo, J. A., Garcia-Gonzalez, M. and Guerrero, M. G. (2007) Outdoor cultivation of microalgae for carotenoid production: current state and perspectives. *Appl. Microbiol. Biotechnol.*, 74(6): 1163–1174.

Dewapriya, P. and Kim, S. K. (2014) Marine microorganisms: an emerging avenue in modern nutraceuticals and functional foods. *Food Res. Int.*, 56: 115–125.

Durmaz, Y., Monteiro, M., Bandarra, N., Gökpinar, Ş. and Işik, O. (2007) The effect of low temperature on fatty acid composition and tocopherols of the red microalga, Porphyridium cruentum. *J. Appl. Phycol.*, 19(3): 223–227.

El-Baky, H. H. A., El Baz, F. K. and El-Baroty, G. S. (2009) Production of phenolic compounds from *Spirulina maxima* microalgae and its protective effects *in vitro* toward hepatotoxicity model. *Afr. J. Pharm. Pharmacol.*, 3(4): 133–139.

Espin, J. C., Garcia-Conesa, M. T. and Tomas-Barberan, F. A. (2007) Nutraceuticals: facts and fiction. *Phytochemistry*, 68(22–24): 2986–3008.

Fabregas, J. and Herrero, C. (1986) Marine microalgae as a potential source of minerals in fish diets. *Aquaculture*, 51, 237–243.

Fabregas, J. and Herrero, C. (1990) Vitamin content of four marine microalgae. Potential use as source of vitamins in nutrition. *J. Indust. Microbiol.*, 5(4): 259–263.

Fajardo, A. R., Cerdán, L. E., Medina, A. R., Fernández, F. G. A., Moreno, P. A. G. and Grima, E. M. (2007) Lipid extraction from the microalga *Phaeodactylum tricornutum. Eur. J. Lipid Sci. Technol.*, 109(2): 120–126.

Fernandez-Sevilla, J. M., Acien Fernandez, F. G. and Molina Grima, E. (2010) Biotechnological production of lutein and its applications. *Appl. Microbiol. Biotechnol.*, 86(1): 27–40.

Francavilla, M., Colaianna, M., Zotti, M., et al. (2012) Extraction, characterization and in vivo neuromodulatory activity of phytosterols from microalga *Dunaliella tertiolecta. Curr. Med. Chem.*, 19(18): 3058–3067.

Gong, Y., Hu, H., Gao, Y., Xu, X. and Gao, H. (2011) Microalgae as platforms for production of recombinant proteins and valuable compounds: progress and prospects. *J. Indust. Microbiol. Biotechnol.*, 38(12): 1879–1890.

Guarnieri, M. and Pienkos, P. (2015) Algal omics: unlocking bioproduct diversity in algae cell factories. *Photosynth. Res.*, 123(3): 255–263.

Guedes, A. C., Amaro, H. M. and Malcata, F. X. (2011a) Microalgae as sources of carotenoids. *Marine Drugs*, 9(4): 625–644.

Guedes, A. C., Amaro, H. M. and Malcata, F. X. (2011b) Microalgae as sources of high added-value compounds – a brief review of recent work. *Biotechnol. Prog.*, 27(3): 597–613.

Guerin, M., Huntley, M. E. and Olaizola, M. (2003) *Haematococcus* astaxanthin: applications for human health and nutrition. *Trends Biotechnol.*, 21(5): 210–216.

Herrero, M., Ibanez, E., Cifuentes, A., Reglero, G. and Santoyo, S. (2006) *Dunaliella salina* microalga pressurized liquid extracts as potential antimicrobials. *J. Food Prot.*, 69(10): 2471–2477.

Himaya, S. W. A., Ngo, D. H., Ryu, B. and Kim, S. K. (2012) An active peptide purified from gastrointestinal enzyme hydrolysate of Pacific cod skin gelatin attenuates angiotensin-1 converting enzyme (ACE) activity and cellular oxidative stress. *Food Chem.*, 132(4): 1872–1882.

Hudek, K., Davis, L. C., Ibbini, J. and Erickson, L. (2014) Commercial products from algae. In: Bajpai, R., Prokop, A. and Zappi, M. (eds) *Algal Biorefineries*. New York: Springer, pp. 275–295.

Huheihel, M., Ishanu, V., Tal, J. and Arad, S. M. (2002) Activity of *Porphyridium* sp. polysaccharide against herpes simplex viruses in vitro and in vivo. *J. Biochem. Biophys. Methods*, 50(2–3): 189–200.

Ibanez, E. and Cifuentes, A. (2013) Benefits of using algae as natural sources of functional ingredients. *J. Sci. Food Agric.*, 93(4): 703–709.

Janssen, C. I. and Kiliaan, A. J. (2014) Long-chain polyunsaturated fatty acids (LCPUFA) from genesis to senescence: the influence of LCPUFA on neural development, aging, and neurodegeneration. *Prog. Lipid Res.*, 53: 1–17.

Ju, Z. Y., Deng, D. F. and Dominy, W. (2012) A defatted microalgae (*Haematococcus pluvialis*) meal as a protein ingredient to partially replace fishmeal in diets of Pacific white shrimp (*Litopenaeus vannamei*, Boone, 1931). *Aquaculture*, 354–355(0): 50–55.

Kim, G. N., Shin, J. G. and Jang, H. D. (2009) Antioxidant and antidiabetic activity of Dangyuja (*Citrus grandis* Osbeck) extract treated with *Aspergillus saitoi*. *Food Chem.*, 117(1): 35–41.

Lavie, C. J., Milani, R. V., Mehra, M. R. and Ventura, H. O. (2009) Omega-3 polyunsaturated fatty acids and cardiovascular diseases. *J. Am. Coll. Cardiol.*, 54(7): 585–594.

Li, H. B., Cheng, K. W., Wong, C. C., Fan, K. W., Chen, F. and Jiang, Y. (2007) Evaluation of antioxidant capacity and total phenolic content of different fractions of selected microalgae. *Food Chem.*, 102(3): 771–776.

Lim, D. K., Garg, S., Timmins, M. et al. (2012) Isolation and evaluation of oil-producing microalgae from subtropical coastal and brackish waters. *PLoS ONE*, 7(7): e40751.

Macías-Sánchez, M. D., Mantell, C., Rodríguez, M. et al. (2007) Supercritical fluid extraction of carotenoids and chlorophyll a from *Synechococcus* sp. *J. Supercrit. Fluids*, 39(3): 323–329.

Marschner, H. (1995) 1 – Introduction, definition, and classification of mineral nutrients. In: *Mineral Nutrition of Higher Plants*, 2nd edn. London: Academic Press, pp. 3–5.

Mayfield, S. P., Manuell, A. L., Chen, S. et al. (2007) *Chlamydomonas reinhardtii* chloroplasts as protein factories. *Curr. Opin. Biotechnol.*, 18(2): 126–133.

Mendiola, J. A., Jaime, L., Santoyo, S. et al. (2007) Screening of functional compounds in supercritical fluid extracts from *Spirulina platensis. Food Chem.*, 102(4): 1357–1367.

Nakajima, Y., Inokuchi, Y., Shimazawa, M., Otsubo, K., Ishibashi, T. and Hara, H. (2008) Astaxanthin, a dietary carotenoid, protects retinal cells against oxidative stress in-vitro and in mice in-vivo. *J. Pharm. Pharmacol.*, 60(10): 1365–1374.

Ozdemir, G., Karabay, N. U., Dalay, M. C. and Pazarbasi, B. (2004) Antibacterial activity of volatile component and various extracts of *Spirulina platensis. Phytother. Res.*, 18(9): 754–757.

Palthur, M. P., Palthur, S. and Chitta, S. K. (2010) Nutraceuticals: a conceptual definition. *Int. J. Pharm. Pharmaceut. Sci.*, 2(3): 19–27.

Plaza, M., Herrero, M., Cifuentes, A. and Ibanez, E. (2009) Innovative natural functional ingredients from microalgae. *J. Agric. Food Chem.*, 57(16): 7159–7170.

Pulz, O. and Gross, W. (2004) Valuable products from biotechnology of microalgae. *Appl. Microbiol. Biotechnol.*, 65(6): 635–648.

Ryckebosch, E., Bruneel, C., Muylaert, K. and Foubert, I. (2012) Microalgae as an alternative source of omega-3 long chain polyunsaturated fatty acids. *Lipid Technol.*, 24(6): 128–130.

Ryu, B., Kang, K. H., Ngo, D. H., Qian, Z. J. and Kim, S. K. (2012) Statistical optimization of microalgae *Pavlova lutheri* cultivation conditions and its fermentation conditions by yeast, *Candida rugopelliculosa. Bioresour. Technol.*, 107: 307–313.

Sharma, K. and Schenk, P. M. (2015) Rapid induction of omega-3 fatty acids (EPA) in *Nannochloropsis* sp. by UV-C radiation. *Biotechnol. Bioeng.*, 112: 1243–1249.

Sheih, I. C., Wu, T. K. and Fang, T. J. (2009) Antioxidant properties of a new antioxidative peptide from algae protein waste hydrolysate in different oxidation systems. *Bioresour. Technol.*, 100(13): 3419–3425.

Shim, J. Y., Shin, H. S., Han, J. G. et al. (2008) Protective effects of *Chlorella vulgaris* on liver toxicity in cadmium-administered rats. *J. Med. Food*, 11(3): 479–485.

Specht, E. A. and Mayfield, S. P. (2014) Algae-based oral recombinant vaccines. *Front. Microbiol.*, 5: 60.

Spolaore, P., Joannis-Cassan, C., Duran, E. and Isambert, A. (2006) Commercial applications of microalgae. *J. Biosci. Bioeng.*, 101(2): 87–96.

Stengel, D. B., Connan, S. and Popper, Z. A. (2011) Algal chemodiversity and bioactivity: sources of natural variability and implications for commercial application. *Biotechnol. Adv.*, 29(5): 483–501.

Sun, M., Qian, K., Su, N., Chang, H., Liu, J. and Shen, G. (2003) Foot-and-mouth disease virus VP1 protein fused with cholera toxin B subunit expressed in *Chlamydomonas reinhardtii* chloroplast. *Biotechnol. Lett.*, 25(13): 1087–1092.

Tokuşoglu, Ö. and Ünal, M. K. (2003) Biomass nutrient profiles of three microalgae: *Spirulina platensis*, *Chlorella vulgaris*, and *Isochrisis galbana*. *J. Food Sci.*, 68(4): 1144–1148.

Tran, M., Henry, R. E., Siefker, D. et al. (2013) Production of anti-cancer immunotoxins in algae: ribosome inactivating proteins as fusion partners. *Biotechnol. Bioeng.*, 110(11): 2826–2835.

Vanthoor-Koopmans, M., Cordoba-Matson, M., Arredondo-Vega, B. et al. (2014). Microalgae and cyanobacteria production for feed and food supplements. In: Guevara-Gonzalez, R. and Torres-Pacheco, I. (eds) *Biosystems Engineering: Biofactories for Food Production in the Century XXI*. New York: Springer International Publishing, pp. 253–275.

Venugopal, V. (2008) *Marine Products for Healthcare: Functional and Bioactive Nutraceutical Compounds from the Ocean*. Boca Raton: CRC Press.

Watanabe, F., Takenaka, S., Kittaka-Katsura, H., Ebara, S. and Miyamoto, E. (2002) Characterization and bioavailability of vitamin B12 compounds from edible algae. *J. Nutr. Sci. Vitaminol. (Tokyo)*, 48(5): 325–331.

Yim, J. H., Son, E., Pyo, S. and Lee, H. K. (2005) Novel sulfated polysaccharide derived from red-tide microalga *Gyrodinium impudicum* strain KG03 with immunostimulating activity in vivo. *Mar. Biotechnol.*, 7(4): 331–338.

2

Functional Foods from Cyanobacteria: An Emerging Source for Functional Food Products of Pharmaceutical Importance

Neha Panjiar[1], Shashank Mishra[1,2]*, Ajar Nath Yadav[3], and Priyanka Verma[4]

[1] *Department of Bio-Engineering, Birla Institute of Technology, Mesra, Ranchi, Jharkhand, India*
[2] *Biotech Park, Lucknow, Uttar Pradesh, India*
[3] *Department of Biotechnology, Akal College of Agriculture, Eternal University, Baru Sahib, India*
[4] *Department of Microbiology, Akal College of Basic Science, Eternal University, Baru Sahib, India*

*Corresponding author e-mail: smishra83@gmail.com

Introduction

Cyanobacteria: A Successful Evolutionary Phylum

Cyanobacteria have emerged as one of the most successful prokaryotic phyla sustained during the course of evolution. They are considered as one of the primitive life forms found on our planet (Dixit and Suseela 2013). Evidence has suggested that eukaryotic photosynthesis in chloroplasts has a cyanobacterial origin, leading to the evolution of aerobic respiration 2.22–2.45 billion years ago (Gothalwal and Bajpai 2012; Schirrmeister et al. 2011). Although the autotrophic mode of nutrition is dominant, a few cynaobacterial species can grow in anaerobic and dark conditions, including *Nostoc* and *Oscillatoria* (Uzair et al. 2012). Certain species of cyanobacteria like *Nostoc* can also fix atmospheric nitrogen (Gothalwal and Bajpai 2012; Yadav et al. 2011). Furthermore, they are also among the few phyla in which complex multicellular structures developed from simple unicellular structures during evolution (Schirrmeister et al. 2011). They exhibit broad morphological diversity, ranging from gram-negative unicellular to multicellular filamentous and colonial forms (Singh et al. 2011). They inhabit various niches, which may be attributed to their evolution towards multicellularity as well as

Microbial Functional Foods and Nutraceuticals, First Edition. Edited by Vijai Kumar Gupta, Helen Treichel, Volha (Olga) Shapaval, Luiz Antonio de Oliveira, and Maria G. Tuohy.
© 2018 John Wiley & Sons Ltd. Published 2018 by John Wiley & Sons Ltd.

their tolerance of environmental stress conditions (Schirrmeister et al. 2011). The various ecological habitats occupied by cyanobacteria include soil, rock, deserts as well as glaciers, freshwater bodies like rivers, lakes, cold springs, hot springs and brackish or marine ecosystems with saline water (Nair and Bhimba 2013).

Cyanobacteria, being one of the major constituents of phytoplankton, provide ample opportunity for exploitation as a producer of secondary metabolites. The morphological, genetic, and ecological diversity of cyanobacteria leads to the production of a wide range of compounds with applications in the food, feed, pharmaceutical, and neutraceutical industries (Tan 2007). Stability of the products of cyanobacterial origin over a wide range of pH and temperature and solubility in water are some of the desirable properties which make these compounds important for research and human welfare. Active cyanobacterial species producing secondary metabolite belong to the order Oscillatoriales (49%), followed by Nostocales (26%), Chroococcales (16%), Pleurocapsales (6%), and Stigonematales (4%) (Gerwick et al. 2008).

Cyanobacteria: A Potential Source of Functional Foods

Conventional food or food ingredients which provide the required amounts of essential nutritional compounds, such as carbohydrates, proteins, fats, vitamins, and minerals, are termed functional foods (El-Sohaimy 2012). They may confer additional health benefits. Functional foods may contain bioactive compounds, which are natural chemicals derived from plants, animals or micro-organisms and are beneficial for human health.

Cyanobacteria have been exploited as a potential food supplement since ancient times. The history of commercial use of *Spirulina maxima* as a food dates back to 1521, when it was harvested from Lake Texcoco and sold in dried form in the markets of today's Mexico City (Farrar 1966). It was also here that the first commercial production of *Spirulina* started in the 1970s. People of the Kanembu tribe living near Lake Chad in Africa have used *Spirulina* as a protein supplement since 1940 (Abdulqader et al. 2000). Evidence also exists for the use of cyanobacteria other than *Spirulina*, for example *Nostoc commune* in Asia, *N. flagelliforme* in China, *N. punctiforme* in China, Mongolia and South America, as food and feed supplements (Facciola 1998; Soeder 1980; Takenaka et al. 1998). *N. punctiforme* is still being marketed in South America as "Lakeplum" (Trainor 1978). Although the market is dominated by the filamentous cyanobacteria, unicellular forms have also created their own niche in the market, like *Aphanotheca sacrum*, which is being consumed in Japan (Fujishiro et al. 2004).

The global population explosion has resulted in the need to look for alternative sustainable sources of food apart from conventional agricultural products. This has promoted interest in the use of functional foods to meet the nutritional demands of the growing human population. Natural products have been used as food supplements for thousands of years. Approximately 80% of the world population depends on traditional methods for primary healthcare, as estimated by the World Health Organization (Dixit and Suseela 2013). Micro-organisms, especially cyanobacteria, are an untapped resource as their secondary metabolites have nutritional or therapeutic values. Commercial exploitation of cyanobacteria since the establishment of human civilization is owing to its different properties which make it a suitable source of functional foods.

Many characteristics make cyanobacteria an attractive alternative for sustainable food production:

- high nutrient content of cyanobacteria, especially *Spirulina*, which is the most commericalized cultivated cyanobacterial species
- worldwide distribution
- need for only small amounts of water for growth; salt water can also be used
- need for smaller amounts of land; unfertile land unsuitable for other types of crops can also be used
- easily digestible
- product stability over a wide pH and temperature range.

According to some statistics, 1 kg of *Spirulina* may replace 1000 kg of assorted fruits and vegetables in terms of nutritional value. Dried *Spirulina* is sold in the United States as a health food, with annual sales of approximately $40 million (Singh et al. 2005).

Realizing the present requirement for extensive research on functional foods from cyanobacteria, many companies have initiated industrial-scale production of cyanobacteria. Culturing, harvesting, processing (drying), and packaging are the crucial steps involved. Currently, more than 70 countries have commercialized products of nutritional importance obtained from cyanobacteria (Gantar and Svircev 2008). Some of the companies are listed in Table 2.1. The demand for large-scale production of cynaobacterial products is increasing, but there are constraints associated with their production which need to be resolved (Figure 2.1). Development of efficient technologies is required to reduce the production cost along with maintenance and improvement of product quality. Potential cyanobacterial strain selection should be focused and research oriented to conserve the valuable germplasm (Emeka et al. 2012). Limitations in terms of water requirement should be considered for possible substitutes like waste water or sea water (Emeka et al. 2012).

Table 2.1 Companies marketing functional food products of cyanobacterial origin.

Country	Companies
USA	Earthrise Farms, Cyanotech Corporation
Myanmar	Myanmar Microalgae Biotechnology Project, Myanmar Spirulina
China	Hainan DIC Microalgae, Nan Pao Resins
India	Ballarpur Industries, EID Parry
Taiwan	Nan Pao Resins
Thailand	Neotech Food, Siam Algae
Cuba	Genix
Chile	Solarium Biotechnology
Canada	Ocean Nutrition
Japan	Nippon Spirulina, Dainippon
Mexico	Spirulina Mexicana
Australia	Panmol
Mongolia	Inner Mongolia Biomedical Engineering
Germany	Blue Biotech
Israel	Koor Foods

CYANOBACTERIA

 Production methods

- Natural environment
- Open ponds with controlled cultivation process
- Photobioreacters

 Applications

- Feed additives
- Pro-vitamin A
- Food pigment reagents
- Dietary supplements
- Bioactive compounds

 Constraints

- Less efficient screening methods
- High content of nucleic acid
- Contamination with weed algal and other microorganisms
- Water requirement in large quantity

 Tentative solutions

- Development of high throughput screening methods
- Conservation of valuable germplasm
- Genetic modification of strains
- Optimization of production process
- Improvement of production technologies for the safety of food products
- Mantaining pure culture of efficient strains

Figure 2.1 General overview of the applications, constraints associated with the production of the cynaobacterial food products and their tentative solutions.

Photobioreactors have also been designed to cultivate and scale up cyanobacterial cultures in optimum conditions (Campo et al. 2007).

Functional Food Ingredients of Cyanobacterial Origin

Carbohydrate and Fibers

Carbohydrates are the primary energy source used by the body. One gram of carbohydrate provides four calories of energy to the body. According to the World Health Organization, at least 55% of the body's total energy requirement should be provided by a carbohydrate source (Nishida and Nocito 2007; www.fao.org/docrep/w8079E/w8079e08. htm). Although fulfillment of energy requirements is the main function of carbohydrates, they also play a significant role in the structure and function of cells, tissues, and organs. Intake of optimum daily amounts of carbohydrate helps prevent fat accumulation in the body. Table 2.2 represents the percentage of carbohydrate present in different cyanobacterial species. Amongst the different cyanobacterial species reviewed by Rajeshwari and Rajashekhar (2011), maximum carbohydrate content was found in *Scytonema bohneri* (28.4%) compared to the average percentage of 15–20% for *Spirulina* species. The protein percentage was less in other cyanobacterial species than *Spirulina* as estimated by Fatma et al. (1994), which make *Spirulina* a favorable candidate for functional food.

Dietary fiber, soluble and insoluble, is a type of complex carbohydrate. Even though it is not a nutrient, it is important as it is not fully digested and hence not assimilated in the body. It constitutes the roughage portion of the diet. Soluble fiber dissolves in water and is essential for lowering blood cholesterol as well as maintaining blood sugar level (El-Sohaimy 2012). Insoluble fiber consumption helps in reducing the diet's caloric value. Dietary fiber therefore helps to maintain a healthy digestive system and is of significant importance for obese and diabetic people (Bolton-Smith and Woodward 1994; El-Sohaimy 2012). Recent research also reveals its use for colon cancer prevention (Zeng et al. 2014). The Food and Agricultural Organization (FAO) of the United Nations has prescribed an average intake of 20–35 g of dietary fiber per day after 2 years of age (Dobbing 1989).

In recent years, prebiotics or the non-digestible oligosaccharide carbohydrates have gained the attention of the scientific community as they help in the maintenance of a beneficial gut microbial population. Tremaroli and Bäckhed (2012) have shown that changes in the composition of gut microbiota can lead to alterations in metabolic capacity in

Table 2.2 Carbohydrate and protein content of common cyanobacterial strains.

Cyanobacteria	Carbohydrate (%)	Protein (%)	References
Oscillatoria limnetica	15.23±0.29	16.93±0.25	Subramanian et al. (2014)
Oscillatoria tenuis	11.05±0.31	11.82±0.11	Subramanian et al. (2014)
Phormidium abronema	8.26±0.19	22.71±0.24	Subramanian et al. (2014)
Lyngbya cryptovaginata	19.26±0.27	25.23±0.32	Subramanian et al. (2014)
Tolypothrix tenuis	16.34±0.32	12.21±0.22	Subramanian et al. (2014)
Spirulina platensis LEB-18	7.5	58	Moreira et al. (2013)
Calothrix fusca	19.5	1.6	Rajeshwari and Rajashekhar (2011)
Gloeocapsa livida	18.0	1.8	
Lyngbya limnetica	18.4	3.1	
Scytonema bohneri	28.4	0.7	
Oscillatoria acuminata	14	6.9	
Oscillatoria calcuttensis	9.6	2.2	
Oscillatoria foreaui	8	7	
Spirulina pacifica	-	58.01	Mišurcová et al. (2010)
Spirulina platensis	-	56.05	

obese people, thereby promoting adiposity and influencing physiological processes like satiety control from the brain and hormonal release from the gut. Fiber content of *Spirulina* is in the range of 5–8% dry weight (Shetty et al. 2006).

Proteins and Peptides

Proteins are structurally complex macromolecules with building blocks of amino acids. They are polypeptides composed of more than 50 amino acids, while peptides have fewer than 50 amino acids. Proteins and peptides have diverse biological activities in the human body ranging from catalysis in the form of enzymes, structural support as a major component of connective tissues, cell movement, being actively involved in body defense mechanisms, body regulation in the form of peptide hormones and growth regulators and in stress response. Deficiency or inadequate intake of calories can cause protein energy malnutrition (PEM), mainly kwashiorkor

which accounts for 6 million deaths annually (http://fnic.nal.usda.gov/
dietary-guidance/dietary-reference-intakes/dri-nutrient-reports).
Therefore an economical diet rich in protein is required worldwide and
especially in underdeveloped and developing countries. The proportion of
the disease burden is is expected to rise to 57% by 2020 (Gershwin and
Belay 2008). Management of the global prevalence of these diseases from
the nutritional point of view is of the utmost importance.

Micro-organisms, being widely distributed in nature, can provide a
global solution. However, high protein content bacteria are very scarce,
like *Cellulomonas* with 80% of their dry weight as protein, but their high
protein content is associated with high nucleic acid content, which is
undesirable from a medical point of view (Litchfield 1983). Metabolism
of nucleic acid results in uric acid formation and accumulation in the
body, leading to pathological conditions like gout. According to the
Protein Advisory Group (PAG) of the United Nations, the daily intake of
nucleic acids for a healthy adult should not exceed 2 g (http://archive.
unu.edu/unupress/food/8F051e/8F051E0d.htm). Cyanobacteria, particularly *Spirulina*, can surpass bacteria like *Cellulomonas* nutritionally
by having a nucleic acid concentration of less than 5% and are thus
advantageous as a future protein supplement (Shetty et al. 2006). As
shown in Table 2.2, *Spirulina* has a higher protein content than other
cyanobacterial species. Crude protein content of *spirulina* is 50–60%,
which is greater than other regular protein foods like egg (47%), whole
soyabean flour (36%), parmesan cheese (36%), wheat germ (27%), peanut
(26%), chicken (24%), and fish (22%) (Ali and Saleh 2012). Studies by Fica
et al. (1984) showed that in Romania, protein supplements from *Spirulina*
helped patients with nutritional deficiency; the patients gained weight
and their health improved. In China, it has been implemented as a baby
food ingredient in baked barley sprouts (Shetty et al. 2006).

Lipids and Fatty Acids

Cyanobacteria contain significant quantities of lipid, with a composition
similar to that of vegetable oils (Singh et al. 2002). Cyanobacterial lipids
generally are esters of fatty acids and glycerol with a chain length of C_{14}
to C_{22}, which may be either saturated or unsaturated (Shetty et al. 2006).
However, cyanobacteria lipids, compared to eukaryotes, usually lack
sterols (Quinn and Williams 1983). Lipid composition of cyanobacterial
species is generally affected by their growth phase, pH, temperature, illu-
minance, carbon dioxide concentration,and nitrogen deficiency (Kaiwan-
arporn et al. 2012; Liu et al. 2014; Loura et al. 1987; Sharathchandra and
Rajashekhar 2011). Some of the filamentous cyanobacteria have been
reported to have large quantities of polyunsaturated fatty acids (25–60%

of the total fatty acid present) and essential fatty acids (Shetty et al. 2006). Lipid of *Synechocystis aquatilis* TISTR8612 is rich in essential fatty acids such as linoleic (C18:2) and linolenic (C18:3) acid, which has medical implications (Kaiwan-arporn et al. 2012). Percentage of lipid on a dry weight basis is 2–3% in *Spirulina*, with linoleic (C18:2) and linolenic (C18:3) acid content of about 17.9% and 24.9% of total fatty acids, respectively (Shetty et al. 2006). The fatty acid composition and high amounts of PUFA make the *Spirulina platensis* lipids a special component (Moreira et al. 2013). PUFA gained interest as it has benefits in alleviating cardiovascular, inflammatory, autoimmune disorder, and diabetes and is good for joints (Riemersma 2001). *Spirulina* lipids also have the advantage of being cholesterol free which is medically suitable for people suffering from obesity, atherosclerosis, and diabetes. They have also been studied for their cholesterol-reducing effects (Shetty et al. 2006).

Minerals and Vitamins

Cyanobateria are rich in minerals and vitamins. Levels of iron (15 mg/10 g of *Spirulina*), calcium (100 mg/10 g of *Spirulina*), magnesium (40 mg/10 g of *Spirulina*), potassium (160 mg/10 g of *Spirulina*) and trace elements are very high in *Spirulina* compared to other readily available foods (Shetty et al. 2006). This has medical implications as it is good for bones, teeth, and blood. Other minerals found in *Spirulina* (per 10 g) are zinc (300 µg), phosphorus (90 mg), copper (120 µg), sodium (60 mg), manganese (500 µg), and selenium (2 µg) (Shetty et al. 2006). *Oscillatoria foreaui* and *Lyngbya limnetica* also contain substantial amounts of minerals, such as 57.65±1.9 and 31.20±2.2 copper, 539.20±0.2 and 379.60±0.5 manganese, 6402.00±0.5 and 5455.70±0.2 iron, 211.20±0.2 and 134.10±1.1 zinc, 9.37±1.2 and 2.20±0.7 nickel, 12,812.00±1.0 and 18,650.00±0.5 µg/mL magnesium respectively (Rajeshwari and Rajashekhar 2011).

Cyanobacteria can also surpass other foods as being the richest source of vitamin B12 (Shetty et al. 2006). It has been shown that intake of only 1 g of *Spirulina* is sufficient to meet the daily requirement for vitamin B12. It contains other vitamins also, including A, B1, B2, B3, B6, E, H, folacin, pantothenic acid, and inositol. Pantothenic acid is present in an amount equivalent to the recommended daily allowance. The quantity of β-carotene (32 µg) present in *Spirulina* is more than the RDA (6 µg) and is 20 times more than that present in carrot (Shetty et al. 2006). In India, a large-scale study on preschool children with vitamin A deficiency showed that they recovered with a *Spirulina* food supplement (Seshadri 1993). Many of these vitamins and minerals have strong antioxidant properties, which help in eliminate toxins and fight disease.

Bioactive Compounds

The oceanic cyanobacteria are an extraordinarily rich source of secondary metabolites (Cardellina and Moore 2010). Cyanobacterial secondary metabolites contain various bioactive molecules including cytotoxic (41%), antitumor (13%), antiviral (4%), and antimicrobial (12%), and other compounds (18%) include antimalarials, antimycotics, multidrug resistance reversers, herbicides, insecticides, algaecides, and immunosuppressive agents (Burja et al. 2001). These natural compounds not only serve directly as drugs, but also are being used as prototypes for the synthesis of new drugs. Therefore, cyanobacterial metabolites continue to be explored for their application in many biological areas and can be an exceptional source of compounds for drug discovery (Bajpai et al. 2010; Nunnery et al. 2010; Singh et al. 2005, 2011).

Anticancerous Activity

Statistics from the American Cancer Society showed that 7.6 million people died worldwide during 2007 from cancer and this will only increase in future, if not controlled properly (http://uk.reuters.com/article/2007/12/17/health-cancer-worlddc-idUKN1633064920071218). Currently available drugs like vinca alkaloids and taxanes can become inactive due to the development of drug-resistant properties by the tumor cells, which is a key reason for failure of chemotherapeutic treatment of cancers.

Screening of cyanobacterial extracts for new anticancer compounds was initiated by Moore (Oregon State University) and Gerwick (University of Hawaii) in the 1990s. Tubulin or actin filaments in eukaryotic cells are targeted by many cyanobacterial bioactive compounds, which make these compunds a potent source of anticancer agents (Jordan and Wilson 1998). The small anticancerous peptides dolastatin 10 and dolastatin 12 were isolated from *Symploca* sp. and *Leptolyngbya* sp. (Catassi et al. 2006; Kalemkerian et al. 1999). Curacin A showed antiproliferative property that has been isolated from *Lyngbya majuscula* (Nagle et al. 1995). It was also artificially synthesized because of its pharmacological importance (Muir et al. 2002).

Antiviral Activity

International spread of lethal viral diseases like HIV, AIDS, dengue, and avian influenza (H5N1 virus) has had serious consequences. The anti-HIV highly active antiretroviral therapy (HAART) is effective in controlling the development of HIV infections but is toxic (Luescher-Mattli 2003). Hence, novel drugs are now needed to fight these lethal diseases. Antiviral compounds isolated from cyanobacteria are usually found to have bioactivity by blocking viral absorption or penetration and inhibiting replication stages of progeny viruses after penetration into cells. Protection of human lymphoblastoid T-cells from the cytopathic effect of HIV infection with an extract of *Lyngbya lagerheimeii* and *Phormidium tenue* was reported by Gustafson et al. (1989). A new class of HIV inhibitor called sulfonic acid, containing glycolipid, was isolated from an extract of cyanobacteria and the compounds were found to be active against the HIV virus. Cyanovirin-N (CVN), a peptide isolated from cyanobacteria, inactivates strains of HIV virus and inhibits cell-to-cell and virus-to-cell fusion (Yang et al. 1997).

Antibacterial Activity

In recent years, multidrug-resistant bacteria causing nosocomial infections such as methicillin-resistant *Staphylococcus aureus*, vancomycin-resistant Enterococci and Amp C β-lactamase producing Enterobacteriaceae have posed therapeutic challenges and are of great concern worldwide (Reinert et al. 2007). In an effort to develop new antibiotics, scientists are screening cyanobacterial extracts for their antibacterial activity (Biondi et al. 2008; Kreitlow et al. 1999; Skulberg 2000) and have found them potentially active against various bacteria. Noscomin from *Nostoc commune* exhibited antibacterial activity against *Bacillus cereus*, *Staphylococcus epidermidis*, and *Escherichia coli* (Jaki et al. 2000). Nostocarboline from *Nostoc* was found to inhibit the growth of other cyanobacteria and green alga (Blom et al. 2006). Hirata et al. (2003) found that Nostocine A isolated from *Nostoc spongiaeforme* inhibited the growth of green algae more strongly than cyanobacteria. Asthana et al. (2009) isolated hapalindole (alkaloids) from *Nostoc* CCC537 and *Fischerella* sp. and found antimicrobial activity against *Mycobacterium tuberculosis* H37Rv, *Staphylococcus aureus* ATCC25923, *Salmonella typhi* MTCC3216, *Pseudomonas aeruginosa* ATCC27853, *E. coli* ATCC25992, and *Enterobacter aerogenes* MTCC2822.

Because of growing bacterial resistance against antibiotics, the search for new active substances with antibacterial activity is urgent and cyanobacteria will be potentially promising candidates.

Antiprotozoal Activity

According to a WHO estimate, more than 1 billion people throughout the world are suffering from tropical diseases caused by *Plasmodium*, *Trypanosoma*, *Leishmania*, *Schistosoma*, and others (Simmons et al. 2008). Failures in the treatment of these diseases, especially in cases of malaria (Lanzer and Rohrbach,2007) and leishmaniasis (Prioto et al. 2007), are due to development of resistance by these protozoa. On the other hand, progress in the advancement of drug discovery programs against these diseases is very slow (Sheifert et al. 2007). In an effort to encourage the development of effective and affordable treatment for these diseases, the Panamanian International Co-operative Biodiversity

Group is screening extracts from terrestrial and marine sources. In addition, the protease inhibitor nostocarboline (Barbaraus et al. 2008), an alkaloid isolated from *Nostoc* sp. 78-12, was found to be active against *Trypanosoma brucei*, *Trypanosoma cruzi*, *Leishmania donovani*, and *Plasmodium falciparum*. Aerucyclamide (Portmann et al. 2008) isolated from *Microcystis aeruginosa* PCC 7806 was also found to be active against *T. brucei*, and the already known aerucyclamide B against *P. falciparum*. Clark et al. (2008) isolated six new acyl proline derivatives, tumonoic acids D–I, from the marine cyanobacterium *Blennothrix cantharidosmum*, among which tumonoic acid I displayed moderate activity in an antimalarial assay.

Protease Inhibition Activity

The discovery of new protease inhibitors may be of great pharmaceutical value. Jaspars and Lawton (1998) described some protease inhibitors of cyanobacterial origin, such as microginins, aeruginosins, and cyanopeptolins. Microginins are used in the treatment of high blood pressure. Serine protease inhibitors like cyanopeptolin are applied in the treatment of conditions such as asthma and viral infections.

Immunomodulatory Activity

Different products prepared from *Spirulina* influence immune systems in various ways such as increasing the phagocytic activity of macrophages, stimulating antibody and cytokine production, increasing accumulation of natural killer cells in tissues, and activating T- and B-cells (Khan et al. 2005). The effect of *Spirulina* in mice was studied and increased phagocytic activity and enhanced antigen production in the test animals were reported (Hayashi et al. 1994). Qureshi and Ali (1996) reported increased phagocytic activity, increased antigen production, and increased natural killer cell-mediated antitumor activity in chicken for the same cyanobacterium (*Spirulina*).

Summary and Future Prospects

Cyanobacteria are photosynthetic bacteria populating diverse habitats worldwide. They are the only prokaryotes capable of fixing carbon dioxide from the atmosphere through photosynthesis, with efficiency higher than vascular plants. Their cultivation is economical, simple and does not directly compete with agricultural crops for water and land. Recent research leading to technical improvements and increased consumer demand has resulted in market expansion for cyanobacterial species and their products. However, their biotechnological potential is still not explored completely and requires exhaustive research for industrial-scale development of its approved functional food products. Functional foods are defined as foods providing health benefits apart from basic nutrition. Cyanobacteria provide an ultimate blend of nutrition in the right quantities as a single food. Although they are a promising source offering diverse functional foods, they are still underexplored as a natural resource.

Commercial exploitation of cyanobacteria is hampered by lack of knowledge of culture techniques, biochemical pathways leading to formation of diverse metabolites and cost-effective production methods. Advances in molecular biology, biochemistry, instrumentation, and computational approaches have directed attention towards the biosynthesis of diverse metabolites from cyanobacteria. One vital factor determining their biosynthesis is the genetic make-up of the producer organisms. Strain improvement through genetic modifications could be used to significantly improve the yield and quality of cyanobacterial bioactive compounds. The combined efforts of molecular biology and genetic engineering have led to the sequencing of over 126 genomes of cyanobacterial strains, thus paving the way for further metabolic engineering techniques to explore their unique biosynthetic information. This may also lead to the development of new strains with better growth rate, higher substrate tolerance and consistent production of biomass aimed at producing designer novel metabolites in the near future.

References

Abdulqader, G., Barsanti, L. and Tredici, M.R. (2000) Harvest of *Arthrospira platensis* from Lake Kossorom (Chad) and its household usage among the Kanembu. *J. Appl. Phycol.*, 12: 493–498.

Ali, S.K. and Saleh, A.M. (2012) *Spirulina* – an overview. *Int. J. Pharm. Pharmaceut. Sci.*, 4(3): 9–15.

Asthana, R.K., Tripathi, M.K., Deepali, A. et al. (2009) Isolation and identification of a new antibacterial entity from the Antarctic cyanobacterium *Nostoc* CCC 537. *J. Appl. Phycol.*, 21: 81–88.

Bajpai, R., Sharma, N.K. and Rai A.K. (2010) Anticancerous/antiviral compounds from cyanobacteria. *Algas*, 43: 4–6.

Barbaraus, D., Kaiser, M., Brun, R. and Gademann, K. (2008) Potent and selective antiplasmodial activity of the cyanobacterial alkaloid nostocarboline and its dimers. *Bioorganic Medicinal Chem. Lett.*, 18: 4413–4415.

Biondi, N., Tredici, M.R., Taton, A. et al. (2008) Cyanobacteria from benthic mats of Antarctic lakes as a new source of bioactivities. *J. Appl. Microbiol.*, 105: 105–115.

Blom, J.F., Brutsch, T., Barbaras, D. et al. (2006) Potent algicides based on the cyanobacterial alkaloid nostocarboline. *Organic Lett.*, 8: 737–740.

Bolton-Smith, C. and Woodward, M. (1994) Dietary composition and fat to sugar ratios in relation to obesity. *Int. J. Obesity*, 18: 820–828.

Burja, A.M., Banaigs, B., Abou-Mansour, E., Burgess, G. and Wright P.C. (2001) Marine cyanobacteria: a prolific source of natural products. *Tetrahedron*, 57: 9347–9377.

Campo, J., Garcia-Gonzalez, M. and Miguel, G.G. (2007) Outdoor cultivation of microalgae for carotenoid production: current state and perspectives. *Appl. Microbiol. Biotechnol.*, 74: 1163–1174.

Cardellina, J.H. and Moore, B.S. (2010) Editorial: Richard E Moore (1933–2007). *J. Nat. Products*, 73: 301–302.

Catassi, A., Cesario, A., Arzani, D. et al. (2006) Characterization of apoptosis induced by marine natural products in non small cell lung cancer A549 cells. *Cell. Molec. Life Sci.*, 63: 2377–2386.

Clark, B.R., Engene, N., Teasdale, M.E. et al. (2008) Natural products chemistry and taxonomy of the marine cyanobacterium. *Blennothrix cantharidosmum. J. Nat. Products*, 71: 1530–1537.

Dixit, R.B. and Suseela, M.R. (2013) Cyanobacteria: potential candidates for drug discovery. *Antonie van Leeuwenhoek*, 103: 947–961.

Dobbing, J. (1989) *Dietary Starches and Sugars in Man: A Comparison*. ILSI Human Nutrition Review Series. New York: Springer.

El-Sohaimy, S.A. (2012) Functional foods and nutraceuticals – modern approach to food science. *World Appl. Sci. J.*, 20(5): 691–708.

Emeka, U., Ndukwe, G.I., Mustapha, K.B. and Ayo, R.I. (2012) Constraints to large scale algae biomass production and utilization. *J. Algal Biomass Util.*, 3(2): 14–32.

Facciola, S. (1998) *Cornucopia II: A Sourcebook of Edible Plants*, 2nd edn. Vista: Kampong Publications.

Farrar, W.V. (1966) Tecuitlatl: a glimpse of Aztec food technology. *Nature*, 21: 1341–1342.

Fatma, T., Sarada, R. and Venkataraman, L.V. (1994) Evaluation of selected strains of *Spirulina* for their constituents. *Phykos*, 33: 89–97.

Fica, V. (1984) Observations on the utilization of *Spirulina* as an adjuvant nutritive factor in treating diseases accompanied by a nutritional deficiency, Clinica 11 Medicala, Spitalui clinic Municipiului Bucuresti. *Med. Interna*, 36(3).

Fujishiro, T., Ogawa, T., Matsuoka, M., Nagahama, K., Takeshima,Y. and Hagiwara, H. (2004) Establishment of a pure culture of the hitherto unicellular cyanobacterium *Aphanothecasacrum*, and phylogenetic position of the organism. *Appl. Environ. Microbiol.*, 70: 3338–3345.

Gantar, M. and Svircev, Z. (2008) Microalgae and cynobacteria : food for thought. *J. Phycol.*, 44: 260–268.

Gershwin, M.E. and Belay, A. (2008) *Spirulina in Human Nutrition and Health*. Boca Raton: CRC Press.

Gerwick, W.H., Coates R.C., Engene, N. et al. (2008) Giant marine cyanobacteria produce exciting potential pharmaceuticals. *Microbe*, 3: 277–284.

Gothalwal, R. and Bajpai, P. (2012) Cynobacteria a comprehensive review. *Int. Res. J. Pharm.*, 3(2): 53–57.

Gustafson, K.R., Cardellina, J.H., Fuller, R.W. et al. (1989) AIDS antiviral sulfolipids from cyanobacteria (blue-green algae). *J. Natl Cancer Inst.*, 81: 1254–1258.

Hayashi, O., Katoh, T. and Okuwaki, Y. (1994) Enhancement of antibody production in mice by dietary *Spirulina platensis*. *J. Nutr. Sci. Vitaminol.*, 40: 431–441.

Hirata, K., Yoshitomi, S., Dwi, S. et al. (2003) Bioactivities of nostocine A produced by a freshwater cyanobacterium *Nostoc spongiaeforme* TISTR 8169. *J. Biosci. Bioeng.*, 95: 512–517.

Jaki, B., Orjala, J., Heilmann, J., Linden, A., Vogler, B. and Sticher, O. (2000) Novel extracellular diterpenoids with biological activity from the cyanobacterium *Nostoc commune*. *J. Nat. Products*, 63: 339–343.

Jaspars, M. and Lawton, L.A. (1998) Cyanobacteria – a novel source for pharmaceuticals. *Curr. Opin. Drug Discov. Dev.*, 1: 77–84.

Jordan, M.A. and Wilson L. (1998) Microtubules and actin filaments: dynamic targets for cancer chemotherapy. *Curr. Opin. Cell Biol.*, 10: 123–130.

Kaiwan-arporn, P., Hai, P.D., Thu, N.T. and Annachhatre, A.P. (2012) Cultivation of cyanobacteria for extraction of lipids. *Biomass Bioenergy*, 44: 142–149.

Kalemkerian, G.P., Ou, X.L., Adil, M.R. et al. (1999) Activity of dolastatin 10 against small-cell lung cancer *in vitro* and *in vivo*: induction of apoptosis and bcl-2 modification. *Cancer Chemother. Pharmacol.*, 43: 507–515.

Khan, Z., Bhadouria, P. and Bisen, P.S. (2005) Nutritional and therapeutic potential of *Spirulina*. *Curr. Pharmaceut. Biotechnol.*, 6: 373–379.

Kreitlow, S., Mundt, S. and Lindequist, U. (1999) Cyanobacteria – a potential source of new biologically active substance. *J. Biotechnol.*, 70: 61–73.

Lanzer, M.R. and Rohrbach, P. (2007) Subcellular pH and Ca2$^+$ in *Plasmodium falciparum*: implications for understanding drug resistance mechanisms. *Curr. Sci.*, 92: 1561–1570.

Litchfield, J.H. (1983) Single–cell proteins. *Science*, 219: 740–746.

Liu, Y., Zhang, J., Nie, H. et al. (2014) Study on variation of lipids during different growth phases of living cyanobacteria using easy ambient sonic-spray ionization mass spectrometry. *Analyt. Chem.*, 86: 7096–7102.

Loura, I.C.D., Dubacq, J.P. and Thomas, J.C. (1987) The effects of nitrogen deficiency on pigments and lipids of cyanobacteria. *Plant Physiol.*, 83: 838–843.

Luescher-Mattli, M. (2003) Algae as a possible source of new antiviral agents. *Curr. Med. Chem. Anti-Infect. Agents*, 2: 219–225.

Mišurcová, L., Kráčmar, S., Klejdus, B. and Vacek, J.(2010) Nitrogen content, dietary fiber, and digestibility in algal food products. *Czech J. Food Sci.*, 28(1): 27–35.

Moore, R.E., Corbett, T.H., Patterson, G.M.L. and Valeriote, F.A. (1996) The search for new antitumor drug from blue green algae. *Curr. Pharmaceut. Design*, 2(3): 17–330.

Moreira, L.M., Ribeiro, A.C., Duarte, F.A., Morais, M.G. and Soares, L.A.S. (2013) *Spirulina platensis* biomass cultivated in Southern Brazil as a source of essential minerals and other nutrients. *Afr. J. Food Sci.*, 7(12): 451–455.

Muir, J.C., Pattenden, G. and Ye, T. (2002) Total synthesis of (+)-curacin A, a novel antimitotic metabolite from a cyanobacterium. *J. Chem. Soc. Perkin Transactions*, 1: 2243–2250.

Nagle, D.G., Geralds, R.S., Yoo, H. and Gerwick, W.H. (1995) Absolute configuration of curacin A, a novel antimitotic agent from the tropical marine cyanobacterium *Lyngbya majuscula*. *Tetrahedron Lett.*, 36: 1189–1192.

Nair, S. and Bhimba, B.V.(2013) Bioactive potency of cynobacteria *Oscillatoria* spp. *Int. J. Pharm. Pharmaceut. Sci.*, 5(2): 611–612.

Nishida, C. and Nocito, F.M. (2007) FAO/WHO Scientific Update on carbohydrates in human nutrition: introduction. *Eur. J. Clin. Nutr.*, 61(1): S1–S4.

Nunnery, J.K., Mevers, E. and Gerwick, W.H. (2010) Biologically active secondary metabolites from marine cyanobacteria. *Curr. Opin. Biotechnol.*, 21: 787–793.

Portmann, C., Blom, J.F., Kaiser, M., Brun, R., Jüttner, F. and Gademann, K. (2008) Isolation of aerucyclamides C and D and structure revision of microcyclamide 7806A, heterocyclic ribosomal peptides from *Microcystis aeruginosa* PCC 7806 and their antiparasite evaluation. *J. Nat. Products*, 71: 1891–1896.

Priotto, G., Kasparian, S., Ngouama, D. et al. (2007) Nifurtimox-eflornithine combination therapy for second-stage Trypanosoma brucei gambiense sleeping sickness: a randomized clinical trial in Congo. *Clin. Infect. Dis.*, 45: 1435–1442.

Qureshi, M. and Ali, R. (1996) *Spirulina platensis* exposure enhances macrophage phagocytic function in cats. *Immunopharmacol. Immunotoxicol.*, 18: 457–463.

Rajeshwari, K.R. and Rajashekhar, M. (2011) Biochemical composition of seven species of cyanobacteria isolated from different aquatic habitats of western ghats, southern India. *Brazil. Arch. Biol. Technol.*, 54(5): 849–857.

Reinert, R., Donald, E.L., Rosi, F.X., Watal, C. and Dowzicky, M. (2007) Antimicrobial susceptibility among organisms from the Asia/Pacific Rim, Europe and Latin and North America collected as part of TEST and the in vitro activity of tigecycline. *J. Antimicrob. Chemother.*, 60: 1018–1029.

Riemersma, R.A. (2001) The demise of the n-6 to n-3 fatty acid ratio. *Eur. J. Lipid Sci. Technol.*, 103: 372–373.

Sallal, A.K., Nimer, N.A. and Radwan, S.S. (1990) Lipid and fatty acid composition of freshwater cyanobacteria. *J. Gen. Microbiol.*, 136: 2043–2048.

Schirrmeister, B.E., Antonelli, A. and Bagheri, H.C.(2011) The origin of multicellularity in cyanobacteria. *BMC Evolution. Biol.*, 11(45): 1–21.

Seshadri, C.V. (1993) *Large-scale nutritional supplementation with Spirulina alga*. All India Project, MCRC, Madras.

Sharathchandra, K. and Rajashekhar, M. (2011) Total lipid and fatty acid composition in some freshwater cyanobacteria. *J. Algal Biomass Util.*, 2(2): 83–97.

Sheifert, K., Lunke, A., Croft, S.L. and Kayser, O. (2007) Antileishmanial structure activity relationships of synthetic phospholipids: *in-vitro* and *in-vivo* activities of selected derivatives. *Antimicrob. Agents Chemother.*, 51: 4525–4528.

Shetty, K., Paliyath, G., Pometto, A. and Robert, E.L. (2006) *Food Biotechnology*, 2nd edn. Boca Raton: CRC Press.

Simmons, L.T., Engene, N., Ureña, L.D. et al. (2008) Viridamides A and B, lipodepsipeptides with antiprotozoal activity from marine cyanobacterium *Oscillatoria nigroviridis*. *J. Nat. Products*, 71: 1544–1550.

Singh, R.K., Tiwari, S.P., Rai, A.K. and Mohapatra. T.M. (2011) Cyanobacteria: an emerging source for drug discovery. *J. Antibiotics*, 64: 401–412.

Singh, S., Kate, B.N. and Banerjee, U.C. (2005) Bioactive compounds from cyanobacteria and microalgae: an overview. *Crit. Rev. Biotechnol.*, 25: 73–95.

Skulberg, O.M. (2000) Microalgae as a source of bioactive molecules: experience from cyanophyte research. *J. Appl. Phycol.*, 12: 341–348.

Soeder, C.J. (1980) *The Scope of Microalgae for Food and Feed*. New York: Elsevier/North-Holland Biomedical Press, pp. 9–20.

Subramanian, G., Anusha, M.B., Bai, A.L.G., Latha, R., Sasikala, J. and Chandrasakaran, R. (2014) Carbohydrate and protein content of five species of marine cyanobacteria from Pamban and Vadakadu (Rameswaram) coastal regions, Tamil Nadu, India. *Int. J. Adv. Interdisciplinary Res.*, 1(6): 18–20.

Takenaka, H., Yamaguchi, Y., Sakaki, S. et al. (1998) Safety evaluation of *Nostoc flagelliforme* (Nostocales, Cyanophyceae) as a potential food. *Food Chem. Toxicol.*, 36: 1073–1077.

Tan, L.T. (2007) Bioactive natural products from marine cyanobacteria for drug discovery. *Phytochemistry*, 68: 954–979.

Trainor, F.R. (1978) *Introductory Phycology*. New York: Wiley and Sons.

Tremaroli, V. and Bäckhed, F. (2012) Functional interactions between the gut microbiota and host metabolism. *Nature*, 489: 242–249.

Uzair, B., Tabassum, S., Rasheed, M. and Rehman, S.F. (2012) Exploring marine cyanobacteria for lead compounds of pharmaceutical importance. *Sci. World J.*, 179(82): 1–10.

Yadav, S., Sinha, R.P., Tyagi, M.B. and Kumar, A. (2011) Cyanobacterial secondary metabolites. *Int. J. Pharma Bio Sci.*, 2(2): 144–167.

Yang, H., Lee, E. and Kim, H. (1997) *Spirulina platensis* inhibits anaphylactic reaction. *Life Sci.*, 61: 1237–1244.

Zeng, H., Lazarova, D.L. and Bordonaro, M. (2014) Mechanisms linking dietary fiber, gut microbiota and colon cancer prevention. *World J. Gastrointest. Oncol.*, 6(2): 41–51.

3

Seaweed Carotenoid, Fucoxanthin, as Functional Food

Nissreen Abu-Ghannam* and Emer Shannon

School of Food Science and Environmental Health, Dublin Institute of Technology, Dublin, Ireland

*Corresponding author e-mail: nissreen.abughannam@dit.ie

Fucoxanthin: Overview and Sources

Fucoxanthin is a light-harvesting carotenoid pigment that occurs in the chloroplasts of the eukaryotic Chromalveolata (phylum Heterokontophyta, class Ochrophyta), including brown macroalgae (Phaeophyceae), and in unicellular microalgae, such as diatoms (Bacillariophyceae) (Cavalier-Smith and Chao 2006). Fucoxanthin is estimated to account for more than 10% of the total production of carotenoids in nature, and is responsible for the brown to yellow color of brown macroalgae (seaweeds) and diatoms, where it masks green chlorophyll *a* and *c* (Hurd et al. 2014; Peng et al. 2011). Fucoxanthin was first isolated in Germany in 1914 from the brown seaweeds, *Dictyota*, *Fucus*, and *Laminaria* (Willstätter and Page 1914). Industrially, Japanese wakame (*Undaria pinnatifida*) is the seaweed most widely utilized for fucoxanthin extraction due to high concentrations of the pigment (≥10%) in the lipid extract (Billakanti et al. 2013). Other marine macroalgae known to contain fucoxanthin include the species *Ascophyllum, Fucus, Laminaria, Pelvetia, Ecklonia, Eisenia, Himanthalia, Sargassum, Saccharina, Ectocarpus, Schytosiphon, Petalonia, Carpophyllum, Hizikia, Padina, Dictyota, Myagropsis, Turbinaria, Cladosiphon, and Cystophora* (Dominguez 2013; Haugan and Liaaen-Jensen 1994; Heo et al. 2010; Jaswir et al. 2011; Kanazawa et al. 2008; Mikami and Hosokawa 2013; Mise et al. 2011; Miyashita and Hosokawa 2007; Sangeetha et al. 2010; Shang et al. 2011; Yan et al. 1999).

Notable fucoxanthin-containing microalgae include the diatoms *Odontella aurita, Phaeodactylum tricornutum, Chaetoceros gracilis, Thalassiosira*

Microbial Functional Foods and Nutraceuticals, First Edition. Edited by Vijai Kumar Gupta, Helen Treichel, Volha (Olga) Shapaval, Luiz Antonio de Oliveira, and Maria G. Tuohy.
© 2018 John Wiley & Sons Ltd. Published 2018 by John Wiley & Sons Ltd.

weissflogii, and *Cyclotella meneghiniana*; the Prymnesiophyceae species *Emiliania huxleyi*, *Pavlova lutheri*, and *Phaeocystis pouchetii*; and the Chrysophyceae species *Pelagococcus subviridis* (Di Valentin et al. 2013; Pyszniak and Gibbs 1992; United States Department of Agriculture 2015; Wright and Jeffrey 1987; Xia et al. 2013). These microscopic algae occur as marine plankton.

Globally, Japan, Korea, and China have the greatest seaweed production, consumption, and most developed fucoxanthin extraction industry (Ryan 2014). Irish coastlines also support the growth of fucoxanthin-containing brown seaweeds, such as bladderwrack (*Fucus vesiculosus*), channeled wrack (*Pelvetia canaliculata*), knotted wrack (*Ascophyllum nodosum*), oarweed (*Laminaria digitata*), sea rod (*Laminaria hyperborea*), sea spaghetti (*Himanthalia elongata*), serrated wrack (*Fucus serratus*), and sugar kelp (*Saccharina latissima*; formerly *Laminaria saccharina*) (Dominguez,2013; Morrissey et al. 2001; Stengel and Dring 1998).

Chemistry of Fucoxanthin

Carotenoids are tetraterpene pigments which occur in plants, algae, and photosynthetic bacteria. Carotenoids composed entirely of hydrogen and carbon belong to the subclass of carotenes, while those that additionally contain oxygen are classed as xanthophylls. Fucoxanthin is a xanthophyll, similar to the plant xanthophylls violaxanthin, neoxanthin, and lutein (Kotake-Nara and Nagao 2011). Xanthophylls share some chemical and physical properties with carotenes, such as lipophilicity and antioxidant activity due to their ability to quench reactive oxygen and nitrogen species (Kim and Chojnacka 2015). However, the presence of oxygen in the hydroxyl and epoxide groups of xanthophylls makes them more polar than carotenes (Landrum 2009). Fucoxanthin's systematic name is (3S,3′S,5R,5′R,6S,6′R,8′R)-3,5′-dihydroxy-8-oxo-6′,7′-didehydro-5,5′,6,6′,7,8-hexahydro-5,6-epoxy-β,β-caroten-3′-yl acetate. It contains an allelic bond, an epoxy group, and six oxygen atoms, as shown in Figure 3.1 (Chemspider

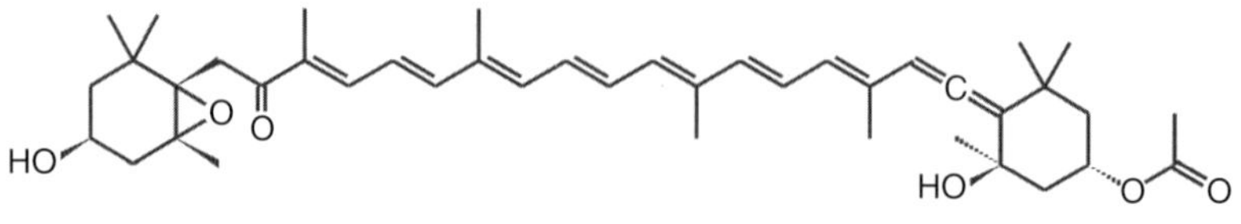

Figure 3.1 Fucoxanthin molecular structure (formula $C_{42}H_{58}O_6$).

2015). In industry, it is most commonly extracted with solvents such as *n*-hexane, methanol, DMSO, ethanol, petroleum ether, diethyl ether, dimethyl ether, acetone, or ethyl acetate, and dried to a powder (Kanda et al. 2014; Kim 2011b).

In algal cells, fucoxanthin is contained in the chloroplasts, within membrane-bound compartments called thylakoids. In the thylakoids, fucoxanthin binds with chlorophyll *a*, *c*, and apoproteins, forming complexes that absorb light in the blue-green region of the spectrum and transfer energy to the alga. At depths of several meters, this is commonly the only spectral wavelength available to marine algae (Kita et al. 2015). Fucoxanthin captures a broader spectrum of light (449–540 nm) than chlorophyll *a* and *c* alone, which increases the efficiency of photosynthesis (Kim et al. 2011; Pyszniak and Gibbs 1992). It also protects the algal cells from damage by reactive oxygen species caused by constant exposure to high levels of oxygen and light in the ocean. Fucoxanthin generally occurs more significantly in the blade of the seaweed thallus, which experiences the greatest light exposure, compared to the stipe and holdfast (Lobban and Wynne 1981).

Fucoxanthin can exist in a *trans* or *cis* configuration. The *trans* isomer is the more chemically stable and potent antioxidant of the two, and comprises ~90% of the fucoxanthin found in nature (Holdt and Kraan 2011; Nakazawa et al. 2009). Fucoxanthin content varies widely amongst macro- and microalgae. For example, the brown seaweed *Fucus serratus* has been reported to contain 0.56 mg/g (dry weight, DW) (Haugan and Liaaen-Jensen 1994) compared to 2.67 mg/g in *Undaria pinnatifida* (DW) (Mori et al. 2004). Most diatoms and other microalgae have a greater fucoxanthin content than brown seaweeds (Kawee-ai et al. 2013), but are less commonly used commercially for extraction due to the necessity for photobioreactors and strict culturing conditions. Xia et al. (2013) reported the fucoxanthin content in eight species of diatoms, ranging from 2.24 mg/g (DW) in *Chaetoceros gracilis* to 18.47 mg/g (DW) in *Odontella aurita*. Seasonal and geographic variations hugely affect content. For example, brown seaweeds harvested from September to March, during the mature phase of the sporophyte, commonly contain higher concentrations of fucoxanthin (Fung et al. 2013; Terasaki et al. 2009). This has been attributed to the upregulation of the xanthophyll, or violaxanthin, cycle pathway in reduced levels of sunlight during the winter. Under environmental stressors such as reduced light exposure, the formation of fucoxanthin from zeaxanthin via the epoxidation of antheraxanthin, violaxanthin, and diadionoxanthin may be accelerated to regulate photosynthetic pathways (Campbell et al. 1999; Goss and Jacob 2010; Mikami and Hosokawa 2013; Ramus et al. 1977).

Current Applications

Despite its discovery and isolation in seaweed over 100 years ago, fucoxanthin has remained somewhat underutilized in food and pharmaceutical applications. Initial studies focused on quantification and extraction methods, structural elucidation, and biosynthetic pathways. It was not until the late 1990s that scientific papers began to emerge on fucoxanthin's potential as a functional food. This was most probably due to growing clinical evidence at the time of the role of antioxidants in the prevention of chronic diseases. Currently, fucoxanthin is available to retail consumers in the form of relatively expensive weight loss supplements, of varying purity and quality, from health stores and online. Unlike other carotenoids, such as β-carotene, which are commonly used as food colorants. 100% pure fucoxanthin is not sold as a bulk food ingredient (Hurst 2002). Its instability due to oxidation and high extraction costs have, to date, been prohibitive. Currently, fucoxanthin is available to food producers as a percentage of various seaweed extracts. Analytical grade fucoxanthin (≥95% pure) is produced for laboratories, but retails at €556 per 50 mg (Sigma-Aldrich 2015a).

Food and Pharmaceutical Regulations

In 2009, the European Food Safety Authority (EFSA) published its scientific opinion on fucoxanthin and the substantiation of health claims related to it and maintenance or achievement of a normal body weight, in the case of *Undaria pinnatifida* thallus extract, under Article 13(1) of Regulation (EC) No 1924/2006. It determined that the extract, as a food constituent, had been sufficiently characterized to be consumed in an amount equivalent to 15 mg pure fucoxanthin per day. However, to date, the EFSA has not accepted that a relationship has been established between its consumption and the maintenance or achievement of a normal body weight (European Food Safety Authority 2009). Other health claims, such as antidiabetic or anticancer effects, relating to fucoxanthin have not yet been evaluated by the EFSA. In the Republic of Ireland, the Health Products Regulatory Authority does not list fucoxanthin as a controlled substance, or include it in the *List of Medicinal Herbs considered acceptable as THMPs*, under the European Traditional Herbal Medicinal Products Directive (2004/24/EC) which came into effect in Ireland in 2007 (Health Products Regulatory Authority 2015).

The clinical evidence supporting fucoxanthin's health benefits has not yet been evaluated by the US Food and Drug Administration. However, fucoxanthin can be sold in the United States as a food supplement with the disclaimer "These statements have not been evaluated by the Food and Drug Administration. This product is not intended to diagnose, treat, cure, or prevent any disease" on the product label (US Food and Drug Administration 2014). In Japan, Food for Specified Health Uses (FOSHU) under the Ministry of Health, Labor and Welfare have not evaluated fucoxanthin as a functional food ingredient or pharmaceutical; however, it can be sold as a food supplement.

Applications in Human Health

Although fucoxanthin has not been evaluated by some food regulatory bodies, in recent years it has been studied clinically for its antioxidant properties, which inhibit free radical damage in cells, reducing the risk of many chronic diseases. It has also been studied for its anticancer, anti-type 2 diabetes, antiobesity, anticholesterol, anti-inflammatory, antiangiogenic, antimalarial, and antihypertensive activities, and for the treatment of Alzheimer's disease (Gammone and d'Orazio 2015; Hosokawa et al. 1999; Ikeda et al. 2003; Kawee-ai et al. 2013; Kim 2011a; Kotake-Nara et al. 2001; Maeda et al. 2007; Rodrigues et al. 2012; Shiratori et al. 2005; Sivagnanam et al. 2015).

Antiobesity Effects

Obesity, type 2 diabetes, metabolic syndrome, and chronic inflammatory diseases are global health epidemics. The World Health Organization estimates that, globally, 2.3 billion people will be overweight and 700 million obese by 2015. To prevent and treat diseases such as these, natural, bioactive, functional compounds such as fucoxanthin are increasingly being studied as an alternative to, or as combination therapy with, orthodox medicines (Peng et al. 2011; Watson, 2014).

In obesity treatment, fucoxanthin has been shown to mediate the induction of uncoupling protein-1 (UCP-1) in abdominal adipose tissue mitochondria in murine studies, leading to the oxidation of fatty acids and heat production, resulting in a reduction in white adipose tissue (d'Orazio et al. 2012; Maeda et al. 2005). Abidov et al. (2010) conducted a double-blind placebo-controlled study of 115 non-diabetic, obese, premenopausal women with a liver fat content above 11% at the Russian Academy of Medical Sciences. A daily supplement of 300 mg brown seaweed extract (species not specified) containing 2.4 mg fucoxanthin, combined with 300 mg pomegranate seed oil, was administered. An olive oil capsule was administered to the placebo group. The treatment group showed a significant increase in resting energy expenditure and mean weight loss of 4.9 kg after 16 weeks.

In Japan, a kombu (*Laminaria japonica*) extract of fucoxanthin (3%) was evaluated for its antimetabolic syndrome effects in human clinical trials. A daily dosage of the extract, equivalent to 0.5–1.0 mg pure fucoxanthin/day, was found to have a significant effect on blood serum parameters related to metabolic syndrome (Oryza 2015). However, no official RDA for fucoxanthin has been established by the World Health Organization.

Topical preparations as vehicles for fucoxanthin in the treatment of obesity have also been reported. Dai et al. (2014) developed a stable microemulsion containing 0.25% pure fucoxanthin using medium chain triglyceride as the oil phase, Tween 80 as a surfactant, and polyethylene glycol 400 as a co-surfactant.

Anticancer Effects

Fucoxanthin's anticancer activity is hypothesized to be due to its ability to induce apoptosis in tumor cells (Nakazawa et al. 2009). Hosokawa et al. (1999) found that fucoxanthin induced apoptosis in human promyelocytic leukemia HL-60 cells by cleaving procaspase-3 and poly-ADP-ribose polymerase. Kim et al. (2010) showed that fucoxanthin induced reactive oxygen species generation, inactivated the Bcl-xL signaling pathway, and induced caspase-3, -7, and poly-ADP-ribose polymerase cleavage, triggering the apoptosis of HL-60 cells, indicating that the generation of reactive oxygen species was a critical target in fucoxanthin-induced apoptosis of these cells. Wang et al. (2014) reported significant growth inhibition of cells in nine human cancer cell lines with extracts of *Undaria pinnatifida* containing fucoxanthin. Satomi and Nishino (2013) reported fucoxanthin extract to have a significant effect on the expression and enzymatic activity of the xenobiotic metabolizing enzymes CYP1A1, CYP1A2 and CYP3A4, which are involved in the activation of pro-carcinogens. The study found that the inhibitory effect of fucoxanthin ($\leq 45\,\mu M$) on these enzymes in human hepatocellular carcinoma HepG2 cells and recombinant human CYPs could also attenuate the action of some anticancer drugs that are normally activated by CYP3A4.

Antidiabetic Effects

Fucoxanthin's antidiabetic activity has been studied in mice with induced type 2 diabetes. It has been found to improve insulin resistance and decrease blood glucose levels mainly via the regulation of cytokine secretions from white adipose tissue (Miyashita et al. 2011). Other studies found that a fucoxanthin-enriched diet promoted the recovery of blood glucose uptake to muscle by the upregulation of GLUT4 mRNA expression. Fucoxanthin has also been shown to affect the peroxisome proliferator-activated receptor γ (PPARγ) and promote gene expression related to lipid metabolism in adipocytes. In cultivated cells, fucoxanthin prevented inflammation and insulin resistance by inhibiting nitric oxide and PGE2 production through the downregulation of iNOS and COX-2 mRNA expression, as well as adipocytokine production in white adipose tissue (Hosokawa et al. 2010; Maeda et al. 2006; Miyashita et al. 2011).

Dietary Antioxidant Effects

The antioxidant capacity of seaweeds has been widely reported (Heffernan et al. 2014). As a dietary antioxidant, fucoxanthin has been shown to improve the antioxidant capacity of blood serum levels in mammals. Fucoxanthin is unusual in that it donates an electron to reactive oxygen species, instead of a proton (hydrogen), as most antioxidants such as ascorbic acid or β-carotene do. Fucoxanthin can also quench reactive oxygen species under hypoxic physiological conditions, unlike the majority of food-derived antioxidants (Nomura et al. 1997; Yan et al. 1999). A high-fat diet has been associated with obesity in humans and other mammals, and has been shown to cause overproduction of reactive oxygen species (Dandona et al. 2005). Reactive oxygen species are known to cause cellular damage, which is implicated in the pathogenesis of diseases such as type 2 diabetes, cardiovascular disease, cancer, and infectious illnesses (Uzun et al. 2004). Ha et al. (2013) reported that fucoxanthin supplementation improved the antioxidant capacity of blood serum levels in obese rats via activation of the nuclear erythroid factor like-2 pathway and its downstream target gene NQO1. A study by Zaragozá et al. (2008) on the antioxidant effect of fucoxanthin extract from *Fucus vesiculosus* found that the extract exhibited increased antioxidant activity in *ex vivo* assays of erythrocytes and plasma, after 4 weeks of daily oral administration in rats. Significant antioxidant activity was also observed in non-cellular systems and in activated RAW 264.7 mouse leukemic monocyte macrophage cell lines. Therefore, supplementation of fucoxanthin may also reduce the risk of oxidative stress in humans.

Fucoxanthin has also been successfully used topically to protect against UV-B-induced cell damage in hairless mice, and in human fibroblast cell lines as an antioxidant against skin aging caused by free radical damage (Heo and Jeon 2009; Urikura et al. 2011).

Toxicity Studies

No toxicity of fucoxanthin extracts has been reported to date, making it a good candidate for functional food use. A number of clinical trials in animal models have shown no significant toxicity with short- or long-term dosage. For example, in Japan, a recently developed functional food ingredient, containing up to 5% fucoxanthin from kombu (*Laminaria japonica*) extract, was evaluated on rats. Toxicity and micronucleus tests were conducted. No toxicity or abnormalities were found after 14 or 90 days. The LD_{50} of the extract (3.0% fucoxanthin) was calculated to be 2000 mg/kg body mass for rats (Oryza 2015). Maeda et al. (2005) supplemented a murine diet with 0.27% fucoxanthin, equivalent to ~0.25 mg/kg body mass per day for 4 weeks, and found no side effects or abnormalities. Kadekaru et al. (2008) conducted a toxicity study on the repeated oral dosing of fucoxanthin (95% purity) to rats for 28 days, and found that it showed no apparent toxicity. Zaragozá et al. (2008) found no ill effects following a 4-week, acute toxicity test in rats, where a daily treatment of 0.0012% pure fucoxanthin extract from *Fucus vesiculosus* was administered. A single dose toxicity study (Beppu et al. 2009) was conducted with doses of 1000 and 2000 mg/kg body mass and a repeated oral dose toxicity study with doses of 500 and 1000 mg/kg for 30 days on purified fucoxanthin (93% purity) in ICR mice. No mortality, abnormalities, abnormal changes in liver, kidney, spleen, or gonadal tissues were found in either study.

Human toxicity studies are required to assess both the toxicity levels and daily dosage for efficacy against any of the disorders discussed above.

Fucoxanthin as a Functional Food: Challenges and Opportunities

Fucoxanthin faces chemical, organoleptic, and bioavailability challenges as a functional food ingredient. Due to its chemical structure, fucoxanthin in its pure form is easily oxidized by high temperatures, low or high pH, UV light, and long storage periods (Kawee-ai et al. 2013; Mise et al. 2011). This may lead to chemical or enzymatic interactions with other ingredients over time. Organoleptic attributes, such as texture, taste, appearance, or smell, may deteriorate as a result of these interactions. Sensitivity to heat poses a problem for bakery products or sauces that require boiling, and foods that contain fruit or probiotic cultures may be too acidic. The natural brown color, savory flavor, and powdery texture of fucoxanthin itself may also make it unsuitable for some food products. The idea of eating seaweed derivatives may be unpalatable to some consumers, even at undetectable levels. This may be encountered in Western cultures where seaweed is rarely part of the diet, and misconceptions of a potential "fishy" or "sea" taste could arise.

Like other non-polar pigments, such as chlorophyll and lycopene, fucoxanthin is insoluble in water. Incorporation into water-based beverages or sauces would require emulsification or dispersion in appropriate colloids (Socaciu 2007). The lipid-soluble nature of fucoxanthin also affects its bioavailability in mammals (Sangeetha et al. 2010). Fucoxanthin is metabolized in the intestine into fucoxanthinol by cholesterol esterase and lipase, then converted to amarouciaxanthin A in the liver (Bagchi and Preuss 2012; Dominguez, 2013). The presence of some form of dietary lipid is required when fucoxanthin is consumed for solubility and absorption (Peng et al. 2011). Another consideration for fucoxanthin as a functional food is the high cost, due to the energy required for the extraction and freeze-drying process (Billakanti et al. 2013) and the lack of an artificial synthesis method for fucoxanthin. Seasonal variations in fucoxanthin nutritional content (Fung et al. 2013) could also affect health claims and nutritional efficacy.

Despite these potential hurdles, fucoxanthin has several intrinsic, beneficial properties. Fucoxanthin, along with other carotenoids such as astaxanthin, is a more powerful antioxidant than many other natural and synthetic antioxidants (Miyashita and Hosokawa 2007). Using chemiluminescence detection, Nishida et al. (2007) reported that fucoxanthin had stronger singlet oxygen-quenching activities than ascorbic acid, α-tocopherol, quercetin, resveratrol, (-)-epigallocatechingallate, lutein, lycopene, gallic acid, pyrocatechol, α-lipoic acid, and the synthetic antioxidant butylated hydroxytoluene. Consumer awareness and interest in naturally antioxidant-rich foods have grown in recent years (Kim 2013; Rodrigues et al. 2012). Seaweed extracts such as carrageenan, alginate, and other hydrocolloids are already accepted by consumers as widely used ingredients in the food, pharmaceutical, and cosmetic industries (Venugopal 2011). Marine-derived functional foods such as fucoxanthin have the potential to be marketed as a more potent and sustainable alternative to many natural and synthetic antioxidants, particularly in countries such as Japan, China, and Korea, where seaweed is a common part of the diet. Fucoxanthin is also Kosher, Halal, and suitable for vegetarians and vegans.

As discussed later in this chapter, little has been reported in peer-reviewed journals regarding the use of fucoxanthin as a functional ingredient. Most fucoxanthin research has focused on medical rather than functional food applications. However, a study conducted by Prabhasankar et al. (2009) at the Central Food Technological Research Institute, Mysore, in India reported the successful addition of fucoxanthin and fucosterol as a constituent of wakame (*Undaria pinnatifida*) powder extract into semolina (wheat)-based pasta. Fucosterol is a structural component of algal lipid membranes which has been shown to have antioxidant, anticancer, antidiabetic, and hepatoprotective properties in animal trials (Alasalvar et al. 2011; Jung et al. 2013; Lee et al. 2003). Prabhasankar et al. (2009) investigated the effect of different percentage additions of wakame powder on the sensory, cooking, nutritional, and biofunctional quality of pasta. Blends of semolina and seaweed were combined by replacement method in ratios of (semolina/wakame, w/w) 100:0, 95:5.0, 90:10, 80:20, and 70:30. In sensory analysis of taste, mouth-feel, appearance, and strand quality, 15 semi-trained panellists, who were regular consumers of wakame, found no significant ($p > 0.05$) or discernible organoleptic differences between the control and the wakame pasta up to 10% total ingredient mass, with acceptance decreasing after 10% up to 20% wakame content. Above 20%, panellists reported saltiness and a seaweed taste. Since wakame is composed of other components such as polysaccharides, proteins, lipids, and minerals, 10% wakame, in this study, equated to 0.04 mg/g (DW) of fucoxanthin and 1.25 mg/g (DW) of fucosterol of the dry ingredient portion. HPLC analysis of the pasta after processing/ kneading, and after cooking (25 g of raw pasta in 250 mL boiling water for 8 min) showed a loss of less than 10% for both fucoxanthin and fucosterol. The authors hypothesized that this remarkable preservation occurred due to stability of fucoxanthin/fucosterol in the protein matrix of gluten. To the authors' knowledge, the stability of fucoxanthin in a food system has not been reported in any other studies. To date, these lab-scale pilots have not been reported as developed to commercial scale.

A recent Irish study reported the successful incorporation of ethanol and water extracts of *Fucus vesiculosus* and *Ascophyllum nodosum*, at 0.25% and 0.50%, into yoghurt and fluid milk (O'Sullivan 2013). The fucoxanthin content of the seaweed extracts was not reported. However, it is probable that both contained fucoxanthin, due to its presence being widely reported in both species and their extracts (Stengel and Dring 1998; Zaragozá et al. 2008). Sensory analysis found that overall acceptability of the yoghurts was governed by appearance and flavor, with the 100% water extract of *A. nodosum* having the greatest panel preference and least yellowness. Overall acceptability of the milk was governed by perception of a fishy flavor. Again, the 100% water extract of *A. nodosum* (0.50% addition) was found to be the most acceptable in terms of taste, and the least green/yellow. *In vitro* antioxidant analysis found no deterioration of antioxidant activity, shelf-life, or pH in the seaweed-supplemented milk and yoghurt formulations.

Fucoxanthin, like many natural food extracts, is widely available wholesale online from many companies, primarily based in China. It is sold in the form of dried seaweed extract, generally from wakame or kombu, with stated percentages of fucoxanthin purity ranging from 10% to 98%. Price ranges widely, from less than $1 to $2000 per gram, as do claims of purity and certification (Alibaba 2015; Kyndt and d'Silva 2013).

However, in Japan, a government-certified functional food ingredient has recently been commercially developed from kombu (*Laminaria japonica*) containing 1–5% fucoxanthin. The powder and oil products contain only cyclodextrin, or triglyceride, in addition to the kombu extract, and natural tocopherol. Stability testing of the products found the fucoxanthin fraction to be thermostable up to 80 °C for 1 hour, and stable in solution from pH 3.0 to 10.0, with the greatest loss being 6% after 1 week at pH 3.0. The addition of the antioxidant preservative tocopherol to the products is most likely the stabilizing factor in this case. The triglyceride and cyclodextrin may also offer a protective matrix for the fucoxanthin, combined with photo-protective, vacuum packaging. The powder and oil products were reported to have been successfully incorporated into beverages, cakes, shortbread, puffed rice biscuits, spreads, and potato snacks. However, the percentage of fucoxanthin extract used was not specified, nor have any sensory evaluation results been published to date (Oryza 2015).

Outside Asia, in the USA, a project was initiated in 2010 by the Research, Education, and Economics Information System of the Department of Agriculture and the National Institute of Food and Agriculture, Auburn University, Alabama, to optimize large-scale fucoxanthin extraction from the diatom *Chaetoceros gracilis*. The effects of the extracts on energy balance in an animal model of obesity are being used to develop a functional food for the public health treatment of obesity to reduce the risk of developing cardiovascular disease, type 2 diabetes, and some forms of cancer. The project intends the functional food industry to be the immediate beneficiary of the study, but also aims to target the wider food industry, nutritionists, scientists, and engineers (United States Department of Agriculture 2015). There have been no publications from the project to date.

This significant project, in the country with the highest global rates of obesity, along with the success of the pasta study and Japan's government-sponsored development of stable oil- and water-soluble fucoxanthin products, is very encouraging for further exploration of fucoxanthin as a functional food ingredient on a large commercial scale.

Approaches to Overcome Adverse Reactions in Functional Food Models

Micro- and Nanoencapsulation

The principal obstacles to the incorporation of fucoxanthin in a food or beverage matrix are water insolubility, pH instability, sensitivity to oxidation, and impaired bioavailability. These properties are an issue with existing food ingredients, such as other carotenoids and polyphenolic compounds, but can be overcome with various technologies and approaches. Microemulsions composed of a water phase, lipid phase, and an amphiphilic compound are widely used to combine lipid solutes, such as carotenoids, into a hydrophilic matrix for food and pharmaceutical purposes (de Campos et al. 2012).

Indrawatia et al. (2015) encapsulated an acetone extract (primarily *trans* fucoxanthin) of Indonesian *Sargassum* species in maltodextrin and Tween 80 (~70:1). The *Sargassum* extract was combined with canola oil and homogenized in the water-based maltodextrin/Tween 80 emulsion. After freeze drying, the fucoxanthin was found to be stable within the microencapsulates for 63 days at 28 °C under inert atmosphere. Encapsulating fucoxanthin in this manner increases its suitability for incorporation into dried food products. Suhendra et al. (2012) succeeded in formulating a clear, stable, oil-in-water microemulsion for fucoxanthin, capable of delivering this hydrophobic antioxidant in aqueous food systems. Virgin coconut oil was used as the lipid phase with a combination of Tween 20, Tween 80, and Span 80 as non-ionic surfactants. The ratios were oil:surfactants (3:17); oil + surfactants:water (35:65); Tween 80:Tween 20:Span 80 (92.0:2.5:5.5). The microemulsion remained stable after exposure to pH 3.5–6.5, 105 °C for 5 hours, and centrifugation at 4500 rpm for 30 minutes. Quan et al. (2013) reported a water-soluble fucoxanthin food application using fucoxanthin-loaded microspheres, composed of a gum arabic/fish gelatin coacervate shell cross-linked by tannic acid, with a solid lipid core of acetyl palmitate and canola oil. Wet and freeze-dried forms of the solid lipid core microspheres were developed successfully. Stability of encapsulated fucoxanthin during long-term storage was significantly increased, as was sustained release in a simulated gastrointestinal environment.

Nanogels have been used extensively in the pharmaceutical industry to protect acid-labile bioactive compounds from the acidic stomach environment before reaching the intestinal tract (Liechty et al. 2010; McClements et al. 2009a). This technology easily translates to functional food delivery. For example, the biological availability and stability of fucoxanthin were reported to have significantly increased after encapsulation with chitosan-sodium-tripolyphosphate-glycolipid nanogels, prepared by ionic gelation (Ravi and Baskaran 2015). The authors were inspired by a study (Gorusupudi and Baskaran 2013) reporting the improved bioavailability of a similar compound, lutein, in mice, by solubilizing the extract first in the glycolipid fraction of wheat germ oil. Apart from protecting the core material (fucoxanthin) against degradation, encapsulation prevents reactions with other ingredients (McClements et al. 2009b), and any seaweed flavor or brown color is prevented from leaching into the food.

Current Trends in Fucoxanthin Research

Most extraction methods currently practiced are based on the utilization of organic acids and solvents to break down the cell walls of seaweed or microalgae. Current trends are leaning towards greener chemical or physical extraction methods.

Green Extraction Technologies

Pressurized Liquid Extraction

Shang et al. (2011) utilized pressurized liquid extraction with ethanol to optimize fucoxanthin yields from the seaweed *Eisenia bicyclis*. A yield of 0.39 mg/g was achieved with 90% ethanol, 110 °C, 1500 psi, for 5 minutes of extraction static time. This was close to the predicted statistical experimental yield of 0.42 mg/g.

Supercritical Carbon Dioxide Extraction

Supercritical carbon dioxide extraction is another green chemical method with potential for commercial seaweed applications. Sivagnanam et al. (2015) extracted fucoxanthin from the brown seaweeds *Sargassum horneri* and *Saccharina japonica* using supercritical CO_2 with ethanol as a co-solvent (SC-CO_2E) for 2 hours at 45 °C in a semi-batch flow extraction process. SC-CO_2E extraction yield of fucoxanthin from *S. horneri* was 0.77 mg/g, compared to only 0.71 mg/g from traditional acetone-methanol extraction. SC-CO_2E extraction fucoxanthin yield from *S. japonica* was 0.41 mg/g, close to the acetone-methanol extraction yield of 0.48 mg/g. An *in vitro* hydrogen peroxide scavenging antioxidant activity assay showed the SC-CO_2E fucoxanthin extracts from *S. horneri* and *S. japonica* had significantly greater activity, or maximal inhibitory concentration (IC_{50} value), than acetone-methanol, hexane, or ethanol extracts. The IC_{50} values of the SC-CO_2E extracted *S. horneri* fucoxanthin (686 µg/mL) and *S. japonica* (600 µg/mL) were significantly greater than that of an ascorbic acid standard (448 µg/mL). SC-CO_2E fucoxanthin extracts from both species were also found to exert angiotensin I-converting enzyme inhibitory effects *in vitro* comparable with those of the acetone-methanol extract.

Quitain et al. (2013) used supercritical carbon dioxide extraction with *Undaria pinnatifida*, combined with a microwave pretreatment, to disrupt the cell membrane. Yields equivalent to 80% of those obtained with traditional solvent-based methods were reported using 40MPa, for 180 minutes, at 40 °C, without producing any chemical waste. Roh et al. (2008) extracted fucoxanthin and polyphenols from *Undaria pinnatifida* using supercritical carbon dioxide, with ethanol as a co-solvent. However, fucoxanthin yields were significantly lower than those obtained with solvent extracts, ranging from 0.00048 to 0.00753 μg/g. Optimum supercritical extraction was achieved after 50 minutes with a flow rate of 28.17 g CO_2/min, 2 mL/min ethanol, at a pressure of 200 bar, and a temperature of 323 K (49.85 °C). Subramanian et al. (2013) optimized a novel, ultrasound-assisted extraction method, using *Sargassum muticum*. Fucoxanthin yields of 0.613 mg/g were reported with optimum parameters of 70% ethanol, at 80 °C, for 10 minutes, at an amplitude of 78 W, with a solid to solvent ratio of 1:5 w/v g/mL.

Microwave-Assisted Extraction

Xiao et al. (2012) developed a rapid (75 minute) microwave-assisted extraction method combined with high-speed counter-current chromatography, with a two-phase solvent system of hexane-ethyl acetate-ethanol-water (5:5:6:4, v/v/v/v). Fucoxanthin yields close to those of traditional methods were obtained from *Laminaria japonica* (0.83 mg/g), *Undaria pinnatifida* (1.09 mg), and *Sargassum fusiforme* (0.20 mg/g). HPLC showed the extracts to have a minimum purity of 90%.

Enzyme Extraction

Fucoxanthin extraction through the application of targeted enzymes, such as cellulases, has the potential to increase yield and safety. The presence of branched, sulfated, or complex polysaccharides, such as alginate and laminarin, in algal cell walls limits the efficiency of classic extraction methods (Kim 2011c; Kim and Chojnacka 2015). Enzymolysis, i.e., the hydrolysis (of cell wall polysaccharides) with enzymes, should, in theory, aid in degradation of the wall and release of the pigment-containing chloroplasts within. Enzyme-assisted extraction has already been used successfully to increase carotenoid yields from terrestrial plants, such as lycopene from tomatoes. Zuorroa et al. (2011) reported an 8–18-fold increase in lycopene yield from tomato processing waste (skins) in pectinase and cellulase pretreated samples, compared to hexane extraction alone. Barzana et al. (2002) reported a similar result for the extraction of carotenoids (primarily lutein) from marigold flowers. Typically, as much as 50% of total carotenoid yield is lost in the traditional drying and hexane extraction of marigold flowers. A pretreatment of Pectinex, Viscozyme, Neutrase, Corolase, and HT-Proteolytic commercial enzymes was used to significantly reduce the volume of hexane required, and completely eliminate the drying and silage steps that cause degradation of the carotenoids. Carotenoid yields of ≥85% were recovered from the marigold flowers in simple stirring vessels. Barzana et al. (2002) hypothesized that if cost were no barrier, the hexane could be omitted entirely, and enzymes alone, with the correct parameters, would achieve the same yield.

In the extraction of algal bioactives such as antioxidants, polysaccharides, carotenoids, and polyphenols, the use of enzymes has shown significant potential as a viable alternative, or addition, to pure solvent methods (Je et al. 2009; Rhein-Knudsen et al. 2015; Wijesinghe and Jeon 2012). Heo et al. (2005) extracted water-soluble antioxidant compounds from *Sargassum, Ecklonia, Ishige,* and *Schytosiphon* using proteases (Neutrase, Protamex, Alcalase, Flavourzyme, and Kojizyme) and carbohydrate-degrading enzymes (Celluclast, Viscozyme, AMG, Ultraflo, and Termamyl). *In vitro* hydrogen peroxide scavenging activity assays showed the majority of the enzyme extracts had significantly greater activity than the commercial synthetic antioxidants α-tocopherol, butylated hydroxyanisole (BHA), and butylated hydroxytoluene (BHT). For example, the hydrogen peroxide scavenging activity of Ultraflo-extracted *Sargassum horneri* was 92.69%, and Kojizyme-extracted *Ishige okamurae* 96.27%, compared to only 50.32% for BHT, 67.37% for BHA, and 64.11% for α-tocopherol.

Ko et al. (2010) significantly increased the antioxidant activity of *Sargassum coreanum* extracts using Celluclast and Neutrase commercial enzymes. Celluclast is a cellulase derived from the fungus *Trichoderma reesei* which hydrolyzes (1,4)-β-D-glucosidic linkages in cellulose and other β-D-glucans, forming cellobiose glucose and other glucose polymers (Sigma-Aldrich 2015b). Neutrase is a metalloprotease derived from *Bacillus amyloliquefaciens.* Similar to thermolysin, it hydrolyzes proteins into peptides (Nagodawithana and Reed 2013). The *S. coreanum* Neutrase and Celluclast extracts had significantly greater DPPH and hydrogen peroxide radical scavenging activities compared to eight other proteases and cellulases studied. The *S. coreanum* extracts were screened for cancer cell inhibition and found to suppress the growth of HL-60 cells through apoptosis (Ko et al. 2012). Ahn et al. (2012) reported obtaining antioxidant-rich extracts from the chlorophyte (green alga) *Enteromorpha prolifera* using the protease Protamex and the carbohydrase mix Viscozyme. The extracts contained up to 8.4 mg/g total flavonoids and 4.5 mg/g total polyphenols. Other *E. prolifera* extracts were produced using Flavourzyme, an exopeptidase that hydrolyzes N-terminal peptide bonds, and Promozyme, a pullulanase (α-dextrin endo-1,6-α-glucosidase) derived from *Bacillus acidopullulyticus* (Mehta et al. 2012). The Flavourzyme and Promozyme extracts were found to have a significant angiotensin-converting enzyme inhibitory effect at concentrations of 1.0 mg/mL. No toxicity was exerted by any of the enzyme-assisted extracts in RAW264.7 cell cytotoxicity tests.

Little has been published on the use of enzymes for fucoxanthin extraction specifically. Billakanti et al. (2013) reported a significant increase (9.3%) in fucoxanthin extraction yields from wakame (*Undaria pinnatifida*) using alginate lyase derived from *Flavobacterium multivorum* as a preprocessing step, followed by dimethyl ether and ethanol extraction, compared to untreated preprocessing. Optimum enzyme pretreatment parameters were found to be 37 °C, for 2 hours, at pH 6.2, 5% (w/v) solids, with 0.05 wt% enzyme using continuous mixing. Centrifugation was used to separate hydrophilic hydrolysis products from the residual seaweed biomass. Alginate lyase, also known as mannuronate lyase, may have succeeded here where other enzymes have little effect due to its ability to catalyze the hydrolysis of alginate in the cell wall into smaller oligosaccharides, aiding extraction of fucoxanthin from the chloroplasts. Alginate lyase cleaves β-(1-4)-D-mannuronic bonds to yield oligosaccharides with 4-deoxy-α-L-erythro-hex-4-enopyranuronosyl groups at the non-reducing terminus (Sigma-Aldrich 2015b). Enzymes such as cellulose, protease or pectinase cannot break the glycosidic bonds specific to the β-D-mannuronate in alginate polysaccharides that occur in macroalgae (Sho et al. 2010).

Currently, this research institute is screening a number of commercially available Irish seaweed species for their fucoxanthin content, using novel extraction technologies. The species under investigation include wild Atlantic wakame (*Alaria esculenta*), bladderwrack (*Fucus vesiculosus*), channeled wrack (*Pelvetia canaliculata*), knotted wrack (*Ascophyllum nodosum*), oarweed (*Laminaria digitata*), sea rod (*Laminaria hyperborea*), sea spaghetti (*Himanthalia elongata*), serrated wrack (*Fucus serratus*), furbellows (*Saccorhiza polyschides*), and sugar kelp (*Saccharina latissima*).

Conclusion

Fucoxanthin is a bioactive compound found in one of the most prolific and sustainable organisms on the planet, alga (Mohamed et al. 2012; Werner et al. 2004). Its efficacy and potential in terms of health applications have been widely reported in clinical studies. However, further human clinical trials are necessary to determine the safety and required daily dosage of fucoxanthin. Technical modifications, such as encapsulation, and sensory trials must be undertaken before fucoxanthin can be successfully utilized as a functional food ingredient. Factors to consider include solubility in the food matrix, organoleptic effects, stability, preservation against oxidation, consumer acceptability, bioavailability, and toxicity risk. The current, prohibitive cost of fucoxanthin must also be addressed. Possible solutions may include the development of more efficient and greener extraction technologies, which require shorter extraction times and less solvent, and have a more specific and higher extraction yield. The sustainability of potential seaweed cultivars must be assessed before large-scale harvesting, to ensure the preservation of this precious marine resource.

References

Abidov, M., Ramazanov, Z., Seifulla, R. and Grachev, S. (2010) The effects of Xanthigen in the weight management of obese premenopausal women with non-alcoholic fatty liver disease and normal liver fat. *Diabet. Obesity Metab.*, 12: 72.

Ahn, C.B., Park, P.J. and Je, J.Y. (2012) Preparation and biological evaluation of enzyme-assisted extracts from edible seaweed (*Enteromorpha prolifera*) as antioxidant, anti-acetylcholinesterase and inhibition of lipopolysaccharide-induced nitric oxide production in murine macrophages. *Int. J. Food Sci. Nutr.*, 63(2): 187–93.

Alasalvar, C., Miyashita, K., Shahidi, F. and Wanasundara, U. (2011) *Handbook of Seafood Quality, Safety and Health Applications*. Hoboken: John Wiley and Sons.

Alibaba (2015) *Fucoxanthin Manufacturers and Suppliers Directory.* Available at: www.alibaba.com/products/F0/extract_fucoxanthin.html? spm=a2700.7724838.0.0.9B9Vbz (accessed 27 April 2017).

Bagchi, D. and Preuss, H.G. (2012) *Obesity: Epidemiology, Pathophysiology, and Prevention*, 2nd edn. Boca Raton: CRC Press.

Barzana, E., Rubio, D., Santamaria, R.I. et al. (2002) Enzyme-mediated solvent extraction of carotenoids from marigold flower (*Tagetes erecta*). *J. Agric. Food Chem.*, 50: 4491–4496.

Beppu, F., Niwano, Y., Tsukui, T., Hosokawa, M. and Miyashita, K. (2009) Single and repeated oral dose toxicity study of fucoxanthin (FX), a marine carotenoid, in mice. *J. Toxicol. Sci.*, 34(5): 501–510.

Billakanti, J.M., Catchpole, O., Fenton, T. and Mitchell, K. (2013) Enzyme-assisted extraction of fucoxanthin and lipids containing polyunsaturated fatty acids from *Undaria pinnatifida* using dimethylether and ethanol. *Process Biochem.*, 48: 1999–2008.

Brodie, J. and Lewis, J. (2007) *Unravelling the Algae: The Past, Present, and Future of Algal Systematics*. Systematics Association Special Volumes. Boca Raton: CRC Press.

Campbell, S.J., Bite, J.S. and Burridge, T.R. (1999) Seasonal patterns in the photosynthetic capacity, tissue pigment and nutrient content of different developmental stages of *Undaria pinnatifida* (Phaeophyta: Laminariales) in Port Phillip Bay, south-eastern Australia. *Botan. Mar.*, 42: 231–241.

Cavalier-Smith, T. and Chao, E. (2006) Phylogeny and megasystematics of phagotrophic heterokonts (kingdom Chromista). *J. Mol. Evol.*, 62(4): 388–420.

Chemspider (2015) *Fucoxanthin*. Available at: http://www.chemspider.com/ Chemical-Structure.21864745.html (accessed 22 April 2017).

Dai, J., Kim, S.M., Shin, I.S. et al. (2014) Preparation and stability of fucoxanthin-loaded microemulsions. *J. Indust. Eng. Chem.*, 20(4): 2103–2110.

Dandona, P., Aljada, A., Chaudhuri, A., Mohanty, P. and Garg, R. (2005) Review. Metabolic syndrome: a comprehensive perspective based on interactions between obesity, diabetes, and inflammation. *Circulation*, 111(11): 1448–1454.

De Campos, V., Ricci-Júnior, E. and Mansur, C. (2012) Nanoemulsions as delivery systems for lipophilic drugs. *J. Nanosci. Nanotechnol.*, 12(3): 2881–2890.

Di Valentin, M., Meneghin, E., Orian, L. et al. (2013) Triplet–triplet energy transfer in fucoxanthin-chlorophyll protein from diatom *Cyclotella meneghiniana*: insights into the structure of the complex. *Biochim. Biophys. Acta Bioenerget.*, 1827(10): 1226–1234.

Dominguez, H. (2013) *Functional Ingredients from Algae for Foods and Nutraceuticals*. Cambridge: Woodhead Publishing.

D'Orazio, N., Gemello, E., Gammone, M.A., de Girolamo, M., Ficoneri, C. and Riccioni, G. (2012) Fucoxanthin: a treasure from the sea. *Marine Drugs*, 3: 604–616.

Englert, G., Bjørnland, T. and Liaaen-Jensen, S. (1990) 1D and 2D NMR study of some allenic carotenoids of the fucoxanthin series. *Magn. Reson. Chem.*, 28: 519–528.

European Food Safety Authority (2009) Scientific opinion on the substantiation of health claims related to *Undaria pinnatifida* (Harvey) Suringar and maintenance or achievement of a normal body weight (ID 2345) pursuant to Article 13(1) of Regulation (EC) No 1924/2006. *EFSA J.*, 7(9): 1302.

Fung, A. (2012) The fucoxanthin content and antioxidant properties of *Undaria pinnatifida* from Marlborough Sound, New Zealand. MAppSc thesis, Auckland University of Technology.

Fung, A., Hamid, N. and Lu, J. (2013) Fucoxanthin content and antioxidant properties of *Undaria pinnatifida*. *Food Chem.*, 136: 1055–1062.

Gammone, M.A. and d'Orazio, N. (2015) Anti-obesity activity of the marine carotenoid fucoxanthin. *Marine Drugs*, 13(4): 2196–2114.

Gorusupudi, A. and Baskaran, V. (2013) Wheat germ oil: a potential facilitator to improve lutein bioavailability in mice. *Nutrition*, 29(5): 790–795.

Goss, R. and Jakob, T. (2010) Review. Regulation and function of xanthophyll cycle-dependent photoprotection in algae. *Photosynth. Res.*, 106(1-2): 103–122.

Ha, A.W., Na, S.J. and Kim, W.K. (2013) Antioxidant effects of fucoxanthin rich powder in rats fed with high fat diet. *Nutr. Res. Pract.*, 7(6): 475–480.

Haugan, J. and Liaaen-Jensen, S. (1994) Algal carotenoids 54. Carotenoids of brown algae (*Phaeophyceae*). *Biochem. Systemat. Ecol.*, 22(1): 31–41.

Health Products Regulatory Authority (2015) *Herbal Medicines*. Available at: www.hpra.ie/homepage/medicines/regulatory-information/medicines-authorisation/herbal-medicines (accessed 22 April 2017).

Heffernan, N., Smyth, T.J., Fitzgerald, R.J., Soler-Vila, A. and Brunton, N. (2014) Antioxidant activity and phenolic content of pressurised liquid and solid–liquid extracts from four Irish origin macroalgae. *Int. J. Food Sci. Technol.*, 49: 1765–1772.

Heo, S.J. and Jeon, Y.J. (2009) Protective effect of fucoxanthin isolated from *Sargassum siliquastrum* on UV-B induced cell damage. *J. Photochem. Photobiol. B*, 95(2): 101–107.

Heo, S.J., Park, E.J., Lee, K.W. and Jeon, Y.J. (2005) Antioxidant activities of enzymatic extracts from brown seaweeds. *Bioresource Technol.*, 96: 1613–1623.

Heo, S.J., Yoon, W., Kim, K. et al. (2010) Evaluation of anti-inflammatory effect of fucoxanthin isolated from brown algae in lipopolysaccharide-stimulated RAW 264.7 macrophages. *Food Chem. Toxicol.*, (8-9): 2045–2051.

Holdt, S. and Kraan, S. (2011) Bioactive compounds in seaweed: functional food applications and legislation. *J. Appl. Phycol.*, 23: 543–598.

Hosokawa, M., Wanezaki, S., Miyauchi, K. et al. (1999) Apoptosis-inducing effect of fucoxanthin on human leukaemia cell line HL-60. *Food Sci. Technol. Res.*, 5: 243–246.

Hosokawa, M., Miyashita, T., Nishikawa, S. et al. (2010) Fucoxanthin regulates adipocytokine mRNA expression in white adipose tissue of diabetic/obese KK-Ay mice. *Arch. Biochem.*, 504: 17–25.

Hurd, C.L., Harrison, P.J., Bischof, K. and Lobban, C.S. (2014) *Seaweed Ecology and Physiology*, 2nd edn. Cambridge: Cambridge University Press.

Hurst, W.J. (2002) *Methods of Analysis for Functional Foods and Nutraceuticals*. Boca Raton: CRC Press.

Ikeda, K., Kitamura, A., Machida, H. et al. (2003) Effect of *Undaria pinnatifida* (Wakame) on the development of cerebrovascular diseases in stroke-prone spontaneously hypertensive rats. *Clin. Exper. Pharmacol. Physiol.*, 30: 44–48.

Indrawatia, R., Sukowijoyoa, H., Indriatmokoa, I., Wijayanti, R.D. and Limantara, L. (2015) Encapsulation of brown seaweed pigment by freeze drying: characterization and its stability during storage. *Procedia Chem.*, 14: 353–360.

Jaswir, I., Noviendri, D., Salleh, H.M., Taher, M. and Miyashita, K. (2011) Isolation of fucoxanthin and fatty acids analysis of *Padina australis* and cytotoxic effect of fucoxanthin on human lung cancer (H1299) cell lines. *Afr. J. Biotechnol.*, 10(81): 18855–18862.

Je, J.Y., Park, P.J., Kim, E.K., Park, J.S., Yoon, H.D., Kim, K.R. and Ahn, C.B. (2009) Antioxidant activity of enzymatic extracts from the brown seaweed *Undaria pinnatifida* by electron spin resonance spectroscopy. *LWT Food Sci. Technol.*, 42: 874–878.

Jung, H.A., Islam, M.N., Lee, C.M. et al. (2013) Kinetics and molecular docking studies of an anti-diabetic complication inhibitor fucosterol from edible brown algae *Eisenia bicyclis* and *Ecklonia stolonifera*. *Chem. Biol. Interact.*, 206(1): 55–62.

Kadekaru, T., Toyama, H.and Yasumoto, T. (2008) Safety evaluation of fucoxanthin purified from *Undaria pinnatifida*. *J. Jpn. Soc. Food Sci.*, 55: 304–308.

Kanazawa, K., Ozaki, Y., Hashimoto, T. et al. (2008) Commercial-scale preparation of biofunctional fucoxanthin from waste parts of brown sea algae *Laminaria japonica*. *Food Sci. Technol. Res.*, 14(6): 573–582.

Kanda, H., Kamo, Y., Machmudah, S., Wahyudiono, E.Y. and Goto, M. (2014) Extraction of fucoxanthin from raw macroalgae excluding drying and cell wall disruption by liquefied dimethyl ether. *Marine Drugs*, 12(5): 2383–2396.

Kawee-ai, A., Kuntiya, A. and Kim, S.M. (2013) Anticholinesterase and antioxidant activities of fucoxanthin purified from the microalga *Phaeodactylum tricornutum*. *Nat. Prod. Commun.*, 8(10): 1381–1386.

Kim, K.N., Heo, S.J., Kang, S.M., Ahn, G. and Jeon, Y.J. (2010) Fucoxanthin induces apoptosis in human leukaemia HL-60 cells through a ROS-mediated Bcl-xL pathway. *Toxicol. in Vitro*, 24(6): 1648–1654.

Kim, S.K. (2011a) *Marine Medicinal Foods: Implications and Application - Macro and Microalgae*. Advances in Food and Nutrition Research, Volume 64. San Diego: Academic Press.

Kim, S.K. (2011b) *Marine Cosmeceuticals: Trends and Prospects*. Boca Raton: CRC Press.

Kim, S.K. (2011c) *Handbook of Marine Macroalgae Biotechnology and Applied Phycology*. Chichester: John Wiley and Sons Ltd.

Kim, S.K. (2013) *Marine Nutraceuticals: Prospects and Perspectives*. Boca Raton: CRC Press.

Kim, S.K. and Chojnacka, K. (2015) *Marine Algae Extracts: Processes, Products, and Applications*. Hoboken: Wiley VCH.

Kim, S.M., Shang, Y.F. and Um, B.H. (2011) A preparative method for isolation of fucoxanthin from *Eisenia bicyclis* by centrifugal partition chromatography. *Phytochem. Anal.*, 22: 322–329.

Kita, S., Fujii, R., Cogdell, R.J. and Hashimoto, H. (2015) Characterization of fucoxanthin aggregates in mesopores of silica gel: electronic absorption and circular dichroism spectroscopies. *J. Photochem. Photobiol. A: Chem.*, 313: 3–8.

Ko, S.C., Kang, S.M., Ahn, G.N., Yang, H.P., Kim, K.N. and Jeon, Y.J. (2010) Antioxidant activity of enzymatic extracts from *Sargassum coreanum*. *J. Korean Soc. Food Sci. Nutr.*, 39(4): 494–499.

Ko, S.C., Lee, S.H., Ahn, G. et al. (2012) Effect of enzyme-assisted extract of *Sargassum coreanum* on induction of apoptosis in HL-60 tumor cells. *J. Appl. Phycol.*, 24(4): 675–684.

Kotake-Nara, E. and Nagao, A. (2011) Absorption and metabolism of xanthophylls. *Marine Drugs*, 9: 1024–1037.

Kotake-Nara, E., Kushiro, M., Zhang, H., Sagawara, T., Miyashita, K. and Nagao, A. (2001) Carotenoids affect proliferation of human prostate cancer cells. *J. Nutr.*, 131: 3303–3306.

Kyndt, J. and d'Silva, A. (2013) *Algae Coloring the Future Green*, 2nd edn. Tucson: Moura Enterprises LLC.

Landrum, J.T. (2009) *Carotenoids: Physical, Chemical, and Biological Functions and Properties.* Boca Raton: CRC Press.

Lee, S., Lee, Y.S., Jung, S.H., Kang, S.S. and Shin, K.H. (2003) Anti-oxidant activities of fucosterol from the marine algae *Pelvetia siliquosa. Arch. Pharmacol. Res.*, 26(9): 719–722.

Liechty, W.B., Kryscio, D.R., Slaughter, B.V. and Peppas, N.A. (2010) Polymers for drug delivery systems. *Annu. Rev. Chem. Biomol. Eng.*, 1:149–173.

Lobban, C.S. and Wynne, M.J. (1981) *The Biology of Seaweeds.* Berkeley: University of California Press.

Maeda, H., Hosokawa, M., Sashima, T., Funayama, K. and Miyashita, K. (2005) Fucoxanthin from edible seaweed, *Undaria pinnatifida*, shows antiobesity effect through UCP1 expression in white adipose tissues. *Biochem. Biophys. Res. Commun.*, 332(2): 392–397.

Maeda, H., Hosokawa, M., Sashima, T., Takahashi, N., Kawada, T. and Miyashita, K. (2006) Fucoxanthin and its metabolite, fucoxanthinol, suppress adipocyte differentiation in 3T3-L1 cells. *Int. J. Mol. Med.*, 18: 147–152.

Maeda, H,. Hosokawa, M., Sashima, T. and Miyashita, K. (2007) Dietary combination of fucoxanthin and fish oil attenuates the weight gain of white adipose tissue and decreases blood glucose in obese/diabetic KK-Ay mice. *J. Agric. Food Chem.*, 55: 7701–7706.

McClements, D.J., Decker, E.A. and Park, Y. (2009a) Controlling lipid bioavailability through physicochemical and structural approaches. *Crit. Rev. Food Sci. Nutr.*, 49: 48–67.

McClements, D.J., Decker, E.A., Park, Y. and Weiss, J. (2009b) Structural design principles for delivery of bioactive components in nutraceuticals and functional foods. *Crit. Rev. Food Sci. Nutr.*, 49: 577–606.

Mehta, B.M., Kamal-Eldin, A. and Iwanski, R.Z. (2012) *Fermentation: Effects on Food Properties*. Boca Raton: CRC Press.

Mikami, K. and Hosokawa, M. (2013) Biosynthetic pathway and health benefits of fucoxanthin, an algae-specific xanthophyll in brown seaweeds. *Int. J. Mol. Sci.*, 14(7): 13763–13768.

Mise, T., Ueda, M. and Yasumoto, T. (2011) Production of fucoxanthin-rich powder from *Cladosiphon okamuranus*. *Adv. J. Food Sci. Technol.*, 3: 73–76.

Miyashita, K. and Hosokawa, M. (2007) Beneficial health effects of seaweed carotenoid, fucoxanthin. In: Barrow, C. and Shahidi, F. (eds) *Marine Nutraceuticals and Functional Foods*. Boca Raton: CRC Press, pp. 297–319.

Miyashita, K., Nishikawa, S., Beppu, F., Tsukui, T., Abe, M. and Hosokawa, M. (2011) The allenic carotenoid fucoxanthin, a novel marine nutraceutical from brown seaweeds. *J. Sci. Food Agric.*, 91(7): 1166–1174.

Mohamed, S., Hashim, S.N. and Rahman, H.A. (2012) Seaweeds: a sustainable functional food for complementary and alternative therapy. *Trends Food Sci. Technol.*, 23(2): 83–96.

Mori, K., Ooi, T., Hiraoka, M. et al. (2004) Fucoxanthin and its metabolites in edible brown algae cultivated in deep seawater. *Marine Drugs*, 2(2): 63–72.

Morrissey, J., Kraan, S. and Guiry, M. (2001) *A Guide to Commercially Important Seaweeds on the Irish Coast*. Dun Laoghaire: Bord Iascaigh Mhara.

Nagodawithana, T. and Reed, G. (2013) *Enzymes in Food Processing*. London: Academic Press.

Nakazawa, Y., Sashima, T., Hosokawa, M. and Miyashita, K. (2009) Comparative evaluation of growth inhibitory effect of stereoisomers of fucoxanthin in human cancer cell lines. *J. Funct. Foods*, 1: 88–97.

Nishida, Y., Yamashita, E. and Miki, W. (2007) Quenching activities of common hydrophilic and lipophilic antioxidants against singlet oxygen using chemiluminescence detection system. *Carotenoid Sci.*, 11: 16–20.

Nomura, T., Kikuchi, M., Kubodera, A. and Kawakami, Y. (1997) Proton-donative antioxidant activity of fucoxanthin with 1,1-diphenyl-2-picrylhydrazyl (DPPH). *Biochem. Mol. Biol. Int.*, 42(2): 361–370.

Olenina, I., Hajdu, S., Edler, L. et al. (2006) Biovolumes and size-classes of phytoplankton in the Baltic Sea HELCOM. *Balt. Sea Environ. Proc.*, 106: 144.

Oryza (2015) *Fucoxanthin: Dietary Ingredient for Prevention of Metabolic Syndrome and Beauty Enhancement*. Available at: www.oryza.co.jp/pdf/english/Fucoxanthin_1.0.pdf.

O'Sullivan, A.M.N. (2013) Cellular and *in-vitro* models to assess antioxidant activities of seaweed extracts and the potential use of the extracts as ingredients. PhD thesis, University College Cork.

Peng, J., Yuan, J.P., Wu, C.F. and Wang, J.H. (2011) Fucoxanthin, a marine carotenoid present in brown seaweeds and diatoms: metabolism and bioactivities relevant to human health. *Marine Drugs*, 9(10): 1806–1828.

Prabhasankar, P., Ganesan, P., Bhaskar, N., Hirose, A., Stephen, N. and Gowda, L.R. (2009) Edible Japanese seaweed, wakame (*Undaria pinnatifida*) as an ingredient in pasta: chemical, functional and structural evaluation. *Food Chem.*, 115(2): 501–508.

Pyszniak, A.M. and Gibbs, S.P. (1992) Immunocytochemical localization of photosystem I and the fucoxanthin-chlorophyll a/c light-harvesting complex in the diatom *Phaeodactylum tricornutum*. *Protoplasma*, 166: 208–217.

Quan, J., Kim, S.M., Pan, C.H. and Chung, D. (2013) Characterization of fucoxanthin-loaded microspheres composed of cetyl palmitate-based solid lipid core and fish gelatin–gum arabic coacervate shell. *Food Res. Int.*, 50(1): 31–37.

Quitain, A.T., Kai, T., Sasaki, M. and Goto, M. (2013) Supercritical carbon dioxide extraction of fucoxanthin from *Undaria pinnatifida*. *J. Agric. Food Chem.*, 61(24): 5792–5797.

Ramus, J., Lemons, F. and Zimmerman, C. (1977) Adaptation of light-harvesting pigments to downwelling light and the consequent photosynthetic performance of the eulittoral rockweeds *Ascophyllum nodosum* and *Fucus vesiculosus*. *Marine Biol.*, 42(4): 293–303.

Ravi, H. and Baskaran, V. (2015) Biodegradable chitosan-glycolipid hybrid nanogels: a novel approach to encapsulate fucoxanthin for improved stability and bioavailability. *Food Hydrocoll.*, 43: 717–725.

Rhein-Knudsen, N., Ale, M.T. and Meyer, A.S. (2015) Seaweed hydrocolloid production: an update on enzyme assisted extraction and modification technologies. *Marine Drugs*, 13(6): 3340–3359.

Rodrigues, E., Lilian, R., Mariutti, B. and Mercadante, A.Z. (2012) Scavenging capacity of marine carotenoids against reactive oxygen and nitrogen species in a membrane-mimicking system. *Marine Drugs*, 10(8): 1784–1798.

Roh, M.K., Uddin, M.S. and Chun, B.S. (2008) Extraction of fucoxanthin and polyphenol from *Undaria pinnatifida* using supercritical carbon dioxide with cosolvent. *Biotechnol. Bioprocess Eng.*, 13: 724–729.

Ryan, R. (2014) *Opportunity for Significant Growth in Seaweed Sales*. Irish Examiner. Available at: www.irishexaminer.com/farming/ opportunity-for-significant-growth-in-seaweed-sales-299294.html (accessed 22 April 2017).

Sangeetha, R.K., Bhaskar, N., Divakar, S. and Baskaran, V. (2010) Bioavailability and metabolism of fucoxanthin in rats: structural characterization of metabolites by LC-MS (APCI). *Mol. Cell. Chem.*, 333(1-2): 299–310.

Satomi, Y. and Nishino, H. (2013) Inhibition of the enzyme activity of cytochrome P450 1A1, 1A2 and 3A4 by fucoxanthin, a marine carotenoid. *Oncol. Lett.*, 6(3): 860–864.

Shang, Y.F., Kim, S.M., Lee, W.J. and Um, B.H. (2011) Pressurized liquid method for fucoxanthin extraction from *Eisenia bicyclis* (Kjellman) Setchell. *J. Biosci. Bioeng.*, 111(2): 237–241.

Shiratori, K., Ohgami, K., Ilieva, I., Jin, X.H., Koyama, Y. and Miyashita, K. (2005) Effects of fucoxanthin on lipopolysaccharide-induced inflammation *in vitro* and *in vivo*. *Exper. Eye Res.*, 81: 422–428.

Sho, A., Sugiyama, Y., Takahashi, T., Hosokawa, M. and Miyashita, K. (2010) Novel functionality of glycolipids from brown seaweeds. Proceedings of the 101st AOCS Annual Meeting and Expo, Phoenix, Arizona, USA, 2010. Available at: www.researchgate.net/ publication/283921155_Novel_functionality_of_glycolipids_from_ brown_seaweeds (accessed 26 April 2017).

Sigma-Aldrich (2015a) *F6932 SIGMA Fucoxanthin Carotenoid Antioxidant*. Available at: www.sigmaaldrich.com/catalog/product/sigma/ f6932?lang=en&region=IE (accessed 26 April 2017).

Sigma-Aldrich (2015b) *Analysis: Enzymes, Kits and Reagents*. Available at: www.sigmaaldrich.com/content/dam/sigma-aldrich/docs/Sigma/General _Information/2/biofiles_issue8.pdf (accessed 26 April 2017).

Sivagnanam, S.P., Yin, S., Choi, J.H. and Park, Y.B. (2015) Biological properties of fucoxanthin in oil recovered from two brown seaweeds using supercritical CO_2 extraction. *Marine Drugs*, 13: 3422–3442.

Socaciu, C. (2007) *Food Colorants: Chemical and Functional Properties*. Boca Raton: CRC Press.

Stengel, D.B. and Dring, M.J. (1998) Seasonal variation in the pigment content and photosynthesis of different thallus regions of *Ascophyllum nodosum* (Fucales, Phaeophyta) in relation to position in the canopy. *Phycologia*, 37(4): 259–268.

Subramanian, J., Baik, O.D. and Singhal, R. (2013) Optimization of ultrasound-assisted extraction of fucoxanthin from dried biomass of *Sargassum muticum*. The Canadian Society for Bioengineering Annual Conference, University of Saskatechewan, Saskatoon, Canada, 7–10 July. Paper No. CSBE13-101. Available at: www.csbe-scgab.ca/docs/ meetings/2013/CSBE13101.pdf (accessed 26 April 2017).

Suhendra, L., Raharjo, S., Hastuti, P. and Hidayat, C. (2012) Formulation and stability of o/w microemulsion as fucoxanthin delivery. *Agritech*, 32(3): 230–239.

Terasaki, M., Hirose, A., Narayan, B., Baba, Y., Kawagoe, C. and Yasui, H. (2009) Evaluation of recoverable functional lipid components of several brown seaweeds (*Phaeophyta*) from Japan with special reference to fucoxanthin and fucosterol contents. *J. Phycol.*, 45(4): 974–980.

United States Department of Agriculture, Research, Education, and Economics Information System (2015) *Optimization of the Extraction of Fucoxanthin and Its Potential as an Antiobesity Functional Food.* Available at: https://portal.nifa.usda.gov/web/crisprojectpages/0225483-optimization-of-the-extraction-of-fucoxanthin-and-its-potential-as-an-antiobesity-functional-food.html (accessed 26 April 2017).

United States Food and Drug Administration (2014) *Code of Federal Regulations Title 21 Part 101, Sec. 101.93 Certain types of statements for dietary supplements. Food Labeling, Subpart F - Specific Requirements for Descriptive Claims That Are Neither Nutrient Content Claims nor Health Claims.* Available at: www.accessdata.fda.gov/scripts/cdrh/cfdocs/cfcfr/CFRSearch.cfm?fr=101.93 (accessed 22 April 2017).

Urikura, I., Sugawara, T. and Hirata, T. (2011) Protective effect of fucoxanthin against UVB-induced skin photoaging in hairless mice. *Biosci. Biotechnol. Biochem.*, 75(4): 757–760.

Uzun, H., Zengin, K., Taskin, M., Aydin, S., Simsek, G. and Dariyerli, N. (2004) Changes in leptin, plasminogen activator factor and oxidative stress in morbidly obese patients following open and laparoscopic Swedish adjustable gastric banding *Obes. Surg.*, 14(5): 659–665.

Venugopal, V. (2011) *Marine Polysaccharides: Food Applications.* Boca Raton: CRC Press.

Wang, S.K., Li, Y., Lindsey White, W. and Lu, J. (2014) Extracts from New Zealand *Undaria pinnatifida* containing fucoxanthin as potential functional biomaterials against cancer *in vitro. J. Funct. Biomaterials*, 5: 29–42.

Watson, L. and Walsh, M. (2012) *A Market Analysis towards the Further Development of Seaweed Aquaculture in Ireland.* Dun Laoghaire: Bord Iascaigh Mhara.

Watson, R. (2014) *Nutrition in the Prevention and Treatment of Abdominal Obesity.* New York: Elsevier.

Watson, R., Preedy, V.R. and Zibadi, S. (2013) *Polyphenols in Human Health and Disease.* San Diego: Academic Press.

Werner, A., Clarke, D. and Kraan, S. (2004) *Strategic Review of the Feasibility of Seaweed Aquaculture in Ireland.* Galway: Marine Institute.

Wijesinghe, W.A. and Jeon, Y.J. (2012) Enzyme-assistant extraction (EAE) of bioactive components: a useful approach for recovery of industrially important metabolites from seaweeds: a review. *Fitoterapia*, 83: 6–12.

Willstätter, R. and Page, H.J. (1914) Chlorophyll. XXIV. The pigments of the brown algae. *Justus Liebig's Annalen der Chemie*, 404: 237–271.

Wright, S.W. and Jeffrey, S.W. (1987) Fucoxanthin pigment markers of marine phytoplankton analysed by HPLC and HPTLC. *Marine Ecol. Progr. Series*, 38: 259–266.

Xia, S., Wang, K., Wan, L., Li, A., Hu, Q. and Zhang, C. (2013) Production, characterization, and antioxidant activity of fucoxanthin from the marine diatom *Odontella aurita*. *Marine Drugs*, 11(7): 2667–2681.

Xiao, X., Si, X., Yuan, Z., Xu, X. and Li, G. (2012) Isolation of fucoxanthin from edible brown algae by microwave-assisted extraction coupled with high-speed countercurrent chromatography. *J. Separation Sci.*, 35: 2313–2317.

Yan, X., Chuda, Y., Suzuki, M. and Nagata, T. (1999) Fucoxanthin as the major antioxidant in *Hizikia fusiformis*, a common edible seaweed. *Biosci. Biotechnol. Biochem.*, 63(3): 605–607.

Zaragozá, M.C., López, D., Sáiz, M.P. et al. (2008) Toxicity and antioxidant activity *in vitro* and *in vivo* of two *Fucus vesiculosu*s extracts. *J. Agric. Food Chem.*, 56(17): 7773–7780.

Zuorroa, A., Fidaleo, M. and Lavecchia, R. (2011) Enzyme-assisted extraction of lycopene from tomato processing waste. *Enzyme Microbial Technol.*, 49: 567–573.

4

Functional Foods from Mushroom

Vinay Basavegowda Raghavendra[1]*, Chandrasekar Venkitasamy[1], Zhongli Pan[1,2], and Chandra Nayak[3]

[1] *Department of Biological and Agricultural Engineering, University of California, Davis, California, USA*
[2] *Healthy Processed Foods Research Unit, USDA-ARS-WRRC, Albany, USA*
[3] *Department of Studies in Biotechnology, Mangalore University, Mangalagangotri, Mangalore, Karnataka, India*

*Corresponding author e-mail: viragh79@gmail.com

Introduction

Food is used not only to satisfy hunger but also to provide nutrition. Nutritious foods prevent nutrition-related diseases. People have become increasingly conscious about their health which has sparked a global market for natural foods consumed as dietary supplements (Siro et al. 2008). All natural foods are functional by virtue of their nutrients but the concept of functional food derives from the observation that some processed foods containing particular ingredients can provide beneficial effects for human health. Functional foods are foods with positive health benefits that exceed those attributable to the nutritional value of the food which promotes optimal health and helps in reducing the risk of many diseases. According to Hasler (1998), "Functional foods are the foods that provide health benefit beyond basic nutrition. Food choices can control our health, knowing that some food can provide specific health benefits." Food products are only considered as functional foods if they provide a basic nutritional impact on the human organism, either improving physical conditions or decreasing the risk of diseases (Siro et al. 2008). "Functional foods may improve health, decrease the risk of some diseases, and could even be used for curing some fever. It was recognized that there is a demand for these products as different demographical studies revealed that the medical service of the aging population is rather expensive" (Ouwehand et al. 2002).

Microbial Functional Foods and Nutraceuticals, First Edition. Edited by Vijai Kumar Gupta, Helen Treichel, Volha (Olga) Shapaval, Luiz Antonio de Oliveira, and Maria G. Tuohy.
© 2018 John Wiley & Sons Ltd. Published 2018 by John Wiley & Sons Ltd.

According to a Canadian definition "The functional foods are similar in appearance to a conventional food consumed as a part of the usual diet that enhance physiological benefits which reduce the risk of chronic diseases beyond basic nutritional functions" (Miles and Chang 1997).

Design and development of functional foods is a scientific challenge relevant to functions that are sensitive to modulation by food components, which are pivotal to maintenance of well-being and health, and that, when altered, may be linked to a change in the risk of a disease. Increasing demands for functional foods have been seen since the introduction of whole foods or food components with ability to prevent cancer, osteoporosis or cardiovascular disease, improve immunity, detoxification, physical performance, weight loss, cognitive function, and the ability to cope with stress, and inhibit inflammation, antioxidants and modulating the effects of hormones (Roberfroid 2000).

Historically, mushrooms were classified among the lower plants in the division of Thallophyta by Linnaeus based on anatomically uncomplicated structural bodies (lack of true roots, true stem, true leaves, true flowers, and true seeds). Modern studies have established that mushroom biota, together with other fungi, have features which are sufficiently distinctive to place them in a separate fungal kingdom, the Mycetaceae (Chang and Miles 2004). More than 14 000 species of mushrooms are included in the Basidiomycetes and Ascomycetes. Only 25% of mushroom species are edible, in various degrees, 50% are inedible and the remaining 25% are poisonous (Chang 1980).

The early civilizations of the Greeks, Egyptians, Romans, Chinese, and Mexicans probably knew about the therapeutic value of mushrooms and used them in different ways. The Greeks believed that mushroom provided health and strength for soldiers and praised mushroom as a delicacy, and the Romans thought mushrooms were the "food of the gods." The Chinese treasured mushrooms as "food of life" and Mexicans used mushroom as a "gift of love" in religious ceremonies (Chang and Miles 2004). The Chinese were the first to cultivate *Auricularia auricula-judae* in AD 600 and *Lentinus edodes* mushrooms between AD 1000 and 1100, followed by *Agaricus bisporus* cultivation in France in 1600 and *Pleurotus bisporus* cultivation in the USA in 1900 (Chang 1999).

In recent years, mushroom has gained a remarkable amount of interest for its delicious taste and nutritional and medicinal value (Crisan and Sands 1978). Mushrooms are not only a source of nutrients but also are a protein-rich food cultivated all over the world as an important agricultural product.

Definition

A mushroom is the fleshy, spore-bearing fruiting body of a fungus (basidiomycete) typically produced above ground on soil or on its food source. Hence the word "mushroom" is most often applied to those fungi (Basidiomycota, Agaricomycetes) that have a stem, cap, and gills on the underside of the cap. These gills produce microscopic spores that help the fungus spread across the ground or its occupant surface (Huang et al. 1985).

According to Chang and Buswell (1996), mushroom is a macro-fungus with a distinctive fruiting body which can be either epigeous or hypogeous and large enough to be seen with the naked eye. Chang and Miles (2004) stated that "Mushrooms need not be Basidiomycetes, nor aerial, nor fleshy, nor edible, but also few Ascomycetes, which can grow underground having fleshy or non-fleshy texture and need not be edible".

Cultivation

Mushrooms have been cultivated for many decades and are now an important agricultural product worldwide. To date, more than 60 mushroom species have been cultivated in East Asian countries, France, and America. Nearly 30 species have been cultivated only in China. Some grow on fresh wood residues (e.g., *Lentinula, Pleurotus, Flammulina, Auricularia, Pholiata, Tremella, Agrocybe, Ganiderma*), some on slightly composted lignocellulosic materials (e.g., *Volvariella, Stropharia, Coprinus*), some on well-composted materials or animal dung (e.g., *Agaricus*), while others grow on soil and humus (e.g., *Lepiota, Leptista, Morchella, Gyromitra*). The most popular species grown all over the world are *L. edodes* and *Pleurotus* spp. The production statistics of 10 major edible mushroom-producing countries are shown in Table 4.1.

Nutritional Value

Moisture

Moisture content of mushrooms is close to 90% of their fresh weight. Some fleshy mushrooms contain moisture in a different range. *Grifolia* and *Agaricus bisporus* contain the highest moisture content of about 92.8–94.8% (Manzi et al. 2001), *L. edodes* has 81.8–90.0% (Mau et al. 2001b), and *Flammulina velutipes* has 87.2–89.1% (Yang et al. 2001). The moisture content varies from species to species based on the time of harvest, cropping conditions, temperature, relative humidity, and the postharvest period.

Table 4.1 Production statistics of 10 major edible mushroom-producing countries in 2014 (FAO, IFAD, WFP 2014).

Country	Production ($1000)	Production (metric tons)
China	921 928	5 150 000
Italy	1 416 342	785 000
USA	700 864	388 450
Netherlands	553 907	307 000
Poland	396 936	220 000
Spain	263 421	146 000
France	210 329	116 574
Iran	158 188	87 675
Canada	147 949	82 000
UK	131 891	73 100

Protein

The crude protein content of edible mushrooms is less than that of animal meat but usually higher than several food products, including milk. In general, mushrooms are considerably higher in protein content, which is twice that of asparagus, cabbage and other vegetables and 12 times higher than oranges and apples, respectively. On dry weight basis, the protein content of mushroom ranges from 19% to 35% whereas the protein content of rice is 7.3%, wheat is 39.1%; and that of soybean and milk is higher than 25%. Edible mushroom contains total proteins of albumins, globulins, natural glutelins, prolamins, and prolamin-like materials. Protein contents of some common edible mushrooms vary depending on species and stage of development. The protein content of common mushroom species is given in Table 4.2.

Table 4.2 Crude protein contents of some popular edible mushrooms.

Mushroom	Protein content (%)
Agaricus bisporus	23.9-34.8
Agaricus campestris	33.2
Boletus edulis	29.7
Pleurotus ostreatus	30.4
Pleurotus sajor-caju	26.6
Volvariella dysplasia	28.5
Volvariella volvacea	25.9

Amino Acids

The nutritional value of mushrooms can be determined based on their content of essential amino acids. The essential amino acid (EAA) index is measured according to dietary protein in terms of an essential amino acid pattern, and the amount of high-quality protein and the nutritional quality of mushroom are calculated using the formula (Crisan and Sands 1978):

$$Nutritional\ index = \frac{EAA\ index\ X\ percentage\ protein}{100}.$$

The EAA index, amino acid scores, and nutritional index values for various mushrooms compared with other foods are shown in Table 4.3.

Table 4.3 shows the rankings based on EAA index, amino acid scores, and nutritional indexes calculated against the FAO reference protein pattern; biological values correlate closely with the EAA indexes. Values for mushrooms represent the mean of the three highest values (high) and lowest values (low) (Huang et al. 1985).

Proteins are made up of over 20 amino acids, out of which only nine (lysine, methionine, tryptophan, threonine, valine, leucine, isoleucine, histidine and phenylamine) are present simultaneously and help in protein synthesis. Animal products provide better and more balanced essential amino acids than plant products. For example, cereal grains lack some important amino acids like lysine, and legumes usually lack methionine and tryptophan, but edible mushrooms contain all nine amino acids which are essential for protein synthesis in humans. According to Bano and Rajarathnam (1988), the most abundant essential amino acid present in mushroom is lysine and very scarce essential amino acids are methionine and tryptophan.

In addition to the nine essential amino acids, some edible mushrooms contain sodium salts of glutamic and aspartic amino acids which give foods a pleasant savory taste known as umami. The taste of the mushroom also depends on the presence of amino acids as shown in Table 4.4. According to chemical analysis, the umami flavor of mushroom was attributed to MSG-like amino acids such as aspartic acid (Asp) and glutamic acid (Glu) and flavor 5′-nucleotides such as 5′-guanosine monophosphate (5′-GMP), 5′-inosine monophosphate (5′-IMP) and 5′-xanthosine monophosphate (5-XMP) (Sommer 2008). The presence of Asp and Glu by themselves gives a sour taste whereas sodium salts of Glu and Asp elicit the umami taste (Fuke and Shimadzu 1993). Umami is also found in various vegetables (eggplant, tomato, potato, cabbage, edible mushroom, carrot, and soybean), sea foods (fish, seaweed, oyster, prawn, crab, sea urchin, clam, and scallop) and meat (beaf, pork, and chicken) (Fuke and Ueda 1996). Glu and Asp amino,acids are found most abundantly in mushrooms compared to all these foods.

Table 4.3 Comparison of nutritive value of mushrooms with other foods.

Essential amino acid index	Food	Amino acid scores	Food	Nutritional indexes	Food
100	Pork, chicken, beef	100	Pork	59	Chicken
99	Milk	98	Chicken, beef	43	Beef
98	Mushrooms (high)	91	Milk	35	Pork
91	Potatoes, kidney beans	89	Mushrooms (high)	31	Soybeans
88	Corn	63	Cabbage	28	Mushrooms (high)
86	Cucumbers	59	Potatoes	26	Spinach
79	Peanuts	53	Peanuts	25	Milk
76	Spinach, soybeans	50	Corn	21	Kidney beans
72	Mushrooms (low), cabbage	46	Kidney beans	20	Peanuts
69	Turnips	42	Cucumbers	17	Cabbage
53	Carrots	33	Turnips	14	Cucumbers
44	Tomatoes	32	Mushrooms (low)	11	Corn
		31	Carrots	10	Turnips
		28	Spinach	09	Potatoes
		23	Soybeans	08	Tomatoes
		18	Tomatoes	06	Carrots
				05	Mushrooms (low)

Source: Crisan and Sands (1978).

Table 4.4 Classification of amino acids of edible mushrooms.

Sl. No.	Characteristics of groups	Amino acids
1	MSG-like or palatable amino acids	Aspartic acid and glutamic acid
2	Sweet amino acids	Alanine, glycine, serine, and threonine
3	Bitter amino acids	Arginine, histidine, isoleucine, and leucine
4	Tasteless amino acids	Lysine and tyrosine

Many researchers have found that the umami taste has been elicited by di-or tripeptides of glutamic acids such as Glu-Glu, Glu-Asp, Glu-Asp-Glu, Glu-Gly-Ser, Gly-Asp, Ala-Glu, Gly-Asp-Gly, Val-Asp-Val, Asp-Leu, and Val-Glu-Leu (Ohyama et al. 1988). The presence of umami was influenced by many factors including the species type, maturity stage, part of mushroom, and quality. Umami amino acid contents in various mushroom species are listed in Table 4.5.

The umami taste of edible mushrooms is the reason for their frequent use as a flavor enhancer in new food product development. Several functional foods have been patented using the umami taste. The glutamic acid content in most edible mushrooms is higher than the aspartic acid content, establishing the fact that Glu is the main umami amino acid in most edible mushrooms. Wild mushroom species have a stronger umami taste than commercial or cultivated ones (Zhang et al. 2013). It is also reported that the presence of glutamate plays an important role in the regulation of gastrointestinal disorders such as gastric problems (functional dyspepsia, chronic atrophic gastritis, and indigestion) (Nakamura et al. 2008). All these health-improving functions of glutamine and glutamate imply that mushrooms can be used not only as nutritious and palatable food, but also as a functional food or as a raw material for functional foods.

Flavored condiments are used as a functional food by adding edible mushroom powder (*A. bisporus*) including sodium glutamate, salt, sugar, maltodextrin, yeast extract, and edible oil (Sun et al. 2007). Soybean sauce was produced by fermenting pine mushroom with soybeans to produce a strong pine mushroom taste without any alteration of the sauce (An 2007). Mushroom powder is used as a functional food in Asian countries, usually fresh *Agaricus bisporus* with sodium glutamate, salt, sugar, yeast extract, flavor nucleotides, and egg yolk powder (Shao and Huang 2009). The dried mushroom slices are mixed with salt, granulated, dried, and sterilized to produce edible mushroom essence spices (Tu 2012).

Table 4.5 Umami taste amino acid content of various mushroom species.

Mushroom	L-Asp	L-Glu	MSG-like	Reference
Agaricus blazei	1.11	3.29	4.4	Tsai et al. 2008
Agaricus bisporus (J.E.Lange) Imbach	1.33	4.71	6.04	Dijkstra and Wiken 1976
Agaricus campestris L.	0.21	34.78	34.99	Beluhan and Ranogajec 2011
Agrocybe cylindracea (DC) Maire	0.94	2.18	3.12	Tsai et al. 2008
Amanita rubescens	2.77	17.53	20.30	Guzman et al. 1997
Auricularia fuscosiccinea (brown)	0.12	0.14	0.26	Mau et al. 1998b
Auricularia fuscosuccinea (white)	0.06	0.16	0.22	Mau et al. 1998b
Auricularia mesenteria	0.07	0.27	0.34	Mau et al. 1998b
Boletus edulis Bull.	0.33	39.09	39.42	Beluhan and Ranogajec 2011
Calocybe gambosa	0.19	25.67	25.86	Beluhan and Ranogajec 2011
Cantharellus cibaricus	0.06	29.99	30.05	Beluhan and Ranogajec 2011
Coriolus versicolor	0.41	0.09	0.50	Mau et al. 2001a
Craterellus cornucopioides	nd	45.85	45.85	Beluhan and Ranogajec 2011
Dictyophoraindusiata	0.31	0.54	0.85	Mau et al. 2001a
Entolomaclypeatum	nd	23.89	23.89	Beluhan and Ranogajec 2011
Flammulina velutipes (Curtis) Singer	0.03	1.54	1.57	Yang et al. 2001
Ganoderma lucidum (Curtis) P. Karst	0.06	0.11	0.17	Mau et al. 2001a
Ganoderma tsugae	0.22	0.06	0.28	Mau et al. 2001a
Hericium erinaceus (Bull.) Pers.	0.50	0.50	1.00	Mau et al. 2001a
Lentinula edodes (Berk.) Pegler	0.41	1.30	1.71	Yang et al. 2001
Macrolepiota procera	0.12	33.65	33.77	Beluhan and Ranogajec 2011

Table 4.5 (Continued)

Mushroom	L-Asp	L-Glu	MSG-like	Reference
Morchellaelata	nd	38.29	38.29	Beluhan and Ranogajec 2011
Pleurotus cystidiosus	0.05	1.16	1.21	Yang et al. 2001
Pleurotus ferulae	0.36	1.40	1.76	Tsai et al. 2009
Pleurotus ostreatus (Jacq. ex Fr.) P. Kumm.	0.17	41.09	1.21	Yang et al. 2001
Ramria flava	1.63	9.42	11.05	Leon-Guzman et al. 1997
Tremella fuciformis Berk.	0.03	0.06	0.09	Mau et al. 2001a
Tricholoma giganteum	0.34	0.34	0.68	Mau et al. 2001a

Solid-state fermentation by *Phellinus linteus* and *Inonotus obliquus* added to cooked rice resulted in a fermented rice product with umami taste having improved nutritional value, and also enhanced the production of enzymes and total antioxidant phenolic products. *Pleurotus eryngii, Cordyceps, Termitomyces,* and *Antrodia camphorata* were used to ferment buckwheat, polished rice, wheat, and rice germ, respectively (Chen et al. 2012). Mushrooms such as *A. salmonea* and *Grifola frondosa* were used to ferment oat and wheat (Huang et al. 2006).

Fat

Mushrooms are considered low-calorie foods as they provide low amounts of fat (Leon-Guzman et al. 1997) ranging from 1.1% to 8.3% on a dry weight basis, with an average fat content of 4.0%. *Lentinula edodes* has the highest fat content of 4.9–8.0% and *Pleurotus* has the lowest crude fat. Generally, mushrooms contain all classes of lipid compounds including monoglycerides, diglycerides, triglycerides, free fatty acids, sterols, esters, and phospholipids.

According to Huang et al. (2006), edible mushroom contains a higher percentage of saponifiable lipids than non-saponifiable lipids. These authors reported that saponifiable lipid ranges from 78.1% in *Auricularia auricula-judae* to 58.8% in *Volvariella volvacea* expressed in terms of percentage of dry weight of lipids. Miles and Chang (2004) reported that high content of saponifiable lipid was due to the presence of provitamin D2 and ergosterol.

The unsaturated fatty acid content of mushroom ranges from 70% to 72% of total fatty acids which is mainly due to the linoleic acid content. Chang (1987) reported that 85.4% of total free fatty acids in *Volvariella volvacea*, 80.5% in *A. bisporus*, and at least 74.1% in *A. auricula-judae* are unsaturated fatty acids. The higher proportion of linoleic acid in edible mushrooms plays a significant role as a functional food. Total fatty acid composition in edible mushroom varies from species to species. Mushroom contains short chain fatty acids with 14 to 18 carbon atoms, including myristic acid (C14), palmitic acid (C16), palmitoleic acid (16:1), stearic acid (C18:0), oleic acid (18:1), and linoleic acid (C18:2).

Vitamins

Edible mushroom are a good source of several vitamins, including thiamine (B1), riboflavin (B2), niacin, biotin, and ascorbic acid (C) (Crisan and Sands 1978). Lau et al. (1985) reported that the thiamin content of *V. volvacea* is 0.35 mg, *Agaricus* species 1.14 mg, *Pleurotus* species 1.16–4.6 mg and L. *edodes* 7.8 mg per 100 g dry weight of mushroom. The niacin content varies from 54.9 mg in *V. volvacea* and 64.9 mg in *A. bisporus* to 46.0–108.7 mg per 100 g dry weight in *Pleurotus* species. The riboflavin content was higher in *A. bisporus* (5.0 mg) and *L.edodes* (4.9 mg) than in *V. volvacea* (1.63-2.98 mg). Lau et al. (1985) reported that vitamin C content in edible mushroom ranges from 7.4 mg to 9.4 mg/100 g dry weight. γ-Ergosterol, a provitamin form of D2, is the common sterol found in edible mushrooms. *V. volvacea* had the highest provitamin D2 content on a dry weight basis (0.47%) followed by *L. edodes* (0.27%) and *A. bisporus* (0.23%) while *Tremella fuciformis* had the least (0.01%). The mature stage of *V. volvacea* had a higher content of provitamin D2 than the egg stage and the higher content of provitamin D2 was found in the cap rather than the stalk. *V. volvacea* contains 0.2% provitamin D2, 0.07% provitamin D4 and 0.07% γ-ergosterol followed by *Agaricus* spp., which contains 0.23% of provitamin D1.

Carbohydrate

Carbohydrate constitutes the major component of mushroom dry matter, usually about 50–60%. This group comprises various compounds like monosaccharides, their derivatives, and oligosaccharides. Mannitol and α-trehalose are the main representatives of the polyols and oligosaccharides which constitute the prevailing component of mushroom dry matter. Crisan and Sands (2003) reported that pentoses, methyl pentoses, and hexoses as well as disaccharides, amino acids, sugars, alcohols, and sugar acids are the constituents of mushroom. Mushroom contains glycogen and chitin as the major polysaccharides which are usually present in animals and not as starch and cellulose as in plants. Chitin is a water-soluble structural N-containing polysaccharide present in 80–90% of dry matter of mushroom cell walls. Chitin content increases with the maturation of several cultivated species and decreases during cooking. Lectin-carbohydrate has been shown to have antitumor and immunomodulatory activities (Wang et al. 2000). The lectin isolated from *Agaricus campestris* has proved to be more effective in hemagglutination (Cheung 2010). Numerous mushroom polysaccharides are considered as a potential source of prebiotics and β-glucans and are used as probiotics (Aida et al. 2009).

β-Glucans

Mesomo et al. (2010) produced β-glucans from the aqueous extraction of *A. blazie*. They analyzed their role in the shelf-life of cheese bread made from *A. blazie* and found that β-glucans were an excellent source of nutrients and had no side effects on the cheese bread formulation. The product was stored for 30 days without any major changes in taste, texture or appearance. Lemos (2009) developed and characterized a product similar to burger from *A. brasiliensis* mushroom with 12% of mushroom. The product was rich in protein (20.31%), carbohydrates (27.84%), dietary fiber (24.47%), ash (6.12%), and lipid (1.60%) when compared with commercial products like ground beef and vegetable burger and had high consumer acceptance. Bassan et al. (2011) developed a gluten-free sponge-like functional food from *A. brasiliensis* mushroom which was proved to have a high level of nutritional value.

Medicinal Properties

Mushrooms are not only sources of nutrients but also have been reported as therapeutic foods, useful in preventing diseases (Silva et al. 2010). They have been proved to enhance the immune system and natural defenses against several diseases. Mushrooms will also help patients receiving radiotherapy and chemotherapy by increasing the number of leukocytes in the blood and enhancing immune function, increasing and improving appetite, reducing pain, stopping hair loss, with potential antioxidant, antidiabetic, hypoglycemic, antitumor, immunomodulator, antimicrobial, antiviral, antiallergic, anti-inflammatory, hepatoprotective, anticancer and genoprotective properties, and general health-improving effect (Figure 4.1) (Lindequist et al. 2005).

Research into medicinal, pharmaceutical, and functional foods during the last few decades suggests that mushrooms have much to offer for human healthcare. In treatments where chemical medicines might have side effects and might not provide a complete remedy, use of mushrooms and their products may augment success. Mushrooms have either good nutritional

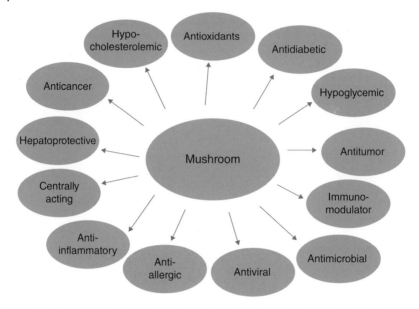

Figure 4.1 Pharmacological properties of edible mushroom. Source: Adapted from Lindequist et al. (2005).

value or medicinal value. Some of them possess both nutritional and medicinal properties and provide health benefits beyond basic nutrition and are therefore considered as functional foods (Rathee et al. 2012).

A database of scientific evidence exists about the specific health effects of mushrooms and their bioactive molecules.

Antitumor Effects

Polysaccharides are the best known and most potent mushroom-derived substances with antitumor and immunomodulating properties (Liu 2002; Mizuno 1995). Antitumor polysaccharides from mushroom have been extensively studied in recent years. These polysaccharides vary in chemical composition, structure, and antitumor activity. Many mushroom polysaccharides are present mainly as glucans with different types of glycosidic linkages. Polysaccharides such as (1-3), (1-6)-β-glucans, and (1-3)-α-glucans are helpful in antitumor action. However, antitumor polysaccharide may have different chemical structures, such as hetero-β-glycan, heteroglycan, β-glucan-protein, and heteroglycan-protein complexes including glycoproteins and triterpenoids. Glucans with higher molecular weight are more effective than glucans of lower molecular weight (Wasser 2002). Mushrooms such as *L. edodes*, *Grifola frondosa*, *Agaricus blazei*, and *Pleurotus* spp., taken as part of the diet, are likely to provide some protection against some diseases, particularly tumors. Polysaccharides from mushrooms do not attack cancer cells directly, but produce antitumor effects by activating T-cells through a thymus-dependent immune mechanism (Konno et al. 2002).

Several researchers have focused their studies on *L. edodes* with the objective of identification and purification of the responsible ingredients against tumor effects. Chihara et al. (1969) reported that water-soluble polysaccharide isolated from *L. edodes* inhibited the growth of mouse sarcoma 180 in albino mice and lead to complete regression at the dose of 1–10 mg/kg body weight. Fujimoto et al. (1991) isolated and characterized antitumor polysaccharide KS-2 composed of xylose-containing polysaccharide and protein from *L. edodes*, which completely suppressed the growth of Ehrlich ascites carcinoma in mice treated with 1–5 mg/kg body weight. Similar results were also obtained by Sugiyama et al. (1995), who identified anticarcinogenic actions of a water-soluble compound from *L. edodes* which suppressed the growth of Ehrlich ascites carcinoma in mice.

Fiber is considered to be an important ingredient in a balanced and healthy diet. According to Kalac (2009), soluble protein of 4–9% and insoluble fiber of 22–30% are present in mushroom. These fibers play an important role for diabetic patients by reducing their daily insulin requirements and stabilizing their blood glucose profile. Further research on mushroom fiber is required to confirm its health benefits. Various biological activities of mushrooms reported by several researchers are shown in Table 4.6.

Silva et al. (2011) developed a natural potential antioxidant ingredient to preserve soybean oil. They studied the antioxidant activity of extracts of mushroom *Agaricus blazei* and proved that mushroom extract is effective in preserving soybean oil. Mushrooms possess many antioxidant properties. Kasuga et al. (1993) studied the antioxidant activity of crude ethanol extract of 150 Japanese mushrooms using peroxide value in the methyl linoleate system and found that the peroxide value of *Suillus* genus mushrooms was 80% lower than the control. They also studied antioxidant activity of polar and non-polar extracts of oogitake, kugitake, and Maitake mushrooms and found that both polar and non-polar extracts had the highest antioxidant activity (Kasuga et al. 1993). Methanolic extracts of *D. indusiata*, *H. erinaceus*, *T. giganteum*, *F. velutipes*, *L. edodes*, and *P. cystidiosus* mushrooms showed inhibitory effect on lipid peroxidase, 1,1-diphenyl-2-picrylhydrazyl (DPPH) radical scavenging, and hydroxyl scavenging and could chelate ferrous ion (Mau et al. 2002).

Cheung (2010) reported that crude methanol and water extracts of the common Chinese edible mushrooms *L. edodes* and *V. volvacea* showed highest antioxidant activity during β-carotene bleaching, DPPH radical

Table 4.6 Biological activity of mushrooms and active constituents.

Mushroom	Compounds	Uses	References
Auricularia auricula-judae (Bull.) J.S.	Dietary fiber	May be used as antioxidant	Tsuchida et al. 1999b
Agaricus blazei	Riboglucan, glucomannan	Cancer, infections, inflammation, allergy, asthama, and diabetes	Cho et al. 1999
Agaricus bisporus (J.E. Lange) Imbach	Fibers, lectins	Hypocholesterolemic and hypoglycemic	Tsai et al. 2007
Agaricus campestris L.	Lectins	Hypoglycemic, ulcers	Ahmad et al. 1984
Agrocybe aegerita	Glucan	Antioxidant	Zhang et al. 2002
Agrocybe cylindracea (DC) Maire	Agrocybin	Anti-inflammatory, antioxidant	Elsayed et al. 2014
Amanita muscaria	Glucan	Antitumor and immunostimulator	Kiho et al. 1992
Auricularia fuscosuccinea (brown)	Glucan	Antiaging, antioxidant, antitumor and anticoagulant	Lo et al. 2012
Auricularia fuscosuccinea (white)	Glucan	Antioxidant	Lin, et al. 2013
Auricularia mesenterica	Glucan	Antimutagenic, antioxidant	Yoon et al. 2003
Auricularia polytricha	Polysaccharides	Antinociceptive, antioxidant	Mau et al. 2001b
Boletus edulis Bull.	Polysaccharides	Antitumor	Mau et al. 2001b
Calocybe gambosa	Polysaccharides	Antimutagenic, anti-inflammatory, antioxidant	Alves et al. 2013

Calvatia gigantea (Batsch ex Pers.) Lloyd	Polysaccharides	Antioxidant	Badshah et al. 2012
Cantharellus cibaricus	Phenols	Antioxidant, antimicrobial	Ramesh and Pattar 2010
Clitocybe maxima	Phenols, polysaccharides	Antihyperglycemic and antioxidant	Liu et al. 2012
Colybia maculata (Alb. &Schwein.) P. Kumm	Purine derivatives	Antiviral and antioxidant	Leonhardt et al. 1987
Coprinus comatus	Polysaccharides, tocopherols	Antioxidant	Han et al. 2006
Coriolus versicolor	Polysaccharide	Anticancer	Patel and Goyal 2012
Craterellus cornucopioides	Glucan	Antioxidant	Barros et al. 2008
Dictyophora indusiata	Fucamannogalacton	Antitumor, antioxidant	Ramesh and Pattar 2010
Flammulina velutipes (Curtis) Singer	Fibers, polysaccharides	Antioxidant, hypocholesterolemic, antiallergic	Fukushima et al. 2000
Ganoderma lucidum (Curtis) P. Karst	Glucans, triterpenes polysaccharide	Hypoglycemic antioxidant and antitumor Antiviral (HIV-1) Antiallergic Anti-inflammatory Antihepatotoxic	Hikino et al. 1985 Lin et al. 1995

(Continued)

Table 4.6 (Continued)

Mushroom	Compounds	Uses	References
Ganoderma pfeifferi Bres.	Polysaccharide, sesquiterpenoid hydroquinones	Antimicrobial, antiviral	Mothana et al. 2000
Ganoderma tsugae	Arabinoglucan	Antitumor and anti-immunomodulatory	Zhang et al. 2007
Grifola frondosa (Dicks.) Gray	Ergosterol (1)	Antioxidant, hypotensive, hypoglycemic, immunotherapy, anti-inflammatory	Lee et al. 2003
Hericium erinaceus (Bull.) Pers.	Phenol-analogous compounds	Antioxidant, ameliorative effect in Alzheimer's dementia	Kim et al. 2011
Hypsizygus marmoreus	Polysaccharide	Antioxidant and antiallergic	Kim et al. 2014
Inonotus hispidus (Bull.) P. Karst.	Phenolic compounds	Antiallergic and antiviral	Ali et al. 1996
Inonotus obliquus Linn.	Polysaccharides	Used as a folk medicine for cancer and stomach diseases, anticancer	Chen et al. 2010
Laricifomes officinalis (Vill.) Kotl. & Pouzar	Phenols, triterpenoids, coumarins, antithrombin	Hypoglycemic, antioxidant, antimicrobial, and antitumor	Molitoris 1994
Lentinula edodes (Berk.) Pegler	Lentinan, oxalic acid	Antioxidant Hypocholesterolemic Immunotherapy Antimicrobial and antiprotozoal	Wang et al. 1996 Lee et al. 2007 Cheung 2009
Macrolepiota procera	Phenols	Antioxidant	Fernandes et al. 2013
Morchella esculenta	Tocopherol	Antioxidant	Kim et al. 2011
Phellinus linteus	Proteoglycan	Anti-inflammatory	Song et al. 2003

Species	Compound	Activity	Reference
Pholiota nameko (T. Ito) S. Ito & S. Imai	Polysaccharide	Antiallergic	Ji et al. 2012
Pleurotus citrinopileatus	Polysaccharide	Antitumor	Zhang et al. 1994
Pleurotus cystidiosus	Tocopherol, arabinogalacton	Antioxidant, antitumor and immunostimulating polysaccharide	Menikpurage et al. 2009
Pleurotus eryngii (DC) Quel.	Glucans	Antiallergic, antiproliferative, antiangiogenic	Sano et al. 2002 Ngai and Ng 2006
Pleurotus ferulae	Polysaccharide	Antihyperlipidemic	Alam et al. 2011
Pleurotus ostreatus (Jacq. ex Fr.) P. Kumm.	Phenolic compound, tocopherol	Antioxidant and hypocholesterolemic	Wang et al. 1996
Polyporus confluens	Glucan	Hypocholesterolemic, antitumor, and immunostimulatory	Sugiyama et al. 1992
Schizophyllum commune Fries	Schizophyllan	Immunotherapy	Hazama et al. 1995
Trametes versicolor (L. Fr.) Quel.	Coriolan, a β-glucan-protein complex	Immunotherapy and hypoglycemic	Hazama et al. 1995
Termitomyces albuminosus	Termitomycesphins G and H	Antioxidant and neuritogenetic	Qu et al. 2012
Tricholoma giganteum	Polysaccharide	Antioxidant, antitumor, and antihypertensive	Mizuno et al. 1995
Volvariella volvacea (Bulliard ex. Fries) Singer	β-glucosidase	Antioxidant, antitumo,r and anti-inflammatory	Cai et al. 1998

scavenging activity, and erythrocyte hemolysis assay. Similar antioxidant properties have also been reported for *A. cylindracea* and *A. aegerita* (Tsai et al. 2007). Inhibition of lipid peroxidation by *L. edodes* and *V. volvacea* showed antioxidant behavior by scavenging free radicals (Cheung and Cheung 2003). These studies confirmed that edible mushrooms have potential as natural antioxidants due to the ability of their phenolics to inhibit lipid oxidation.

Wang et al. (1996) successfully isolated and characterized two novel structural isomers, asiaticusin A and asiaticusin B, from the fruiting body of *A. asiaticus* using spectrometry, mass spectrometry (MS), and nuclear magnetic resonance. Cheung (2010) identified a polyphenolic compound, epigallocatechin 3-gallate, through liquid chromatography mass spectrometry (LCMS) analysis from *P. tuber-regium* mushroom that showed potential antioxidant activity. In addition, the gas chromatography mass spectrometry (GCMS) method revealed the presence of several phenolics, including cinnamic acid, hydroxybenzoic acid, protocatechuic acid, and caffeic acid, from *A. bisporus* and *L. edodes* (Mattila et al. 2001).

The ability of mushroom-derived preparations (MDPs) to prevent oxidative damage to cellular DNA has been evaluated using the single-cell gel electrophoresis assay (Wang et al. 2011). MDPs obtained from nine common mushrooms including *A. bisporus* and *Ganoderma lucidum* varied in their ability to protect against oxidative DNA (Fukushima et al. 2000). These findings indicated that some edible mushrooms consist of biologically active compounds which have possible commercial value as dietary supplements for offsetting adverse biological effects associated with coronary heart disease, cancer, and age-related neurodegenerative diseases. They might also facilitate the development of treatments for the repair of indiscriminate cellular DNA damage that occurs during certain forms of chemotherapy and radiotherapy (Hazama et al. 1995). The edible mushroom *G. frondosa* contains ergosterol(1), ergosta-4-6-8(14), 22-tetraen-3-one and 1-oleoyl-2-linoleoyl-3-palmitoylglycerol, which inhibit cyclo-oxygenase I and II activity (Zhang et al. 2002).

There is no doubt that mushroom-based products can serve as superior dietary supplements. In China several mushrooms, particularly *Ganoderma*, have been used as dietary supplements for the past 2000 years. *Ganoderma* contains several compounds including tetero-glucans, lectins, terpenoids, steroids, nucleic acids, and immunomodulatory proteins such as Ling Zhi-8 and it could be used as an anticancer, antitumor, antiviral, and antibacterial agent. Synergistic effects of several components in an extract are responsible for its therapeutic value and it can be tentatively concluded that mushroom products are of multifunctional value.

Conclusion

Mushrooms are defined as "a macro fungus with distinctive fruiting bodies that could be hypogeous or epigeous, large enough to be seen by naked eyes and to be picked by hands." The Basidiomycetes and some species of Ascomycetes are categorized as mushrooms. Mushrooms constitute 22 000 known species which are widely available and about 10% of them have been explored for their nutritional value and health benefits. Only a few species of mushrooms are edible. Edible mushrooms are appreciated not only for their texture and flavor but also for their chemical and nutritional characteristics. Many Asian countries traditionally use wild edible mushrooms as nutritional foods and medicine. Mushrooms have higher protein contents and minerals and contain less fat but are rich in B vitamins, vitamin D, vitamin K, and sometimes vitamins A and C. Mushrooms are not only sources of nutrients but also have been reported as therapeutic foods, useful in preventing diseases such as hypertension, diabetes, hypercholesterolemia, and cancer. Certain mushroom species have antitumor, antiviral, antithrombotic, and immunomodulating properties. A limited number of mushroom species have the potential to lower elevated blood sugar levels. The health beneficial characteristics of mushrooms are mainly due to the presence of dietary fiber, in particular chitin and β-glucans. Therefore, it would be useful to carry out more research on mushrooms with a view to identifying active ingredients which contribute health-promoting functional food characteristics and also nutraceutical and disease-preventing functions.

References

Ahmad, N., Bansal, A.K. and Kidwai, J.R. (1984) Effect of PHA-B fraction of Agaricus bisporus lectin on insulin release and 45Ca2C uptake by islet of Langerhans *in vitro*. *Acta Diabetol.*, 21: 63–70.

Aida, F.M., Shuhaimi, M., Yazid, M. and Maaruf, A.G. (2009) Mushroom as a potential source of prebiotics: a review. *Trends Food Sci. Technol.*, 20(11): 567–575.

Alam, N., Yoon, K.N. and Lee, T.S. (2011) Antihyperlipidemic activities of Pleurotus ferulae on biochemical and histological function in hypercholesterolemic rats. *J. Res. Med. Sci.*, 16(6): 776.

Ali, N.A., Jansen, R., Pilgrim, H., Liberra, K. and Lindequist, U. (1996) Hispolon, a yellow pigment from Inonotus hispidus. *Phytochemistry*, 41(3): 927–929.

Alves, M.J., Ferreira, I.C., Froufe, H.J., Abreu, R.M.V., Martins, A. and Pintado, M. (2013) Antimicrobial activity of phenolic compounds identified in wild mushrooms, SAR analysis and docking studies. *J. Appl. Microbiol.*, 115(2): 346–357.

An, Y.H. (2007) Method for producing soybean paste containing pine mushroom. Google Patent US20090053363.

Badshah, H., Qureshi, R.A., Khan, J. et al. (2012) Pharmacological screening of Morchella esculenta (L.) Pers., Calvatia gigantea (Batsch ex Pers.) Lloyd and Astraeus hygrometricus Pers., mushroom collected from South Waziristan (FATA). *Chemistry.* 45–65.

Bano, Z. and Rajarathnam, S. (1988) Pleurotus mushrooms. Part II. Chemical composition, nutritional value, post-harvest physiology, preservation, and role as human food. *Crit. Rev. Food Sci. Nutr.*, 27: 87–158.

Barros, L., Cruz, T., Baptista, P., Estevinho, L.M. and Ferreira, I.C. (2008) Wild and commercial mushrooms as source of nutrients and nutraceuticals. *Food Chem. Toxicol.*, 46(8): 2742–2747.

Bassan, J.C., Ferreira, G.A.O., Bueno, M. and Escouto, L.F.S. (2011)_ Physical characteristics and sensory type in gluten-free cake sponge cake with mushroom *Agaricus brasiliensis. Rev. Alimentus*, 1(1): 273–287.

Beluhan, S. and Ranogajec, A. (2011) Chemical composition and nonvolatile components of Croatian wild edible mushrooms. *Food Chem.*, 124(3): 1076–1082.

Cai, Y.J., Buswell, J.A. and Chang, S.T. (1998) β-Glucosidase components of the cellulolytic system of the edible straw mushroom, Volvariella volvacea. *Enzyme Microb. Technol.*, 22(2): 122–129.

Chang, S.T. (1980) Mushroom production in Southeast Asia. *Mushroom Newsl. Trop.*, 1(2): 18–22.

Chang, S.T. (1999) World production of cultivated edible and medicinal mushrooms in 1997 with emphasis on Lentinus edodes (Berk.) Sing, in China. *Int. J. Med. Mushrooms*, 1(4): 43–48.

Chang, S.T. and Buswell, J.A. (1996) Mushroom nutriceuticals. *World J. Microb. Biot.*, 12: 473–476.

Chang, S.T. and Miles, P.G. (1987) Historical record of the early cultivation of *Lentinus* in China. *Mushroom J. Trop.*, 7: 47–52.

Chang, S.T. and Miles, P.G. (1997) Mushroom biology: concise basics and current developments. *World Scientific*, 144.

Chang, S.T. and Miles, P.G. (2004) *Mushrooms: Cultivation, Nutritional Value, Medicinal Effect, and Environmental Impact*. Boca Raton: CRC Press, pp. 27–38.

Chang, S.T., Buswell, J.A. and Chiu, S.W. (1993) *Mushroom Biology and Mushroom Products*. Hong Kong: Chinese University Press.

Chen, S.Y., Ho, K.J., Liang, C.H., Tsai, C.H., Huang, L.Y. and Mau, J.L. (2012) Preparation of culinary-medicinal king oyster mushroom Pleurotus eryngii fermented products with high ergothioneine content and their taste quality. *Int. J. Med. Mushrooms*, 14: 85–93.

Chen, Y., Gu, X., Huang, S.Q., Li, J., Wang, X. and Tang, J. (2010) Optimization of ultrasonic/microwave assisted extraction (UMAE) of polysaccharides from Inonotus obliquus and evaluation of its anti-tumor activities. *Int. J. Biol. Macromol.*, 46(4): 429–435.

Cheung, L.M. and Cheung, C.K. (2003) Antioxidant activity and total phenolics of edible mushroom extracts. *Food Chem.*, 81: 249–265.

Cheung, P.C.K. (2009) Nutritional value and health benefits of mushrooms. In: Cheung, P.C.K. (ed.) *Mushrooms as Functional Foods*. Chichester: Wiley, pp. 71–109.

Cheung, P.C.K. (2010) Mushroom and health. *Nutr. Bull.*, 35: 292–299.

Chihara, G., Maeda, Y., Sasaki, T. and Fukuoka, F. (1969) Inhibition of mouse sarcoma 180 by polysaccharides from *Lentinus edodes* (Berk.). *Nature*, 222: 687–688.

Cho, S.M., Park, J.S., Kim, K.P., Cha, D.Y., Kim, H.M. and Yoo, I.D. (1999) Chemical features and purification of immunostimulating polysaccharides from the fruit bodies of *Agaricus blazei*. *Korean J. Mycol.*, 27: 170–174.

Chung, M.J., Chung, C.K., Jeong, Y. and Ham, S.S. (2010) Anticancer activity of subfractions containing pure compounds of Chaga mushroom (*Inonotus obliquus*) extract in human cancer cells and in Balbc/c mice bearing Sarcoma-180 cells. *Nutr. Res. Pract.*, 4(3): 177–182.

Crisan, E.V. and Sands, A. (1978) *Nutritional Value*. New York: Academic Press, pp. 137–168.

Dijkstra, F. and Wiken, T. (1976) Studies on mushroom flavours. 1. Organoleptic significance of constituents of the cultivated mushroom, Agaricus bisporus. *Zeitschrift fur Lebensmittel- Untersuchung und-Forschung*, 160: 255–262.

El-Mekkawy, S., Meselhy, M.R., Nakamura, N., Tezuka, Y., Hattori, M. and Kakiuchi, N. (1998) Anti-HIV-1 and anti- HIV-1-protease substances from *Ganoderma lucidum*. *Phytochemistry*, 49: 1651– 1657.

Elsayed, E.A., El Enshasy, H., Wadaan, M.A. and Aziz, R. (2014) Mushrooms: a potential natural source of anti-inflammatory compounds for medical applications. *Mediat. Inflamm.*, 3(2): 24–36.

FAO, IFAD, WFP (2014) *The State of Food Insecurity in the World.* 1(1): 24–32.

Fernandes, A., Barros, L., Barreira, J.C. et al. (2013) Effects of different processing technologies on chemical and antioxidant parameters of Macrolepiota procera wild mushroom. *LWT-Food Sci. Technol.*, 54(2): 493–499.

Fortes, R.C. and Novaes, M.R.C. (2006) Effects of dietary supplementation with Agaricales mushrooms and other fungi in medicinal therapy against cancer. *Rev. Bras. Cancerol.*, 52(4): 363–371.

Fujimoto, S., Furue, H., Kimura, T., Kondo, T., Orita, K. and Taguchi, T. (1991) Clinical outcome of postoperative adjuvant immunochemotherapy with sizofiran for patients with resectable gastric cancer-a randomized controlled study. *Eur. J. Cancer*, 27: 1114–1118.

Fuke, S. and Shimizu, T. (1993) Sensory and preference aspects of umami. *Trends Food Sci. Technol.*, 4(8): 246–251.

Fuke, S. and Ueda, Y. (1996) Interactions between umami and other flavor characteristics. *Trends Food Sci. Technol.*, 7(12): 407–411.

Fukushima, M., Nakano, M., Morii, Y., Ohashi, T., Fujiwara, Y. and Sonoyama, K. (2000) Hepatic LDL receptor mRNA is increased by dietary mushroom (*Agaricus bisporus*) fiber and sugar beet fiber. *J. Nutr.*, 130: 2151–2156.

Han, C., Yuan, J., Wang, Y. and Li, L. (2006) Hypoglycemic activity of fermented mushroom of Coprinus comatus rich in vanadium. *J. Trace Elements Med. Biol.*, 20(3):191–196.

Hasler, C.M. (1998) Functional foods: their role in disease prevention and health promotion. *Food Technol.*, 52: 63–70.

Hazama, S., Oka, M., Yoshino, S., Iizuka, N. and Kyamamoto, W. (1995) Clinical effects and immunological analysis of intraabdominal and intrapleural injection of lentinan for malignant ascites and pleural effusion of gastric carcinoma. *Cancer Chemother. Pharmacol.*, 22: 1595–1597.

Hikino, H., Konno, C., Mirin, Y. and Hayashi, T. (1985) Isolation and hypoglycemic activity of ganoderans A and B, glycans of Ganoderma lucidum fruit bodies. *Planta Med.*, 51(4): 339–340.

Huang, B.H., Yung, K.H. and Chang, S.T. (1985) The sterol composition of *Volvariella volvacea* and other edible mushrooms. *Mycologia*, 77: 959–963.

Huang, S.J., Tsai, S.Y., Lee, Y.L. and Mau, J.L. (2006) Nonvolatile taste components of fruit bodies and mycelia of Cordyceps militaris. *LWT–Food Sci. Technol.*, 39(6): 577–583.

Ji, H.F., Zhang, L.W., Zhang, H.Y., Li, G.L. and Yang, M.D. (2012) Antioxidant activities of extracts from Pholiota Nameko. *Adv.Materials Res.*, 343: 457–462.

Jones, P.J. (2002) Clinical nutrition: 7. Functional foods – more than just nutrition. *Can. Med. Assoc. J.*, 166(12): 1555–1563.

Kalac, P. (2009) Chemical composition and nutritional values of European species of wild growing mushrooms: a review. *Food Chem.*, 113: 9–16.

Kasuga, A., Aoyagi, Y. and Sugahara, T. (1993) Antioxidative activities of several mushroom extracts. *J. Japan. Soc. Food Sci. Technol.*, 40: 56–63.

Kiho, T., Nagai, K., Ukai, S. and Haga, M. (1992) Structure and antitumor activity of a branched $(1 \rightarrow 3)$-β-d-glucan from the alkaline extract of Amanita muscaria. *Carbohydrate Res.*, 224: 237–243.

Kim, J.A., Lau, E., Tay, D. and de Blanco, E.J.C. (2011) Antioxidant and NF-κB inhibitory constituents isolated from Morchella esculenta. *Nat. Prod. Res.*, 25(15): 1412–1417.

Kim, S.P., Kang, M.Y., Kim, J.H., Nam, S.H. and Friedman, M. (2011) Composition and mechanism of antitumor effects of Hericium erinaceus mushroom extracts in tumor-bearing mice. *J. Agric. Food Chem.*, 59(18): 9861–9869.

Kim, T., Park, K., Jung, H. S., Kong, W.S., Jeon, D. and Lee, S.H. (2014) Evaluation of anti-atopic dermatitis activity of Hypsizigus marmoreus extract. *Phytother. Res.*, 28(10): 1539–1546.

Konno, S., Aynehchi, S., Dolin, D.J., Schwartz, A.M., Choudhury, M.S. and Tazakin, H.N. (2002) Anticancer and hypoglycaemic effects of polysaccharides in edible and medicinal Maitake mushroom [*Grifola frondosa* (Dicks.:Fr.) S.F.Gray]. *Int. J. Med. Mushrooms*, 4: 185–195.

Lau, O.W., Shiu, K.K. and Chang, S.T. (1985) Determination of ascorbic acid in vegetables and fruits by differential pulse polarography. *J. Sci. Food Agric.*, 36: 733–739.

Lee, B.C., Bae, J.T., Pyo, H.B. et al. (2003) Biological activities of the polysaccharides produced from submerged culture of the edible Basidiomycete Grifola frondosa. *Enzyme Microb. Technol.*, 32(5): 574–581.

Lemos, F.M.R. (2009) Preparation and characterization of the product similar to burgers mushroom *Agaricus brasiliensis*. Thesis, Federal University of Parana.

Leon-Guzman, M.F., Silva, I. and Lopez, M.G. (1997) Proximate chemical composition, free amino acid contents, and free fatty acid contents of some wild edible mushrooms from Queretaro, *Mexico. J. Agric. Food Chem.*, 45(11): 4329–4332.

Leonhardt, K., Anke, T. and Hillen-Maske, E. (1987) 6-Methylpurine, 6-methyl-9-β-D-ribofuranosylpurine, and 6-hydroxymethyl-9-β-D-ribofuranosylpurine as antiviral metabolites of Collybia maculata (Basidiomycetes). *Zeitschrift für Naturforschung C*, 42(4): 420–424.

Lin, J.M., Lin, C.C., Chen, M.F., Ujiie, T. and Takada, A. (1995) Radical scavenger and antihepatotoxic activity of Ganoderma formosanum, Ganoderma lucidum and Ganoderma neo-japonicum. *J. Ethnopharmacol.*, 47(1): 33–41.

Lin, W.Y., Yang, M.J., Hung, L.T. and Lin, L.C. (2013) Antioxidant properties of methanol extract of a new commercial gelatinous mushrooms (white variety of Auricularia fuscosuccinea) of Taiwan. *Afr. J. Biotechnol.*, 12: 6210–6221.

Lindequist, U., Niedermeyer, T.H.J. and Julich, W.D. (2005) The pharmacological potential of mushrooms. *Evidence-Based Comp. Altern. Med.*, 2: 285–299.

Liu, J. (2002) Biologically active substances from mushrooms in Yunnan, China. *Heterocycles*, 57: 157–167.

Liu, Y.T., Sun, J., Luo, Z.Y. et al. (2012) Chemical composition of five wild edible mushrooms collected from Southwest China and their antihyperglycemic and antioxidant activity. *Food Chem. Toxicol.*, 50(5): 1238–1244.

Lo, Y.C., Lin, S.Y., Ulziijargal, E. et al. (2012) Comparative study of contents of several bioactive components in fruiting bodies and mycelia of culinary-medicinal mushrooms. *Int. J. Med. Mushrooms*, 14(4): 85–97.

Manzi, P., Aguzzi, A. and Pizzoferrato, L. (2001) Nutritional value of mushrooms widely consumed in Italy. *Food Chem.*, 73: 321–325.

Mattila, P., Konko, K., Euvola, M., Pihlava, J., Astola, J. and Vahteristo, L. (2001) Contents of vitamins, mineral elements and some phenolic compound in cultivated mushrooms. *J. Agric. Food Chem.*, 42: 2449–2453.

Mau, J.L., Wu, K.T., Wu, Y.H. and Lin, Y.P. (1998a) Nonvolatile taste components of ear mushrooms. *J. Agric. Food Chem.*, 46(11): 4583–4586.

Mau, J.L., Lin, Y.P., Chen, P.T., Wu, Y.H. and Peng, J.T. (1998b) Flavor compounds in king oyster mushrooms *Pleurotus eryngii*. *J. Agric. Food Chem.*, 46: 4587–4591.

Mau, J.L., Lin, H.C. and Chen, C.C. (2001a) Non-volatile components of several medicinal mushrooms. *Food Res. Int.*, 34: 521–526.

Mau, J.L., Lin, H.C. and Chen, C.C. (2001b) Non-volatile taste components of several medicinal mushrooms. *Food Res. Int.*, 34: 521–526.

Mau, J.L., Lin, H.C. and Chen, C.C. (2002) Antioxidant properties of several medicinal mushrooms. *J. Agric. Food Chem.*, 50(21): 6072–6077.

Menikpurage, I.P., Abeytunga, D.T.U., Jacobsen, N.E. and Wijesundara, R.L.C. (2009) An oxidized ergosterol from Pleurotus cystidiosus active against anthracnose causing colletotrichumgloeosporioides. *Mycopathologia*, 167(3): 155.

Menikpurage, I.P., Soysa, S.S. and Abeytunga, D.T.U. (2012) Antioxidant activity and cytotoxicity of the edible mushroom, Pleurotus cystidiosus against Hep-2 carcinoma cells. *J. Natl Sci Found Sri Lanka*, 40: 2–18.

Mesomo, M.C., Helms, K.M., Zuim, D.R., Visentainer, J.V., Costa, S.M.G. and Pintro, P.T.M. (2010) Evaluation of shelf life of cheese added to the residue of the extract Agaricus blazei Murrill. *Amb. Rev. Sector Agric. Environ. Sci.*, 6(3).

Miles, P.G. and Chang, S.T. (1997) *Mushroom Biology: Concise Basics and current Developments*. Singapore: World Scientific.

Miles, P. G. and Chang, S.T. (2004) *Mushrooms: Cultivation, Nutritional Value, Medicinal Effect, and Environmental Impact*. Boca Raton: CRC Press.

Mizuno, T. (1995) Bioactive biomolecules of mushrooms: food function and medicinal effect of mushroom fungi. *Food Rev. Int.*, 11: 7–12.

Mizuno, T. (1999) The extraction and development of antitumoractive polysaccharides from medicinal mushrooms in Japan (review). *Int. J.Med. Mushrooms*, 1: 9–30.

Mizuno, T., Kinoshita, T., Zhuang, C., Ito, H. and Mayuzumi, Y. (1995) Antitumor-active heteroglycans from Niohshimeji mushroom, Tricholoma giganteum. *Biosci. Biotechnol. Biochem.*, 59(4): 568–571.

Molitoris, H.P. (1994) Mushrooms in medicine. *Folia Microbiol.*, 39(2): 91–98.

Mothana, R.A., Jansen, R., Jülich, W.D. and Lindequist, U. (2000) Ganomycins A and B, new antimicrobial farnesyl hydroquinones from the basidiomycete Ganoderma pfeifferi. *J. Nat. Prod.*, 63(3): 416–418.

Nakamura, E., Torii, K. and Uneyama, H. (2008) Physiological roles of dietary free glutamate in gastrointestinal functions. *Biol. Pharmaceut. Bull.*, 31(10): 1841–1843.

Ngai, P.H. and Ng, T.B. (2006) A hemolysin from the mushroom Pleurotus eryngii. *Appl. Microbiol. Biotechnol.*, 72(6): 1185–1191.

Ohyama, S., Ishibashi, N., Tamura, M., Nishizaki, H. and Okai, H. (1988) Synthesis of bitter peptides composed of aspartic acid and glutamic acid. *Agric. Biol. Chem.*, 52(3): 871–872.

Ouwehand, A.C., Salminen, S. and Isolauri, E. (2002) Probiotics: an overview of beneficial effects. *Antonie Van Leeuwenhoek*, 82(1-4): 279–289.

Patel, S. and Goyal, A. (2012) Recent developments in mushrooms as anti-cancer therapeutics: a review. *3 Biotech*, 2(1): 1–15.

Qu, Y., Sun, K., Gao, L. et al. (2012) Termitomycesphins G and H, additional cerebrosides from the edible Chinese mushroom Termitomyces albuminosus. *Biosci. Biotechnol. Biochem.*, 76(4): 791–793.

Ramesh, C. and Pattar, M.G. (2010) Antimicrobial properties, antioxidant activity and bioactive compounds from six wild edible mushrooms of western ghats of Karnataka, India. *Pharmacog. Res.*, 2(2): 107.

Rathee, S., Rathee, D., Rathee D., Kumar, V. and Rathee, P. (2012) Mushroom as therapeutic agents. *Brazil. J. Pharmacog.*, 22(2): 459–474.

Roberfroid, M.B. (2000) Concepts and strategy of functional food science: the European perspective. *Am. J. Clin. Nutr.*, 71: 1660–1664.

Sano, M., Yoshino, K., Matsuzawa, T. and Ikekawa, T. (2002). Inhibitory effects of edible higher basidiomycetes mushroom extracts on mouse type IV allergy. *Int. J. Med. Mushrooms*, 4(1).

Shao,W. and Huang, Y. (2009) A composite mushroom flavor seasoning and its preparation method. Chinese Patent CN200910063604.X.

Silva, M.C.S., Nozuka, J., Oliveira, P.V. et al. (2010) *In vivo* bioavailablity of selenium in enriched *Pleurotus ostreatus* mushrooms. *Metallomic*, 2: 162–166.

Siro, I., Kapolna, E., Kapolna, B. and Lugasi, A. (2008) Functional food. Product development, marketing and consumer acceptance – a review. *Appetite*, 51: 456–467.

Sommer, I. (2008) Effect of gamma irradiation on selected compounds of fresh mushrooms. MSc thesis, Wien: Uniwien, 0105517.

Song, Y.S., Kim, S.H., Sa, J.H., Jin, C., Lim, C.J. and Park, E.H. (2003) Anti-angiogenic, antioxidant and xanthine oxidase inhibition activities of the mushroom Phellinus linteus. *J. Ethnopharmacol.*, 88(1): 113–116.

Stojkovic, D., Reis, F.S., Barros, L. et al. (2013) Nutrients and non-nutrients composition and bioactivity of wild and cultivated Coprinus comatus (OF Müll.) Pers. *Food Chem. Toxicol.*, 59: 289–296.

Sugiyama, K., Saeki, S., Tanaka, A., Yoshida, S., Sakamoto, H. and Ishiguro, Y. (1992) Hypocholesterolemic activity of Ningyotake (Polyporus confluens) mushroom in rats. *J. Jpn Soc. Nutri. Food Sci.*, 45(3): 265–270.

Sugiyama, K., Akachi, T. and Yamakawa, A. (1995) Hypocholesterolemic action of eritadenine is mediated by a modification of hepatic phospholipid metabolism in rats. *J. Nutr.*, 125: 2134–2144.

Sun, J.Y., Sun, M.Q., He, C.S, and Zhao, J. (2007) A mushroom refined condiments and its manufacturing method. Chinese Patent CN101011135.

Tsai, S., Tsai, H. and Mau, J. (2007) Nonvolatile taste components of fruit bodies and mycelia of shaggy ink cap Mushroom Coprinus comatus (OF Mull.: Fr.) Pers. (Agaricomycetideae). *Int. J. Med. Mushrooms*, 9: 47–55.

Tsai, S.Y., Tsai, H.L. and Mau, J.L. (2008) Non-volatile taste components of Agaricus blazei, Agrocybe cylindracea and Boletus edulis. *Food Chem.*, 107: 977–983.

Tsai, S.Y., Huang, S.J., Lo, S.H., Wu, T.P., Lian, P.Y. and Mau, J.L. (2009) Flavour components and antioxidant properties of several cultivated mushrooms. *Food Chem.*, 113(2): 578–584.

Tsuchida, H., Mizuno, M., Taniguchi, Y., Ito, H., Kawade, M. and Akasaka, K. (1999) Glucomannan separated from *Agaricus blazei* mushroom culture and antitumor agent containing as active ingredient. Japanese Patent 11-080206.

Tu, H.M. (2012) A method for preparing mushrooms essence and its application in seasoning production. Chinese Patent CN201210116730.9.

Wang, H., Gao, J. and Ng, T.B. (2000) A new lectin with highly potent antiheptoma and antisarcoma activities from the oyster mushroom *Pleurotus ostreatus*. *Biochem. Biophys. Res. Commun.*, 275: 810–816.

Wang, H.X., Liu, W.K., Ng. T.B. and Chang S.T. (1996) The immunomodulatory and antitumor activities of lectins from the mushroom. *Immunopharmacognosy*, 31: 205–211.

Wang, Z.M., Peng, X., Lee, K.L.D., Tang, J.C.O., Cheung, P.C.K. and Wu, J.Y. (2011) Structural characterisation and immunomodulatory property of an acidic polysaccharide from mycelial culture of Cordyceps sinensis fungus Cs-HK1. *Food Chem.*, 125(2): 637–643.

Wasser, S.P. (2002) Medicinal mushrooms as a source of antitumor and immunomodulating polysaccharides. *Appl. Microbiol. Biot.*, 60: 258–274.

Yang, J.H., Lin, H.C. and Mau, J.L. (2001) Non-volatile taste components of several commercial mushrooms. *Food Chem.*, 72: 465–471.

Yang, J.H., Lin, H.C. and Mau, J.L. (2002) Antioxidant properties of several commercial mushrooms. *Food Chem.*, 77: 229–237.

Yoon, S.J., Yu, M.A., Pyun, Y.R. et al. (2003) The nontoxic mushroom Auricularia auricula contains a polysaccharide with anticoagulant activity mediated by antithrombin. *Thromb. Res.*, 112(3): 151–158.

Zhang, J., Wang, G., Li, H. et al. (1994) Antitumor polysaccharides from a Chinese mushroom,"yuhuangmo" the fruiting body of Pleurotus citrinopileatus. *Biosci Biotechnol. Biochem.*, 58(7): 1195–1201.

Zhang, M., Cui, S.W., Cheung, P.C.K. and Wang, Q. (2007) Antitumor polysaccharides from mushrooms: a review on their isolation process, structural characteristics and antitumor activity. *Trends Food Sci. Technol.*, 18(1): 4–19.

Zhang, Y., Mills, G. and Nair, M.G. (2002) Cyclooxygenase inhibitory and antioxidant compounds from the mycelia of the edible mushroom *Grifola frondosa*. *J. Agric. Food Chem.*, 50: 7581–7585.

Zhang, Y., Venkitasamy, C., Pan, Z. and Wang, W. (2013) Recent developments on umami ingredients of edible mushrooms–a review. *Trends Food Sci. Technol.*, 33(2): 78–92.

5

Microbial Production of Organic Acids

Ram Naraian* and Simpal Kumari

Department of Biotechnology, Mushroom Training and Research Centre, Faculty of Science, Veer Bahadur Singh Purvanchal University, Jaunpur (UP), India

*Corresponding author e-mail: ramnarain_itrc@rediffmail.com

Introduction

Organic acids are chemical compounds widely distributed in nature as normal constituents of plants or animal tissues. Microbial production of organic acids is a promising approach (Sauer et al. 2008). It is well established that almost all organisms produce organic acids of low molecular weight.

Organic acids represent a rising chemical segment in which several bio-based compounds such as fumaric, propionic, and itaconic acids are synthesized (Jang et al. 2012). These compounds are formed through the microbial fermentation of carbohydrates and related substrates. Organic acids constitute a key group among the building-block chemicals that can be produced by microbial processes (Sauer et al. 2008). Several organic acids are produced by natural or genetically engineered micro-organisms such as bacteria, fungi, and yeast. During fermentation, many anaerobic and facultatively anaerobic micro-organisms produce organic acids including lactic, succinic, acetic, citric, butyric, and propionic acids (Gottschalk 1985). The gram-positive facultative anaerobes grow on sugars and produce organic acids (Nishimura et al. 2007; Schaffer and Burkovski 2005; Takeno et al. 2007). It seems surprising that so many fungi have a high efficiency of organic acid production at high concentrations. Filamentous fungi are commercially used for production of different organic acids. Yeasts are also used for the production of most acids.

Organic acids have been used for many years in the food, chemical, agriculture, and pharmaceutical industries. Production of many organic acids via economical processes will create new markets by providing

Microbial Functional Foods and Nutraceuticals, First Edition. Edited by Vijai Kumar Gupta, Helen Treichel, Volha (Olga) Shapaval, Luiz Antonio de Oliveira, and Maria G. Tuohy.
© 2018 John Wiley & Sons Ltd. Published 2018 by John Wiley & Sons Ltd.

new opportunities for the chemical industry (Sauer et al. 2008). The chemical industries use organic acids as basic compounds for a wide variety of polymer and solvent production processes. In addition to food safety, supplementation of foods with organic acids imparts flavor and antioxidant activity and maintains organoleptic properties (Gurtler and Mai 2014).

Types of Organic Acid

Organic acids differ on the basis of the involvement of carbon, hydrogen, and oxygen elements. Major types of organic acid produced by microbial activity and discussed below are citric acid, succinic acid, lactic acid, itaconic acid, lactobionic acid, gluconic acid, fumaric acid, propionic acid, and acetic acid (Figure 5.1).

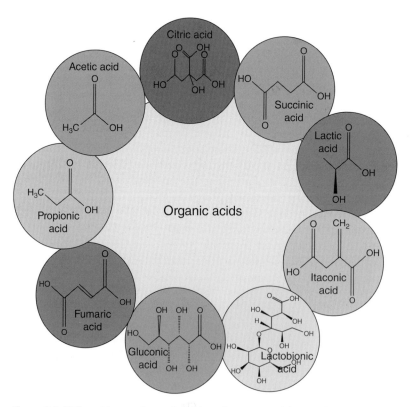

Figure 5.1 Different types of organic acids and summary of their molecular structures.

Citric Acid

Citric acid ($C_6H_8O_7$, 2-hydroxy-1, 2, 3-propane tricarboxylic acid) (CA) is one of the most versatile and widely used organic acids (Figure 5.2). This is as colorless translucent crystals, odorless, with a strongly acid taste (Franz 1991).

Citric acid was first isolated in 1874, from lemon juice. However, the presence of citric acid was initially reported by Wehmer in 1923 as a byproduct of calcium oxalate produced by *Penicillium glaucum* (Vandenberghe et al. 1999). Furthermore, Currie (1917) mentioned the process of industrial production using *Aspergillus niger* in a sugar-based medium. This technique finally became the method of choice for commercial production, mainly due to its economic advantage and biological production over chemical synthesis. The production of CA by bacteria, yeast, and filamentous fungi is well studied.

Several species of bacteria such as *Arthrobacter paraffinens* (Kroya Fermentation Industry 1970), *Bacillus licheniformis* (Sardinas,1972), and *Corynebacterium* sp. (Fukuda et al. 1970) have been studied for citric acid production. Kapoor et al. (1983) reported that these bacteria were efficiently grown in medium containing glucose, urea, calcium, and ammonium sulfate. Later, *Aerobacter, Pseudomonus, Micrococcus, Bacillus, Brevibacterium, Corynebacterium,* and *Arthrobacter* were evaluated (Kapoor et al. 1983) in media containing isocitric acid which was later converted into citric acid (Table 5.1).

Barbesgaard et al. (1992) reported various yeast-based citrate production processes. Yeasts such as *Saccharomycopsis lipolytica* (Ikeno et al. 1975; Wojtatowicz et al. 1993), *Candida tropicalis* (Kapelli et al. 1978), *C. oleophila* (Ishi et al. 1972), *C. guilliermondii* (Gutierrez et al. 1993), *C. citroformans* (Uchio et al. 1975), *C. parapsilosis* (Omar and Rehm 1980), *C. lipolytica* (Miall and Parker 1975; Rane and Sims 1993), *Hansenula anamola* (Oh et al. 1973) and *Yarrowia lipolytica* have been tested for citric acid production using diverse raw materials (Ikeno et al. 1975; Kubicek et al. 1986). Furthermore, Suzuki et al. (1974) also reported some yeasts including *Brettanomyces, Debaromyces, Kloeckers, Torulopsis,* and *Pichia* capable of citric acid production (see Table 5.1).

Citric acid production is the oldest microbial process. Several authors (Hang and Woodams 1987; Roukas 1991; Lu et al. 1997; Pintado et al. 1998; Vandenberghe et al. 1999) reported sugar-based citric acid production by the filamentous fungus *Aspergillus niger*. In several studies many species

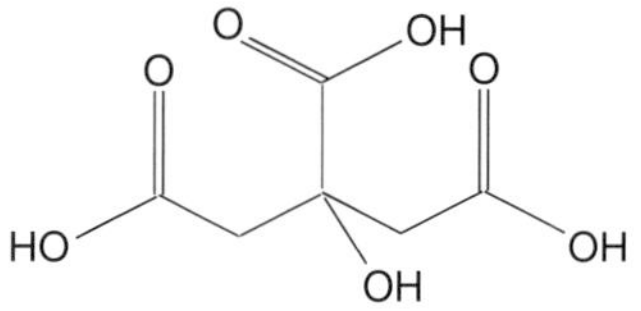

Figure 5.2 Citric acid.

Table 5.1 Summary of studies on citric acid production using three major classes of micro-organisms (bacteria, yeast, and filamentous fungi).

Class of micro-organism	Strains/isolates	Reference
Bacteria	*Arthrobacter paraffinens* *Bacillus licheniformis* *Corynebacterium* sp. *Aerobacter* *Pseudomonas* *Micrococcus* *Bacillus* *Brevibacterium* *Arthrobacter*	Kroya Fermentation Industry 1970 Sardinas 1972 Fukuda et al. 1970 Kapoor et al. 1983
Yeast	*Saccharomycopsis lipolytica* *Yarrowia lipolytica* *Candida tropicalis* *C. oleophila* *C. guilliermondii* *C. citroformans* *C. parapsilosis* *C. lipolytica* *Hansenula anomola* *Brettanomyces* *Debaromyces* *Torulopsis* *Kloeckers* *Pichia*	Ikeno et al. 1975; Wojtatowicz et al. 1993 Ikeno et al. 1975; Kubicek et al. 1986 Kapelli et al. 1978 Ishi et al. 1972 Gutierrez et al. 1993 Uchio et al. 1975 Omar and Rehm 1980 Miall and Parker 1975; Rane and Sims 1993 Oh et al. 1973 Suzuki et al. 1974
Filamentous fungi	*Aspergillus niger* *Aspergillus aculeatus* *A. awamori* *A. carbonarius* *A. foetidus* *A. phoenicis* *A. wentii* *Penicillium janthinellum*	Hang and Woodams 1987 El Dein and Emaish 1979 Grewal and Kalra 1995 El Dein and Emaish 1979 Tran et al. 1998 Yokoya et al. 1992 Karow and Waksman 1947 Grewal and Kalra 1995

of fungi were subsequently reported including *Aspergillus aculeatus* (El Dein and Emaish 1979), *A. carbonarius* (El Dein and Emaish 1979), *A. awamori* (Grewal and Kalra 1995), *A. foetidus* (Tran et al. 1998), *A. wentii* (Karow and Waksman 1947), *A. fonsecaeus, A. phoenicis*

(Yokoya et al. 1992), and *Penicillium janthinellum* (Grewal and Kalra 1995) (see Table 5.1).

Citric acid is used in many industrial applications, in variable ways, e.g., as a flavoring agent, acid for vegetable oils and fats, as a stabilizer and preservative in the food and pharmaceutical industries. It eliminates haze due to trace metals, and prevents color and flavor deterioration. It also inhibits turbidity of wine and adjusts pH. It inverts sucrose, prevents oxidation, and produces darker color in candies, jams, jellies, and dairy products as an antioxidant and emulsifier in cheese and cream (Franz 1991). It is also used in cosmetic products as an antioxidant. It is used for the treatment of boiler water, metal plating, detergents, tanning, and textile processes.

Succinic Acid

Succinic acid (SA) is a diprotic, dicarboxylic acid with the molecular formula $C_4H_6O_4$ and structural formula $HOOC-(CH_2)_2-COOH$ (Zeikus et al. 1999) (Figure 5.3). It is a white, odorless solid and well-known intermediate metabolite of the tricarboxylic acid cycle (TCA).

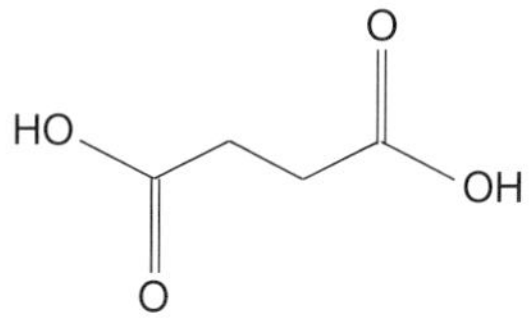

Figure 5.3 Succinic acid.

Georgius Agricola (1546) reported the first purification of succinic acid from amber so it is also called amber acid (Song and Lee 2006). Currently succinic acid is chemically produced from butane through maleic anhydride and petroleum-based raw materials.

The first microbial production of succinic acid was the engineering of mixed acid fermentation of *Escherichia coli* using glucose (Chatterjee et al. 2001). The *Actinobacillus succinogenes* bacterium was investigated for the production of succinic acid using glucose as a carbon source (Li et al. 2011; Xi et al. 2012). In several similar studies *Anaerobiospirillum succiniciproducens* produced succinic acid using glycerol as a carbon source (Jabalquinto et al. 2004; Lee et al. 2001). Lee et al. (2003) reported production of succinic acid by *Mannheimia succiniciproducens* using whey and corn steep liquor. Strains including *Bacillus fragilis* (Beauprez et al. 2010), *Corynebacterium glutamicum* (Okino et al. 2005), and genetically engineered *E. coli* (Khan et al. 2009) were also found to produce succinic acid (Table 5.2).

The production of succinic acid is industrially exploited by fermentation through host organisms such as *Saccharomyces cerevisiae* (Raab and Lang 2011). However, succinic acid production by other yeasts, such as *Candida brumptii* (Sato et al. 1972), *C. zeylanoides* (Kamzolova et al. 2009), *C. catenulata* (Kamzolova et al. 2009), and

Table 5.2 Summary of studies on succinic acid production using three major classes of micro-organisms (bacteria, yeast, and filamentous fungi).

Class of micro-organism	Strains/isolates	Reference
Bacteria	*Actinobacillus succinogenes*	Li et al. 2011, Xi et al. 2012
	Anaerobiospirillum succiniciproducens	Lee et al. 2001; Jabalquinto et al. 2004
	Mannheimia succiniciproducens	Lee et al. 2003
		Beauprez et al. 2010
	Bacillus fragilis	Okino et al. 2005
	Corynebacterium glutamicum	Kahn et al. 2009;
	Escherichia coli	Chatterjee et al. 2001
Yeast	*Saccharomyces cerevisiae*	Andreas et al. 2011
	Candida brumptii	Sato et al. 1972
	Candida zeylanoides	Mandeva et al. 1981;
	Candida catenulata	Kamzolova et al. 2009
	Yarrowia lipolytica	Kamzolova et al. 2009
		Yuzbashev et al. 2010
Filamentous fungi	*Penicillium simplicissimum*	Gallmetzer et al. 2002
	Fusarium sp.	Foster 1949
	Aspergillus sp.	Bercovitz et al. 1990

Yarrowia lipolytica (Yuzbashev et al. 2010), is also well established (see Table 5.2).

Gallmetzer et al. (2002) reported that the filamentous fungus *Penicillium simplicissimum* accumulates succinic acid naturally. A few other species such as *Fusarium* (Foster 1949) and *Aspergillus* (Bercovitz et al. 1990) are also common producers (see Table 5.2).

Succinic acid is widely used as a flavoring agent in foods and beverages, in dyes, insecticides, perfumes, lacquers, in vehicle water cooling systems and the manufacture of clothing, paint, inks, and fibers. Chen et al. (2012) reported increased demand for succinic acid in the synthesis of biodegradable polymers such as polybutyrate succinate (PBS), polyamides, and various green solvents. It is also used in therapeutic medicines (Kondrashova 2002; Kondrashova et al. 2005).

Lactic Acid

Lactic acid (LA) is an organic compound (2-hydroxypropanoic acid) with chemical formula $C_3H_6O_3$ and structural formula $CH_3CHOHCOOH$ (Figure 5.4). It is a colorless or white to light yellow hydroxyl acid in either solid or powder form. Lactic acid has an asymmetric carbon atom and is present in two optically active D (-) and L (+) forms (Senthuran et al. 1997).

Originally, lactic acid was recognized in sour milk. However, in 1857, Pasteur discovered that it is not a milk component but a fermentation product generated by certain micro-organisms (Benninga 1990). Fremy (1881) produced lactic acid by fermentation and it was first used for industrial production (Narayanan et al. 2004).

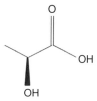

Figure 5.4 Lactic acid.

Microbial production of lactic acid is well established using the bacteria *Carnobacterium funditum* and *Carnobacterium alterfunditum* (Franzmann et al. 1991). *Lactococcus* sp. is widely used for lactic acid production by sucrose inversion (Akerberg et al. 1998; Roissart 1994). In general, lactic acid is produced from lactose using *Lactobacillus plantarum* (Wenge and Matthews 1999) and *Enterococcus faecium* (Nolasco-Hipolito et al. 2012). Similarly, *Leuconostoc mesenteroides* (Fitzpatrick and Keeffe 2001), a lactic acid bacterium, produces optically pure lactic acid using different carbon and nitrogen sources. A few other bacteria including *Escherichia*, *Bacillus*, and *Kluyveromyces* have also been reported to produce lactic acid (Maas et al. 2008). Garde et al. (2002) used a mixed culture of *Lactobacillus pentosus* and *L. brevis* to produce lactic acid from wheat straw hemicelluloses (Table 5.3).

Moreover, lactic acid is produced by *Saccharomyces cerevisiae* in medium containing glucose at low pH (Dequin and Barre 1994). There are several genetically engineered yeasts, such as *Kluyveromyces lactis* (Bianchi et al. 2001), *Pichia stipitis* (Llmen et al. 2007), *Candida boidinii* (Osawa et al. 2009), *Candida utilis* (Ikushima et al. 2009), *Torulaspora delbruekii* (Porro et al. 1999) and *Kluyveromyces marxianus* (Pecota et al. 2007), which have been studied for producing lactic acid (see Table 5.3).

The filamentous fungus *Rhizopus* under aerobic conditions produced lactic acid using glucose. *Rhizopus oryzae* produces lactic acid and requires a mineral medium composition (Bai et al. 2003; Lockwood et al. 1936). Similarly, *R. arrhizus* can convert starch directly to L-lactic acid due to its amylolytic enzyme activity (Wee et al. 2006) (see Table 5.3).

Lactic acid has many potential applications in industry (Datta and Henry 2006), as a descaling agent, pH regulator, neutralizer, chiral intermediate, solvent, cleaning agent, slow acid, metal complexing agent, and antimicrobial agent (Datta et al. 1995). Lactic acid has been utilized by the food industry as an additive for preservation, flavor, mineral fortification, baked goods, pickled vegetables, and beverages. It is frequently used in cosmetics as a moisturizer, pH regulator, skin lightener, and skin

Table 5.3 Summary of studies on lactic acid production using three major classes of micro-organisms (bacteria, yeast, and filamentous fungi).

Class of micro-organism	Strains/isolates	Reference
Bacteria	*Carnobacterium funditum* *Carnobacterium alterfunditum*	Franzmann et al. 1991
	Lactobacillus plantarum	Wenge 1999
	Lactococcus sp.	Roissart,1994; Akerberg et al. 1998
	Leuconostoc mesenteroides	Fitzpatrick and Keeffe 2001
	Enterococcus faecium	Nolasco-Hipolito et al. 2012
	Escherichia	Maas et al. 2008
	Bacillus	
	Kluyveromyces	
	Lactobacillus pentosus	Garde et al. 2002
	Lactobacillus brevis	
Yeast	*Saccharomyces cerevisiae*	Dequin and Barre 1994
	Kluyveromyces lactis	Bianchi et al. 2001
	Pichia stipites	Llmen et al. 2007
	Candida boidinii	Osawa et al. 2009
	Candida utilis	Ikushima et al. 2009
	Torulaspora delbruekii	Porro et al. 1999
	Kluyveromyces marxianus	Pecota et al. 2007
Filamentous fungi	*Rhizopus oryzae* *R. arrhizus*	Lockwood et al. 1936; Bai et al. 2003 Wee et al. 2006

hydrater. Use of lactic acid is also reported in a wide variety of mineral preparations including tablets, prostheses, surgical sutures, and controlled drug delivery systems (Narayanan et al. 2004).

Itaconic Acid

Itaconic acid (IA) (methylene butanedioic acid) has the chemical formula $C_5H_6O_4$ and is commonly known as methylene succinic acid (Figure 5.5). It is a colorless or white crystalline unsaturated dicarbonic acid (Tate 1981). It is stable at acidic, neutral, and middle basic conditions under moderate temperatures (Tate 1981).

Itaconic acid was first discovered by Baup (1837) as a thermal decomposition product of citric acid. Later, Kinoshita (1932) found itaconic acid from *Aspergillus itaconicus*, which was isolated from dried salted plums. Pfeifer et al. (1952) first reported its large-scale industrial production.

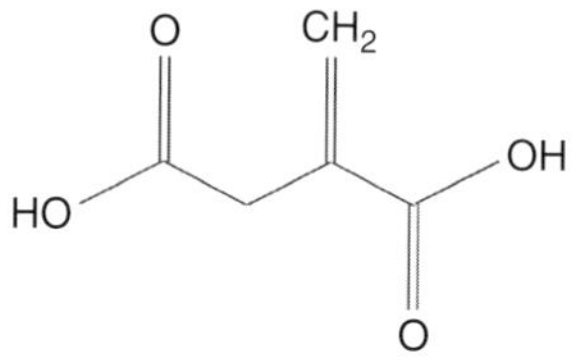

Figure 5.5 Itaconic acid.

The fermentation of glucose by yeast belonging to the genus *Candida* produced itaconic acid (Tabuchi et al. 1981). Willke and Vorlop (2001) also reported itaconic acid production by the *Candida* mutant strain. *Rhodotorula* sp. (Kawamura et al. 1981) and *Pseudozyma antarctica* (William et al. 2006) also produce itaconic acid from glucose and other sugars under nitrogen-limited growth conditions (Table 5.4).

Itaconic acid was produced from the osmophilic filamentous fungus *Aspergillus itaconicus* on sugar growth medium (Kinoshita 1932). A few species of the pathogenic fungal genus *Ustilago* are also known to produce itaconic acid during fermentation (Willke and Vorlop 2001). In later reports, a few other fungal strains, mainly *Aspergillus terreus* (Calam et al. 1939), *Ustilago zeae* (Haskins et al. 1955), *Helicobasidium mompa* (Araki et al. 1957), and mold fungus *Ustilago maydis* (Martinez-Espinoza et al. 2002) efficiently produced itaconic acid (see Table 5.4).

This acid supports large industrial applications such as synthesis of lattices, fibers, polyesters, plastics, artificial glass and preparation of bioactive compounds for the agriculture, pharmacy, and medicine sectors (Kin et al. 1998). Itaconic acid is also used in detergents, cleaners, shampoos, pharmaceuticals, and herbicides (Lancashire 1969) and to improve

Table 5.4 Summary of studies on itaconic acid production using two major classes of micro-organisms (yeast and filamentous fungi).

Class of micro-organism	Strains/isolates	Reference
Yeast	*Candida* sp.	Tabuchi et al. 1981
	Rhodotorula sp.	Kawamura et al. 1981
	Pseudozyma antarctica	William et al. 2006
Filamentous fungi	*Aspergillus itaconicus*	Kinoshita 1932
	Aspergillus terreus	Calam et al. 1939
	Ustilago zeae	Haskins et al. 1955
	Helicobasidium mompa	Araki et al. 1957
	Ustilago maydis	Martinez-Espinoza et al. 2002

adhesives of paper, cellophane, etc. (Pitzl et al. 1951). It may remove unsaturated polyester resins, paint guns and lines, chopper guns, polyure-thane foam, most paints and inks. Applications of IA have been extended to biomedical fields, such as dental and ophthalmic uses, and it has high potential for use in sustained drug release during ocular delivery.

Lactobionic Acid

Lactobionic acid (LBA) is a sugar acid, in the form of a white powder, with the chemical formula $C_{12}H_{22}O_{12}$ (Figure 5.6). It is also called 4-*O*-β-galacto-pyranosyl-D-gluconic acid and is soluble in water, and slightly soluble in organic compounds such as glacial acetic acid, ethanol, and methanol.

Lactobionic acid was first synthesized by Fischer and Meyer (1889) after oxidation of lactose with bromine. Later, several processes includ-ing chemical, electrochemical, biocatalytic and heterogeneous catalytic oxidations were introduced for LBA production. LBA is successfully produced by bacteria and filamentous fungi only.

The production of LBA was first recognized in *Pseudomonas* species. Several bacterial strains produce LBA from aqueous lactose solutions, such as *Pseudomonas graveolens* (Stodola and Lockwood 1947), *P. fragi*, *P. mucidolens*, *P. myxogenes*, and *P. taetrolen* (Stodola and Lockwood 1947). LBA production was also reported through *P. quercito-pyrogallica*, *P. calco-acetica*, and *P. aromatica*, which convert lactose into LBA (Masuo et al. 1952). Futhermore, several other bacterial species such as *Acetobacter orientalis* (Kiryu et al. 2012), *Burkholderia cepacia* (Murakami et al. 2006), *Halobacterium saccharovorum* (Tomlinson et al. 1978), *Bacterium anitratum* (Villecourt and Blachere 1955), *Paraconiothyrium* sp. (Murakami et al. 2008), *Pseudomonas* sp. LS13-1 (Yasumitsu et al. 2000), and *Zymomonas mobilis* (Satory et al. 1997) have been investigated for LBA production by lactose oxidizing activity (Table 5.5).

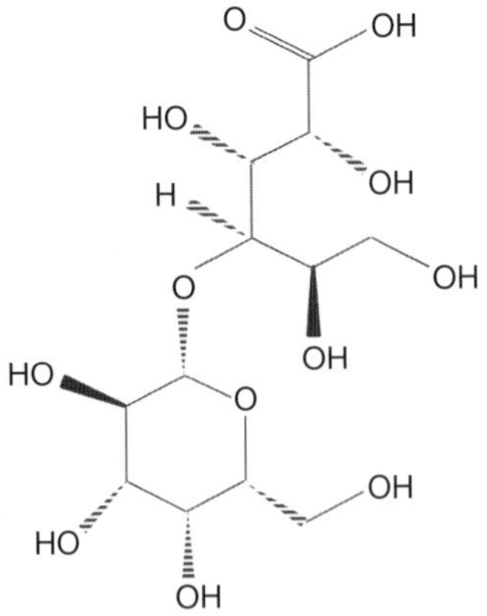

The production of LBA by filamen-tous fungi has also been studied (Bucek et al. 1956). In one study, production of LBA from lactose by *Penicillium chrys-ogenum* (Cort et al. 1956) was reported in shake flask cultures (see Table 5.5).

Lactobionic acid has many applica-tions in the cosmetics (Green et al. 2009; West 2004), pharmaceutical

Figure 5.6 Lactobionic acid.

Table 5.5 Summary of studies on lactobionic acid production using two major classes of micro-organisms (bacteria and filamentous fungi).

Class of micro-organism	Strains/isolates	Reference
Bacteria	*Pseudomonas* sp.	Yasumitsu et al. 2000
	Pseudomonas fragi	Stodola and Lockwood 1947
	P. mucidolens	
	P. myxogenes	
	P. quercito-pyrogallica	Masuo et al. 1952
	P. calco-acetica	
	P. aromatica	
	Acetobacter orientalis	Kiryu et al. 2012
	Burkholderia cepacia	Murakami et al. 2006
	Halobacterium saccharovorum	Tomlinson et al. 1978
	Bacterium anitratum	Villecourt and Blachere 1955
	Paraconiothyrium sp.	Murakami et al. 2008
	Zymomonas mobilis	Satory et al. 1997
Filamentous fungi	*Penicillium chrysogenum*	Cort et al. 1956

(Belzer et al. 1992), food and chemical industries (Gerling 1998). Several brands of cosmetics are currently employing LBA as a key component of novel antiaging products, claiming to reduce photoaging and wrinkles (Grimes et al. 2004), and as a keratinization agent and regenerative skincare product (Green et al. 2009).

Gluconic Acid (Sugar Acid)

Gluconic acid is an organic compound with the molecular formula $C_6 H_{12} O_7$ and structural formula $H O C H_2 (C H O H)_4 C O O H$ (Figure 5.7). It is charactristically a non-corrosive, non-volatile, non-toxic, and mild organic acid with many applications.

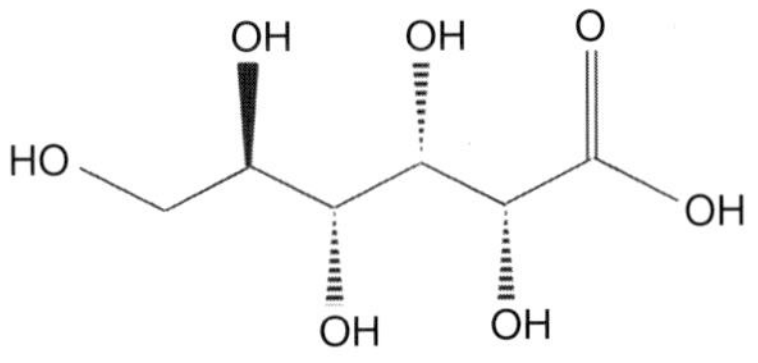

Figure 5.7 Gluconic acid.

Hlasiwetz and Habermann first discovered gluconic acid (Rohr et al. 1983). Bacteria are capable of producing sugar acid, as first reported by Boutroux (1880).

Production of gluconic acid has been predominantly studied with different bacterial species such as *Pseudomonas ovalis* (Bull et al. 1970), *Acetobacter methanolicus* (Loffhagen and Babel 1993), and *A. suboxydans* (Currie and Carter 1930) using diverse media. A few other genera were also found to produce gluconic acid by using glucose, including *Zymomonas mobilis* (Kim and Kim 1992), *Acetobacter diazotrophicus* (Attwood et al. 1991), *Gluconobacter oxydans* (Velizarov and Beschkov 1994), *G. suboxydans* (Shiraishi et al. 1989), *Azospirillum brasiliense* (Rodriguez et al. 2004), etc. (Table 5.6). Gluconic acid production is also possible with yeasts such as *Aureobasidium pullulans* (Anastassiadis et al. 2005; Perwozwansky 1939) (see Table 5.6).

Conversion of glucose to gluconic acid by filamentous fungi is catalyzed by the enzyme glucose oxidase. Various fungal species including *Aspergillus niger* (Moyer 1940), *Penicillium funiculosum* (Kundu and Das 1984), *P. luteum purpurogenum* (May et al. 1927), *P. variabile* (Petruccioli et al. 1994), and *P. amagasakiense* (Kusai et al. 1960) produce gluconic

Table 5.6 Summary of studies on gluconic acid production using three major classes of microorganisms (bacteria, yeast, and filamentous fungi).

Class of micro-organism	Strains/isolates	Reference
Bacteria	*Pseudomonas ovalis*	Bull et al. 1970
	Acetobacter methanolicus	Loffhagen et al. 1993
	Acetobacter suboxydans	Currie and Carter 1930
	Zymomonas mobilis	Kim et al. 1992
	Acetobacter diazotrophicus	Attwood et al. 1991
	Gluconobacter oxydans	Velizarov et al. 1994
	Gluconobacter suboxydans	Shiraishi et al. 1989
	Azospirillum brasiliense	Rodriguez et al. 2004
Yeast	*Aureobasidium pullulans*	Perwozwansky, 1939; Anastassiadis et al. 2005
Filamentous fungi	*Aspergillus niger*	Moyer et al. 1940
	Penicillium funiculosum	Kundu et al. 1984
	P. variabile	Petruccioli et al. 1994
	P. amagasakiense	Kusai et al. 1960
	P. purpurescens	Temash and Olama 1999
	P. luteum purpurogenum	May et al. 1927

acid. Gluconic acid production through *Penicillium purpurescens* using banana waste has been reported (Temash and Olama 1999).

Gluconic acid has wide applications in the food industries as a flavoring agent in syrups and in reducing fat absorption. It is also used in meat and dairy products, particularly in baked goods as a component of the leavening agent (Znad et al. 2003). Gluconic acid is commonly used in chemical industries for bottle washing, washing painted walls and removal of metal carbonate precipitates without causing corrosion.Textile industries also exploit GA where it prevents the deposition of iron and for desizing polyester and polyamide fabrics. It also finds application as an additive to cement for increasing the strength of water resistance. Cleaning compounds such as mouthwashes, common cold treatments, and wound healing compounds use GA as an ingredient.

Fumaric Acid

Fumaric acid 9FA) is a white solid crystal with no odor and slight acidic taste, with the molecular formula $C_4H_4O_4$ (Goto et al. 1998) (Figure 5.8). It is an important intermediate in the citrate cycle (Roa Engel et al. 2008). At higher temperatures, it forms traces of maleic anhydride.

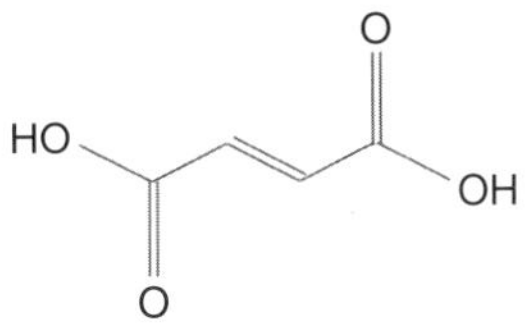

Figure 5.8 Fumaric acid.

Fumaric acid was initially extracted from plants belonging to the genus *Fumaria*. Ehrlich (1911) first discovered fumaric acid production from fungal fermentation of sugar using *R. nigricans*.

The bacteria able to produce maleate isomerase are also capable of synthesizing fumaric acid: *Pseudomonas* sp. (Otsuka 1961), *Alcaligenes faecalis* (Takamura et al. 1969) and *Pseudomonas fluorescens* (Scher and Lennarz 1969). Similarly *P. alcaligenes* has high rates of maleic acid conversion into fumaric acid (Ichikawa et al. 2003; Nakajima Kambe et al. 1997). Song et al. (2013) reported on genetically engineered *E. coli* strains for fumaric acid production (Table 5.7).

Fumaric acid production was reported by yeasts such as *Schizosaccharomyces pombe* (Saayman et al. 2000), *Candida utilis* (Saayman et al. 2000), *Candida sphaerica* (Corte-Real et al. 1989), and *Hansenula anomala* (Corte-Real and Leao 1990) which can convert substrates into fumaric acid (see Table 5.7).

Fumaric acid accumulation has been studied in many filamentous fungi of the *Rhizopus* genus, such as *Rhizopus formosa* (Foster and Waksman 1939), *R. oryzae* (Peleg et al. 1989), *R. nigricans* (Ehrlich 1911), and *R. arrhizus* (Kautola and Linko 1989).

Table 5.7 Summary of studies on fumaric acid production using three major classes of micro-organisms (bacteria, yeast, and filamentous fungi).

Class of micro-organism	Strains/isolates	Reference
Bacteria	*Pseudomonas* sp.	Otsuka 1961
	Alcaligenes faecalis	Takamura et al. 1969
	Pseudomonas alcaligenes	Nakajima Kambe et al.
	Pseudomonas fluorescens	1997; Ichikawa et al. 2003
	E. coli mutant	Scher and Lennarz 1969
		Song et al. 2013
Yeast	*Schizosaccharomyces pombe*	Saayman et al. 2000
	Candida utilis	
	Candida sphaerica	Corte-Real et al. 1989
	Hansenula anomala	Corte-Real and Leao 1990
Filamentous fungi	*Rhizopus nigricans*	Ehrlich 1911
	R. arrhizus	Kautola and Linko 1989
	R. oryzae	Peleg et al. 1989
	R. formosa	Foster and Waksman 1939

Fumaric acid has extensive applications in the food, beverage, and pharmaceutical industries. It is currently used in many food products, including corn and wheat tortillas, refrigerated biscuit doughs, sourdough, rye breads, gelatin desserts, gelling aids, pie fillings, fruit juice, nutraceutical drinks, and wines. Its use as a supplement in cattle feed also leads to a large reduction in methane emission (Roa Engel et al. 2008). Ferrous fumarate is used to treat iron deficiencies like anemia.

Propionic Acid

Propionic acid (PA) is colorless liquid with a rancid odor and $C_3H_6O_2$ molecular formula (Figure 5.9). It is a major intermediate between formic and acetic acid which is miscible in water, but can be removed from water by adding salt. Propionic acid was first discovered by Johann Gottlieb (1844) during the degradation of sugar (Haque et al. 2009).

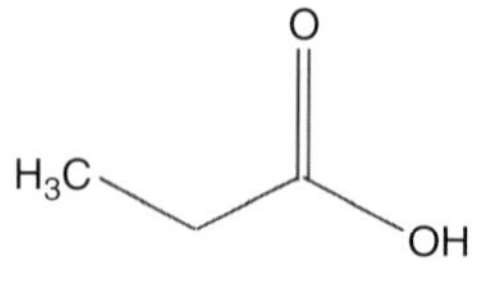

Figure 5.9 Propionic acid.

Available reports suggest that this acid is produced only by bacterial agents such as

Table 5.8 Summary of studies on propionic acid production using one major class of micro-organism (bacteria).

Bacterial strains/isolates	Reference
Clostridium propionicum	Cardon and Barker 1947
Propionibacterium freudenreichii	Playne 1985
Propionibacterium acidipropionici	Steven et al. 1991
Propionibacterium jensenii	Zhuge et al. 2013

Propionibacterium freudenreichii (Playne 1985), *P. acidipropionici* (Steven et al. 1991), and *Clostridium propionicum* (Cardon and Barker 1947). These species can produce PA as a metabolic end product, through the fermentation of lactate. Recently, large-scale production of propionic acid was undertaken using genetically engineered *Propionibacterium jensenii* ATCC 4868 from glycerol (Zhuge et al. 2013) (Table 5.8).

Propionic acid is used in the production of cellulose acetate-based polymers (Playne 1985). It is widely used in the food industry as an artificial flavor and preservative (yeast and mold inhibitor). Its ammonium salt is employed as a common ingredient in animal feed. It is also used in the manufacturing of herbicides, pesticides, pharmaceuticals, perfumes, and plastics (Boyaval and Corre 1995).

Acetic Acid

Acetic acid ($C_2H_4O_2$) (AA), the most familiar edible organic acid systematically named ethanoic acid, is an organic compound commonly called vinegar (Figure 5.10). It is a colorless liquid called glacial acetic acid when in undiluted form. In general, acetic acid occurs naturally in trace amounts in fruits, such as pears. Acetic acid was first produced in 1794 in Germany using grapes. The production of acetic acid predominantly depends on the species and efficiency of micro-organisms. A variety of micro-organisms including bacteria, yeasts, and a few filamentous fungi are able to produce this organic acid.

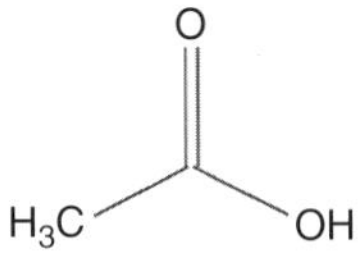

Figure 5.10 Acetic acid.

The literature contains several reports of bacteria producing acetic acid. The most relevant bacteria are gram negative, obligate aerobic, ellipsoidal or cylindrical, found alone, in pairs or in clusters and chains (de Ley et al. 1984). These are able to oxidize ethanol into acetic acid. *Acetobacter* is the oldest classified genus known for production of vinegar which was reclassified by Yamada et al. (1997) as *Gluconacetobacter*

(Yamada et al. 2012). Based on their ability to oxidize acetate and lactate, bacteria are grouped into two particular genera called *Acetobacter* and *Gluconacetobacter* (de Ley et al. 1984; Yamada et al. 1976). Vegas et al. (2010) reported that bacterial species like *Komagataeibacter europaeus*, *Acetobacter hansenii*, and *Acetobacter xylinus* are involved in the production of wine vinegar (Table 5.9).

Spoiling of bottled wine is usually noted as the production of haze, turbidity, and acetic acid by a few genera of yeasts (Sponholz 1993). The production of AA was reported by using different species of yeast such as *Dekkera anomala* (Geros et al. 2000), *Saccharomyces cerevisiae* (Postma et al. 1989), and *D. bruxellensis* (Carrascosa et al. 1981) grown in concentrations of glucose. In a similar study acetic acid production occurred via

Table 5.9 Summary of studies on acetic acid production using three major classes of micro-organisms (bacteria, yeast, and filamentous fungi).

Class of micro-organism	Strains/isolates	Reference
Bacteria	*Komagataeibacter europaeus*	Vegas et al. 2010 Hidalgo et al. 2012, 2013
	Acetobacter hansenii	
	A. xylinus	
	A. malorum	
	A. aceti	
	A. polyoxogenes	
	A. obodiens	
	A. europaeus	
	A. lovaniensis	Gullo et al. 2009
Yeast	*Dekkera bruxellensis*	Carrascosa et al. 1981
	D. anomala	Geros et al. 2000
	Brettanomyces intermedius NRRL Y-2394	Castro-Martinez et al. 2005
	B.custersianus	
	B. clausenni	
	B. bruxellensis	
	Saccharomyces cerevisiae	(Postma et al. 1989)
Filamentous fungi	*Fusarium oxysporum*	Singh et al. 1992
	Polyporus anceps	Perlman et al. 1949

different yeast *Dekkera* and *Brettanomyces* spp. using either glucose or ethanol as a carbon source. Castro-Martinez et al. (2005) reported that acetic acid can be produced directly from glucose by using different strains of yeasts *Brettanomyces bruxellensis* YB-5164, *B. intermedius* NRRL Y-2394, *B. custersianus*, and *B. clausenni*. *A. pasteurianus* seems to be the most common in wine vinegars, although other *Acetobacter* spp. also occur, such as *A. malorum*, *A. aceti*, *A. polyoxogenes*, *A. obodiens*, and *A. europaeus* (Hidalgo et al. 2012, 2013) (see Table 5.9).

The production of acetic acid by filamentous fungi has also been reported. Kumar et al. (1991) screened various strains of *Fusarium oxysporum* and *Monilia brunnae* for their ability to produce acetic acid. *Fusarium oxysporum* DSM 841 was selected as a potential strain producing acetic acid and ethanol. Singh et al. (1992) reported involvement of *Fusarium oxysporum* in acetic acid production. Perlman (1949) mentioned that *Polyporus anceps* is capable of utilizing substrate to produce acetic acid as a major product.

Figure 5.11 Different organic acids used for broad spectrum applications in the field of agriculture, food, pharmaceutical, cosmetics and chemical industries.

Because of its pungent odor and taste, acetic acid has an important place in our kitchens, laboratories, and food industries such as in the preparation of pickles, chutney, salad creams, mayonnaise, dressings, and sauces.

Conclusion

Based on the broad applicability of the organic acids (Figure 5.11), it can be concluded that bacteria, yeasts, and filamentous fungi are the responsible bioagents converting substrate into beneficial acids. The exploitation of these micro-organisms at a large scale should be conducted to achieve a better yield of organic acids for easy and cheap availability. Furthermore, screening of efficient isolates and subsequent efforts regarding strain improvement should be facilitated.

References

Akerberg C., K. Hofvendahl, G. Zacchi and B. Hahn-Hagerdal (1998) Modeling the influence of pH, temperature, glucose and lactic acid concentration on the kinetics of lactic acid production by *Lactococcus lactis* sp. *lactis* ATCC 19435 in whole-wheat flour. *Appl. Microbiol. Biotechnol.*, 49: 682–690.

Anastassiadis S., A. Aivasidis, C. Wandrey and H.J. Rehm (2005) Process optimization of continuous gluconic acid fermentation by isolated yeast-like strains of *Aureobasidium pullulans. Biotechnol. Bioeng.*, 91: 494–501.

Araki T., Y. Yamazaki and N. Suzuki (1957) Studies on the violet root rot sweet potatoes caused by *Helicobasidium mompa. Bull. Natl. Inst. Agric. Sci. Jpn. Ser.*, 8: 53–59.

Attwood M., J.P. van Dijken and J.T. Pronk (1991) Glucose metabolism and gluconic acid production by *Acetobacter diazotrophicus. J. Ferment. Bioeng.*, 72: 101–105.

Bai D.M., M.Z. Jia, X.M. Zhao et al (2003) L(+)-Lactic acid production by pellet form *Rhizopus oryzae* R1021 in a stirred tank fermentor. *Chem. Eng. Sci.*, 58: 785–791.

Barbesgaard P., H.P. Heldt-Hansen and B. Diderichsen (1992) On the safety of *Aspergillus oryzae*: a review. *Appl. Microbiol. Biotechnol.*, 36: 569–572.

Baup S. (1837) Ueber eine neue Pyrogen-Citronensaure und über Benennung der Pyrogen-Säuren uberhaupt. *Ann. Chim. Phys.*, 19: 29–38.

Beauprez J.J., M. de Mey and W.K. Soetaert (2010) Microbial succinic acid production: natural versus metabolic engineered producers. *Process Biochem.*, 45: 1103–1114.

Belzer F.O., A.M. d'Alessandro, R.M. Hoffmann, S.J. Knechtle, A. Reed and J.D. Pirsch (1992) The use of UW solution in clinical transplantation. A 4-year experience. *Ann. Surg.*, 215: 579–585.

Benninga H (1990) *A History of Lactic Acid Making: A Chapter in the History of Biotechnology.* Dordrecht: Kluwer Academic Publishers.

Bercovitz A., Y. Peleg, E. Battat, J.S. Rokem and I. Goldberg (1990) Localization of pyruvate carboxylase in organic acid producing *Aspergillus* strains. *Appl. Environ. Microbiol.*, 56: 1594– 1597.

Bianchi M.M., L. Brambilla, F. Protani, C.L. Liu, J. Lievense and D. Porro (2001) Efficient homolactic fermentation by *Kluyveromyces lactis* strains defective in pyruvate utilization and transformed with the heterologous LDH gene. *Appl. Environ. Microbiol.*, 67: 5621–5625.

Boutroux L. (1880) Physiological chemistry. About a new glucose fermentation. *C.R. Hebd. Seances* Acad. Sci., 91: 236–238.

Boyaval P. and C. Corre (1995) Production of propionic acid. *Lait.*, 75: 453–461.

Bucek W., W.M. Connors, W.M. Cort and H.R. Roberts (1956) Evidence for the formation and utilization of lactobionic acid by *Penicillium chrysogenum. Arch. Biochem. Biophys.*, 63: 477–478.

Bull D.N. and L.L. Kempe (1970) Kinetics of the conversion of glucose to gluconic acid by *Pseudomonas ovalis. Biotechnol. Bioeng.*, 12: 273–290.

Calam C.T., A.E. Oxford and H. Raistrick (1939) Studies in the biochemistry of micro organisms. LXIII. Itaconic acid, a metabolic product of a strain of *Aspergillus terreus* Thom. *Biochem. J.*, 33: 1488– 1495.

Cardon B.P. and H.A. Barker (1947) Amino acid fermentations by *Clostridium propionicum* and *Diplococcus glycinophilus. Arch. Biochem.*, 12: 165–180.

Carrascosa J.M., M.D. Viguera, I. Nunez de Castro, W.A. Schet Ters (1981) Metatabolism of acetaldehyde and Custers effect in the yeast. *Antonie van Leeuwenhoek*, 47: 209–215.

Castro-Martinez C., B.I. Escudero-Abarca, J. Gomez-Rodriguez, P.M. Hayward-Jones and M.G. Aguilar-Uscanga (2005) Effect of physical factors on acetic acid production in *Brettanomyces* strain. *J. Food Process Eng.*, 28: 133–143.

Chatterjee R., C.S. Millard, K. Champion, D.P. Clark and M.I. Donnelly (2001) Mutation of the ptsG gene results in increased production of succinate in fermentation of glucose by *Escherichia coli. Appl. Environ. Microbiol.*, 67: 148–154.

Chen Z., L. Xiao, W. Liu, D. Liu, Y.Y. Xiao and J. Chen (2012) Novel materials which possess the ability to target liver cells. *Expert Opin. Drug Deliv.*, 9: 649–656.

Cort W.M., W.M. Connors, H.R. Roberts and W. Bucek (1956) Evidence for the formation and utilization of lactobionic acid by *Penicillium chrysogenum. Arch. Biochem. Biophys.*, 63: 477–478.

Corte-Real M. and C. Leao (1990) Transport of malic acid and other carboxylic acids in the yeast *Hansenula anomala. Appl. Environ. Microbiol.*, 56: 1109–1113.

Corte-Real M., C. Leao and N. van Uden (1989) Transport of L-malic acid and other dicarboxylic acids in the yeast *Candida sphaerica. Appl. Microbiol. Biotechnol.*, 31: 551–555.

Currie J.N. (1917) Citric acid fermentation. *J. Biol. Chem.*, 31: 15–37.

Currie J.N. and R.H. Carter (1930) Gluconic acid. US Patent 1,896,811.

Datta R. and M. Henry (2006) Lactic acid: recent advances in products, processes and technologies – a review. *J. Chem. Technol. Biotechnol.*, 81: 1119–1129.

Datta R., S.P. Tsai, P. Bonsignore, S.H. Moon and J.R. Frank (1995) Technological and economic potential of poly (lactic acid) and lactic acid derivatives. *FEMS Microbiol. Rev.*, 16: 221–231.

De Ley J., J. Swings and F. Gossele (1984) Genus I *Acetobacter* Beijerinck. In: Kreig N.R. and J.C. Holt (eds) *Bergey's Manual of Systematic Bacteriology.* Baltimore: Williams and Wilkins, pp. 268–274.

Dequin S. and P. Barre (1994) Mixed lactic acid alcoholic fermentation by *Saccharomyces cerevisiae* expressing the *Lactobacillus casei* (L)-LDH. *Biotechnology*, 12: 173–177.

Ehrlich F. (1911) Formation of fumeric acid by means of molds. *Ber. Dtsch. Chem. Ges.*, 44: 3737–3742.

El Dein S.M.N. and G.M.I. Emaish (1979) Effect of various conditions on production of citric acid from molasses in presence of potassium ferrocyanide by *Aspergillus aculeatus* and *A. carbonarius. Indian J. Exp. Biol.*, 17: 105–106.

Fischer E. and J. Meyer (1889) Oxydation des Milchzuckers. *Ber. Deut. Chem. Ges.*, 22: 361–364.

Fitzpatrick J.J. and U.O. Keeffe (2001) Influence of whey protein hydrolyzate addition to whey permeate batch fermentations for producing lactic acid. *Proc. Biochem.*, 37: 183–186.

Foster J.W. (1949) *Chemical Activities of Fungi.* New York: Academic Press.

Foster J.W. and S.A. Waksman (1939) The production of fumaric acid by molds belonging to the genus *Rhizopus. J. Am. Chem. Soc.*, 61: 127–135.

Franz A., W. Burgstaller and F. Schinner (1991) Leaching with *Penicillium simplicisszmum*: influence of metals and buffers on proton extrusion and citric acid production. *Appl. Environ. Microbiol.*, 57: 769–774.

Franzmann P. D., P. Hopfl, N. Weiss and B. J. Tindall (1991) Psychrotrophic, lactic acid-producing bacteria from anoxic waters in Ace Lake, Antarctica: *Carnobacterium funditum* sp. nov. and *Carnobacterium alterfunditum* sp. nov. *Arch. Microbiol.*, 156: 255–262.

Fukuda H., T. Susuki, Y. Sumino and S. Akiyama (1970) Microbial preparation of citric acid. *German Patent* 2: 003–221.

Gallmetzer M., J. Meraner and W. Burgstaller (2002) Succinate synthesis and excretion by *Penicillium simplicissimum* under aerobic and anaerobic conditions. *FEMS Microbiol. Lett.*, 210: 221–225.

Garde A., G. Jonsson, A.S. Schmidt and B.K. Ahring (2002) Lactic acid production from wheat straw hemicellulose hydrolysate by *Lactobacillus pentosus* and *Lactobacillus brevis*. *Biores. Technol.*, 81: 217–223.

Georgius Agricola (1546) *Book IV, De natura fossilium. H. frobenius and N.episcopius.* Basel, Switzerland.

Gerling K.G. (1998) Large scale production of lactobionic acid use and new applications. *Int. Dairy Fed.*, 9804: 251–261.

Geros H., M.M. Azevedo and F. Cassio (2000) Biochemical studies on the production of acetic acid by the yeast *Dekkera anomala*. *Food Technol. Biotechnol.*, 38: 59–62.

Goto M., T. Nara, I. Tokumaru et al. (1998) Method of producing fumaric acid. US Patent 5,783,428.

Gottschalk G. (1985) *Bacterial Metabolism*, 2nd edn. Berlin: Springer.

Green B.A., R.J. Yu and E.J. van Scott (2009) Clinical and cosmeceutical uses of hydroxyacids. *Clin. Dermatol.*, 9: 495–501.

Grewal H.S. and K.L. Kalra (1995) Fungal production of citric acid. *Biotechnol. Adv.*, 13: 209–234.

Grimes P.E., B.A. Green, R.H. Wildnauer and B.L. Edison (2004) The use of polyhydroxy acids (PHAs) in photoaged skin. *Cutis*, 73: 3–13.

Gullo, M., de Vero, L. and Giudici, P. (2009) Succession of selected strains of Acetobacter pasteurianus and other acetic acid bacteria in traditional balsamic vinegar. *Appl. Environ. Microbiol.*, 75(8): 2585–2589.

Gurtler J.B. and T.L. Mai (2014) Traditional preservatives: organic acids. *Encycl. Food Microbiol.*, 3: 119–130.

Gutierrez N.A., I.A. Mckay, C.E. French, J. Brooks and I.S. Maddox (1993) Repression of galactose utilization by glucose in the citrate producing yeast *Candida guilliermondii*. *J. Ind. Microbiol.*, 11: 143–146.

Hang Y.D. and E. E. Woodams (1987) Microbial production of citric acid by solid state fermentation of kiwifruit peel. *J. Food Sci.*, 52: 226–227.

Haque M.N., R. Chowdhury, K M.S. Islam and M.A. Akbar (2009) Propionic acid is an alternative to antibiotics in poultry diet. *Bang. J. Anim. Sci.*, 38: 115–122.

Haskins R.H., J.A. Thorn and B. Boothroyd (1955) Biochemistry of the Ustilaginales. XI. Metabolic products of *Ustilago zeae* in submerged culture. *Can. J. Microbiol.*, 1: 749–756.

Hidalgo C., E. Mateo, A. Mas and M.J. Torija (2012) Identification of yeast and acetic acid bacteria isolated from the fermentation and acetification of persimmon. *Food Microbiol.*, 30: 98–104.

Hidalgo C., E. Mateo, A. Mas and M.J. Torija (2013) Effect of inoculation on strawberry fermentation and acetification processes using native strains of yeast and acetic acid bacteria. *Food Microbiol.*, 34: 88–94.

Ichikawa S., T. Iino, S. Sato, T. Nakahara and S. Mukataka (2003) Improvement of production rate and yield of fumaric acid from maleic acid by heat treatment of *Pseudomonas alcaligenes* strain XD-1. *Biochem. Eng. J.*, 13: 7–13.

Ikeno Y., Y.M., Masuda, K. Tanno, I. Oomori and N. Takahashi (1975) Citric acid production from various raw materials by yeasts. *J. Ferment.Technol.*, 53: 752–756.

Ikushima S., T. Fujii, O. Kobayashi, S. Yoshida and A. Yoshida (2009) Genetic engineering of *Candida utilis* yeast for efficient production of L-lactic acid. *Biosci. Biotechnol. Biochem.*, 73: 1818–1824.

Ishi K., Y. Nakajima and T. Iwakura (1972) Citric acid by fermenetation of waste glucose. *German Patent* 2: 157–847.

Jabalquinto A.M., F.D. Gonzales-Nilo, M. Laivenieks, M. Cabezas, J. G. Zeikus and E. Cardemil (2004) *Anaerobiospirillum succiniciproducens* phosphoenolpyruvate carboxykinase, mutagenesis at metal site 1. *Biochemie*, 86: 47–51.

Jang Y.S., B. Kim, J.H. Shin et al. (2012) Bio-based production of C2–C6 platform chemicals. *Biotechnol. Bioeng.*,109: 2437–2459.

Kamzolova S.V., A.I. Yusupova, E.G. Dedyukhina, T.I. Chistyakova, T.M. Kozyreva and I.G. Morgunov (2009) Succinic acid synthesis by ethanol grown yeast. *Food Technol. Biotechnol.*, 47: 144–152.

Kapelli O., M. Muller and A. Fiechter (1978) Chemical and structural alterations at cell surface of *Candida tropicalis*, induced by hydrocarbon substrate. *J. Bacteriol.*, 133: 952–958.

Kapoor K.K., K. Chaudhary and P. Tauro (1983) Citric acid. In: Reed G. (ed.) *Prescott and Dunn's Industrial Microbiology*. Basingstoke: Macmillan, pp. 709–747.

Karow E.O. and S.A. Waksman (1947) Production of citric acid in submerged culture. *Ind. Eng. Chem.*, 39: 821–825.

Kautola H. and Y.Y. Linko (1989) Fumaric acid production from xylose by immobilized *Rhizopus arrhizus* cells. *Appl. Microbiol. Biotechnol.*, 31: 448–452.

Kawamura D., M. Furuhashi, O. Saito and H. Matsui (1981) Production of itaconic acid by fermentation. Japanese Patent 56,137,893.

Khan M.K., P.K. Mishra, U. Kumar, M. Mishra, T. Sharma and V. Permar (2009) Production of succinic acid of related microbial strain. Glob. *J. Biotechnol. Biochem.*, 4: 47–50.

Kim D.M. and H.S. Kim (1992) Continuous production of gluconic acid and sorbitol from Jerusalem artichoke and glucose using an oxidoreductase of *Zymomonas mobilis* and inulinase. *Biotechnol. Bioeng.*, 39: 336–342.

Kin R., T. Sai and S. So (1998) Itaconate copolymer with quadratic nonlinear optical characteristic. Japanese Patent 10,293,331.

Kinoshita K. (1932) Uber die Production von Itaconsaure and Mannit durch einen neuen Schimmelpiz *Aspergillus itacnicus*. *Acta Phytochim.*, 5: 271–287.

Kiryu K., K. Yamauchi, A. Masuyama et al. (2012) Optimization of lactobionic acid production by *Acetobacter orientalis* isolated from Caucasian fermented milk Caspian sea yogurt. *Biosci. Biotechnol. Biochem.*, 76: 361–363.

Kondrashova M.N. (1991) Interaction of carbonic acid reamination and oxidation processes at various functional states of tissue. *Biokhimiia (Moscow)*, 56: 388–405.

Kondrashova M.N. (2002) Hormone-similar action of succinic acid (in Russian). *Vopr. Biol. Med. Pharmac. Chem.*, 1: 7–15.

Kondrashova M.N., E.L. Maevsky, E.V. Grigorenko et al. (2005) The interaction of Krebs cycle with sympathetic and parasympathetic nervous system. *Appl. Microbiol. Biotechnol.*, 83: 1027–1034.

Kroya Fermentation Industry (1970) Citric acid prepared by fermentation. British Patent 1,187,610.

Kubicek C.P. and M. Rohr (1986) Citric acid fermentation. *Crit. Rev. Biotechnol.*, 3: 331–373.

Kumar P.K.R., A. Singh and K. Schugerl (1991) Formation of acetic acid from Cellulosic substrates. *Appl. Microbiol. Biotechnol.*, 34: 570–572.

Kundu, P. and A. Das (1984) Utilization of cheap carbohydrate sources for production of calcium gluconate by *Penicillium funiculosum* mutant MN 238. *Ind. J. Exper. Biol.*, 23: 279–281.

Kusai K., I. Sekuzu, B. Hagihara, K. Okunuki, S. Yamauchi and M. Nakai (1960) Crystallisation of glucose oxidase from *Penicillium amagasakiense. Biochim. Biophys. Acta*, 40: 555–557.

Lancashire E (1969) Soap compositions having improved curd dispersing properties. US Patent 3: 454–500.

Lee P.C., W.G. Lee, S.Y. Lee and H.N. Chang (2001) Succinic acid production with reduced byproduct formation in the fermentation of *Anaerobiospirillum succiniciproducens* using glycerol as a carbon source. *Biotechnol. Bioeng.*, 72: 41–48.

Lee P.C., S.Y. Lee, S.H. Hong, H.N. Chang and S.C. Park (2003) Biological conversion of wood hydrolysate to succinic acid by *Anaerobiospirillum succiniciproducens. Biotechnol. Lett.*, 25: 111–114.

Li J., X.Y. Zheng, X.J. Fang et al. (2011) A complete industrial system for economical succinic acid production by *Actinobacillus succinogenes. Biores. Technol.*, 102: 6147–6152.

Llmen M., K. Koivuranta, L. Ruohonen, P. Suominen and M. Penttila (2007) Efficient production of L-lactic acid from xylose by *Pichia stipitis. Appl. Environ. Microbiol.*, 73: 117–123.

Lockwood L.B., G.E. Ward and O.E. May (1936) The physiology of *Rhizopus oryzae. J. Agric. Res.*, 53: 849–857.

Loffhagen N. and W. Babel (1993) The influence of energy deficiency – imposing conditions on the capacities of *Acetobacter methanolicus* to oxidize glucose and to produce gluconic acid. *Acta Biotechnol.*, 13: 251–256.

Lu M., J.D. Brooks and I.S. Maddox (1997) Citric acid production by solid-state fermentation in a packed bed reactor using *Aspergillus niger. Enzyme Microbiol. Technol.*, 21: 392–397.

Maas R.H., R.R. Bakker, M.L. Jansen, D. Visser, E. de Jong and G. Eggink (2008) Lactic acid production from limetreated wheat straw by *Bacillus coagulans*: neutralization of acid by fed-batch addition of alkaline substrate. *Appl. Microbiol. Biotechnol.*, 78: 751–758.

Mandeva R.D., I.T. Ermakova and A.B. Lozinov (1981) Metabolite excretion by yeasts of the genus *Candida* in media lacking sources N, P, S, or Mg and having different carbon sources. *Mikrobiologiia*, 50: 62–68.

Martinez-Espinoza A.D., M.D. Garcia-Pedrajas and S.E. Gold (2002) The ustilaginales as plant pests and model systems. *Fungal Genet. Biol.*, 35: 1–20.

Masuo E., F. Shima and K. Yoshida (1952) Oxidation of carbohydrates by *Pseudomonas*. II. Formation of lactobionic acid from lactose. *Ann. Rep. Shionogi. Res. Lab.*, 1: 136–141.

May O.E., H.T. Herrick, G. Thom and N.B. Church (1927) The production of gluconic acid by *Penicillium luteum purpurogenum* Group I. *J. Biol. Chem.*, 75: 417–422.

Miall M. and G.F. Parker (1975) Continuous preparation of citric acid by *Candida lipolytica*. German Patent 2,429,224.

Moyer A.J. (1940) Process for gluconic acid production. US Patent 2,351,500.

Murakami H., A. Seko, M. Azumi et al. (2006) Microbial conversion of lactose to lactobionic acid by resting cells of *Burkholderia cepacia* No. 24. *J. Appl. Glycosci.*, 53: 7–11.

Murakami H., T. Kiryu, T. Kiso and H. Nakano (2008) Production of calcium lactobionate by a lactose-oxidizing enzyme from *Paraconiothyrium* sp. KD-3. *J. Appl. Glycosci.*, 55: 127–132.

Nakajima Kambe T., T. Nozue, M. Mukouyama and T. Nakahara (1997) Bioconversion of maleic acid to fumaric acid by *Pseudomonas alcaligenes* strain XD-1. *J. Ferment. Bioeng.*, 84: 165–168.

Narayanan N., P.K. Roychoudhury and A. Srivastava (2004) L (Ð) lactic acid fermentation and its product polymerization. *Electron. J. Biotechnol.*, 7: 167–178.

Nishimura T., A.A. Vertes, Y. Shinoda, M. Inui and H. Yukawa (2007) Anaerobic growth of *Corynebacterium glutamicum* using nitrate as a terminal electron acceptor. *Appl. Microbiol. Biotechnol.*, 75: 889–897.

Nolasco-Hipolito C., O.C. Zarrabal, R.M. Kamaldin, L. Teck-Yee, S. Lihan and K.B. Bujang (2012) Lactic acid production by *Enteroccocus faecium* in liquefied sago starch. *AMB Express*, 2: 1–9.

Oh M.J., Y.J. Park and S.K. Lee (1973) Citric acid production by *Hansenula anamola*. *Hanguk Sikpum Kawahakhoe Chi*, 5: 215–223.

Okino S., M. Inui and H. Yukawa (2005) Production of organic acids by *Corynebacterium glutamicum* under oxygen deprivation. *Appl. Microbiol. Biotechnol.*, 68: 475–480.

Omar S.H. and H.J. Rehm (1980) Physiology and metabolism of two alkane oxidizing and oxidizing and citric acid producing strains of *Candida parapsilosis. Eur. J. Appl. Microbiol. Biotechnol.*, 11: 42–49.

Osawa F., T. Fujii, T. Nishida et al. (2009) Efficient production of L-lactic acid by Crabtree-negative yeast *Candida boidinii. Yeast*, 26: 485–496.

Otsuka K. (1961) Cis-trans isomerase – isomerisation from maleic acid to fumaric acid. *Agr. Biol. Chem.*, 25: 726–730.

Pecota D.C., V. Rajgarhia and N.A. da Silva (2007) Sequential gene integration for the engineering of *Kluyveromyces marxianus. J. Biotechnol.*, 127: 408–416.

Peleg Y., E. Battat, M.C. Scrutton and I. Goldberg (1989) Isoenzyme pattern and subcellular localization of enzymes involved in fumaric acid accumulation by *Rhizopus oryzae. Appl. Microbiol. Biotechnol.*, 32: 334–339.

Perlman D. (1949) Studies on the growth and metabolism of *Polyporus anceps* in submerged culture. *Am. J. Bot.*, 36: 180–184.

Perwozwansky W.W. (1939) Formation of gluconic acid during the oxidation of glucose by bacteria. *Microbiology (USSR)*, 8: 149–159.

Petruccioli M., P. Piccioni, M. Fenice and F. Federici (1994) Glucose oxidase, catalase and gluconic acid production by immobilized mycelium of *Penicillium variabile* P16. *Biotechnol. Lett.*, 16: 939–942.

Pfeifer V.F., C. Vojnovich and E.N. Heger (1952) Itaconic acid by fermentation with *Aspergillus terreus. Ind. Eng. Chem.*, 44: 2975–2980.

Pintado J., A. Torrado, M.P. Gonzalez and M.A. Murado (1998) Optimization of nutrient concentration for citric acid production by solid state culture of *Aspergillus niger* on polyurethane foams. *Enzyme Microbiol. Technol.*, 23: 149–156.

Pitzl G. (1951) Vinylidene chloride interpolymer as a coating for regenerated cellulose film. US Patent 2,570-478.

Playne M.J. (1985) Propionic and butyric acids. In: Moo-Young, M. (ed.) *Comprehensive Biotechnology: The Principles, Applications and Regulations of Biotechnology in Industry, Agriculture and Medicine.* Oxford: Pergamon, pp. 731–759.

Porro D., M.M. Bianchi, L. Brambilla et al. (1999) Replacement of a metabolic pathway for large scale production of lactic acid from engineered yeasts. *Appl. Environ. Microbiol.*, 65: 4211–4215.

Postma E., C. Verduyn, W.A. Scheffers and J.P. van Dijken (1989) Enzymatic analysis of the Crabtree effect in glucose limited chemostat cultures of *Saccharomyces cerevisiae. Appl. Environ. Microbiol.*, 55: 468–477.

Raab A.M. and C. Lang (2011) Oxidative versus reductive succinic acid production in the yeast *Saccharomyces cerevisiae. Bioeng. Bugs.*, 2: 120–123.

Rane K.D. and K.A. Sims (1993) Production of citric acid by *Candida lipolytica* Y-1095: effect of glucose concentration on yield and productivity. *Enzyme Microbiol. Technol.*, 15: 646–651.

Roa Engel C.A., A.J.J. Straathof and T.W. Zijlmans (2008) Fumaric acid production by fermentation. *Appl. Microbiol. Biotechnol.*, 78: 379–389.

Rodriguez H., T. Gonzalez, I. Goire and Y. Bashan (2004) Gluconic acid production and phosphate solubilisation by the plant growth promoting bacterium *Azospirillum* spp. *Naturwissenschaften*, 91: 552–555.

Rohr M., C.P. Kubicek and J. Kominek (1983) Gluconic acid. *Biotechnology*, 3: 455–465.

Roissart H. and F.M. Luquet (1994) *Bacteries Lactiques: Aspects Fondamentaux et Technologiques.* Paris: Lorica.

Roukas T. (1991) Production of citric acid from beet molasses by immobilized cells of *Aspergillus niger. J. Food Sci.*, 56: 878–880.

Saayman M., H.J.J. van Vuuren, W.H. van Zyl and M. Viljoen-Bloom (2000) Differential uptake of fumarate by *Candida utilis* and *Schizosaccharomyces pombe. Appl. Microbiol. Biotechnol.*, 54: 792–798.

Sardinas J.L. (1972) Fermentative production of citric acid. French Patent 2,113,668.

Sato M., T. Nakahara and K. Yamada (1972) Fermentative production of succinic acid from n-paraffin. *Agric. Biol. Chem.*, 36: 1969–1974.

Satory M., M. Furlinger, D. Haltrich, K.D. Kulbe, F. Pittner and B. Nidetzky (1997) Continuous enzymatic production of lactobionic acid using glucose-fructose oxidoreductase in an ultrafiltration membrane reactor. *Biotechnol. Lett.*, 19: 1205–1208.

Sauer M., D. Porro, D. Mattanovich and P. Branduardi (2008) Microbial production of organic acids: expanding the markets. *Trends Biotechnol.*, 26: 100–108.

Schaffer S. and A. Burkovski (2005) Proteomics. In: Eggeling L and M. Bott (eds) *Handbook of Corynebacterium glutamicum.* Boca Raton: CRC Press, pp. 99–118.

Scher M. and W.J. Lennarz (1969) Maleate isomerase. *J. Biol. Chem.*, 244: 1878–1872.

Senthuran A., V. Senthuran, B. Mattiasson and R. Kaul (1997) Lactic acid fermentation in a recycle batch reactor using immoblized *Lactobacillus casei. Biotechnol. Bioeng.*, 55: 841–853.

Shiraishi F., K. Kawakami, S. Kono, A. Tamura, S. Tsuruta and K. Kusunoki (1989) Characterisation of production of free gluconic acid by *Gluconobacter suboxydans* adsorbed on ceramic honeycomb monolith. *Biotechnol. Bioeng.*, 33: 1413–1418.

Singh A., P.K.R. Kumar and K. Schugerl (1992) Bioconversion of cellulosic materials to ethanol by filamentous fungi. *Adv. Biochem. Eng. Biotechnol.*, 45: 29–55.

Song C.W., D.I. Kim, S. Choi, J.W. Jang and S.Y. Lee (2013) Metabolic engineering of *Escherichia coli* for the production of fumaric acid. *Biotechnol. Bioeng.*, 110: 2025–2034.

Song H. and S.Y. Lee (2006) Production of succinic acid by bacterial fermentation. *Enzyme Microbiol. Technol.*, 39: 352–361.

Sponholz W.R. (1993) Wine spoilage by microorganisms. In: Fleet G.H. (ed.) *Wine Microbiology and Biotechnology*. Amsterdam: Harwood Academic, pp. 395–420.

Steven A.W. and A.G. Bonita (1991) Propionic acid production by a propionic acid tolerant strain of *Propionibacterium acidipropionici* in batch and semicontinuous fermentation. *Appl. Environ. Microbiol.*, 57: 2821–2828.

Stodola F.H. and B.L. Lockwood (1947) The oxidation of lactose and maltose to bionic acids by *Pseudomonas. J. Biol. Chem.*, 171: 213–221.

Suzuki T., Y. Sumino, S. Akiyam and H. Fukuda (1974) Method for producing citric acid. US Patemt 3,801,455.

Tabuchi T. (1981) Itaconic acid fermentation by a yeast belonging to the genus *Candida. Agric. Biol. Chem.*, 45: 475–479.

Takamura Y., T. Takamura, M. Soejima and T. Uemura (1969) Studies on induced synthesis of maleate cis-trans isomerase by malonate. Purification and properties of maleate cis-trans isomerase indiced by malonate. *Agric. Biol. Chem.*, 33: 718–728.

Takeno S., J. Ohnishi, T. Komatsu, T. Masaki, K. Sen and M. Ikeda (2007) Growth and potential for amino acid production by nitrate respiration in *Corynebacterium glutamicum. Appl. Microbiol. Biotechnol.*, 75: 1173–1182.

Tate B.E. (1981) Itaconic acid and derivatives. In: Mark H.F., D.F. Othmer, C.G. Overberger and G.T. Seaborg (eds) *Encyclopedia of Chemical Technology*. New York: Wiley, pp. 865–873.

Temash S. and Z.A. Olama (1999) Optimization of glucose oxides and gluconic acid production by *Penicillium purpurescens*using banana waste under solid state fermentation. *J. Med. Res.*, 20: 54–62.

Tomlinson G.A., M.P. Strohm and L.I. Hochstein (1978) The metabolism of carbohydrates by extremely halophilic bacteria: the identification of lactobionic acid as a product of lactose metabolism by *Halobacterium saccharovorum. Can. J. Microbiol.*, 24: 898–903.

Tran C.T., L.I. Sly and D.A. Mitchell (1998) Selection of a strain of *Aspergillus* for the production of citric acid from pineapple waste in solid state fermentation. *World J. Microbiol. Biotechnol.*, 14: 399–404.

Uchio R., I. Maeyashiki, K. Kikuchi and Y. Hirose (1975) Citric acid production by yeast. Japanese Patent 76,63,059.

Vandenberghe L.P.S., C.R. Soccol A. Pandey, and J.M. Lebeault (1999) Review: Microbial production of citric acid. *Braz. Arch. Biol. Technol.*, 42: 263–276.

Vegas C., E. Mateo and A. Gonzalez (2010) Population dynamics of acetic acid bacteria during traditional wine vinegar production. *Int. J. Food Microbiol.*, 138: 130–136.

Velizarov S. and V. Beschkov (1994) Production of free gluconic acid by cells of *Gluconobacter oxydans. Biotechnol. Lett.*, 16: 715–720.

Villecourt P. and H. Blachere (1955) Oxidation of lactose in lactobionic acid by pathogenic or saprophytic bacteria (*Bacterium anitratum*). *Ann. Inst. Pasteur (Paris)*, 88: 523–526.

Wee Y.J., J.N. Kim and H.W. Ryu (2006) Biotechnological production of lactic acid and its recent applications. *Food Technol. Biotechnol.*, 442: 163–172.

Wenge F. and A.P. Mathews (1999) Lactic acid production from lactose by *Lactobacillus plantarum*: kinetic model and effects of pH, substrate, and oxygen. *Biochem. Engin. J.*, 3: 163–170.

West D. (2004) AHA wrinkle creams come of age. *Chem. Week*, 166: 20.

William E.L., P.K. Cletus and M.K. Tsung (2006) Production of itaconic acid by *Pseudozyma Antarctica* NRRL Y-7808 under nitrogen-limited growth conditions. *Enz. Microbiol. Technol.*, 39: 824–827.

Willke T. and K.D. Vorlop (2001) Biotechnological production of itaconic acid. *Appl. Microbiol. Biotechnol.*, 56: 289–295.

Wojtatowicz M., G.L. Marchin and L.E. Erickson (1993) Attempts to improve strain A-10 of *Yarrow lipolytica* for citric acid production from n-paraffins. *Proc. Biochem.*, 28: 453–460.

Xi Y.L., K.Q. Chen, R. Xu et al. (2012) Effect of biotin and a similar compound on succinic acid fermentation by *Actinobacillus succinogenes* in a chemically defined medium. *Biochem. Eng. J.*, 69: 87–92.

Yamada Y., Y. Okada and K. Kondo (1976) Isolation and characterization of polarly flagellated intermediate strains in acetic acid bacteria. *J. Gen. Appl. Microbiol.*, 22: 237–245.

Yamada Y., K.I. Hoshino and T. Ishikawa (1997) The phylogeny of acetic acid bacteria based on the partial sequences of 16S ribosomal RNA: the elevation of the subgenus *Gluconoacetobacter* to the generic level. *Biosci. Biotechnol. Biochem.*, 61: 1244–1251.

Yamada Y., P. Yukpan and H.T.L. Vu (2012) Description of *Komagataeibacter* gen. nov.with proposals of new combinations (Acetobacteraceae). *J. Gen. Appl. Microbiol.*, 58: 397–404.

Yasumitsu M., T. Ooi and S. Kinoshita (2000) Production of lactobionic acid from whey by *Pseudomonas* sp. LS13-1. *Biotechnol. Lett.*, 22: 427–430.

Yokoya F. (1992) *Citric acid production. In: Industrial Fermentation Series.* Brazil: Campinas.

Yuzbashev T.V., E.Y. Yuzbasheva, T.I. Sobolevskaya et al. (2010) Production of succinic acid at low pH by a recombinant strain of the aerobic yeast *Yarrowia lipolytica. Biotechnol. Bioeng.*, 7: 673–682.

Zeikus J.G., M.K. Jain and P. Elankovan (1999) Biotechnology of succinic acid production and markets for derived industrial products. *Appl. Microbiol. Biotechnol.*, 51: 545–552.

Zhuge X., L. Liu, H.D. Shin et al. (2013) Development of a *Propionibacterium, Escherichia coli* shuttle vector for metabolic engineering of *Propionibacterium jensenii*, an efficient producer of propionic acid. *Appl. Environ. Microbiol.*, 79: 4595–4602.

Znad A., J. Markos and V. Bales (2003) Production of gluconic acid by *Aspergillus niger* growth and non-growth conditions. *Food Technol.*, 34: 1078–1080.

6

Microbes as a Source for the Production of Food Ingredients

Charu Gupta* and Dhan Prakash

Amity Institute for Herbal Research and Studies, Amity University UP, Noida, India

*Corresponding author e-mail: charumicro@gmail.com

Introduction

Bacteria, yeast, and microalgae can act as catalysts for the production of food ingredients, enzymes, and nutraceuticals (Hugenholtz and Smid 2002). With the current trend towards natural ingredients, there is renewed interest in microbial flavors, colors, and bio-processing using enzymes. Microbial production of substances such as organic acids, enzymes, proteins, vitamins, antibiotics, and hydrocolloids also remains important (Dufossé 2009). Lactic acid bacteria, in particular *Lactococcus lactis*, have been demonstrated to be ideal cell factories for the production of these important nutraceuticals. Developments in the genetic engineering of food-grade micro-organisms mean that the production of certain nutraceuticals can be enhanced or newly induced through overexpression and/or disruption of relevant metabolic genes (Bronzwaer 2008).

Microbes are used in the production of food flavors, carotenoids, flavonoids and terpenoids, enzymes, organic acids for use in food, viable probiotic cells, bacteriocins as food preservatives, amino acids and their derivatives for use in foods, nutraceuticals and medications, microbial polysaccharides for use in food, production of xylitol and other polyols, prebiotic oligosaccharides, polyunsaturated fatty acids, microalgae as sources of food ingredients and nutraceuticals, and production of vitamins.

Recently scientists have found that bacteria can also be a rich source of terpenes, an important ingredient in drugs and food additives. New research at Brown University, published in the Proceedings of the

Microbial Functional Foods and Nutraceuticals, First Edition. Edited by Vijai Kumar Gupta, Helen Treichel, Volha (Olga) Shapaval, Luiz Antonio de Oliveira, and Maria G. Tuohy.
© 2018 John Wiley & Sons Ltd. Published 2018 by John Wiley & Sons Ltd.

National Academy of Sciences, shows that the genetic capacity of bacteria to make terpenes is widespread (www.biotecnika.org/content/ december-2014/bacteria-found-be-rich-source-terpenes-important-ingredient-drugs-food-additiv?page=0,1). Using a specialized technique to sift through genomic databases for a variety of bacteria, the researchers found 262 gene sequences that likely code for terpene synthases, an enzyme that catalyzes the production of terpenes. The researchers then used several of those enzymes to isolate 13 previously unidentified bacterial terpenes. Genetically engineered *Streptomyces* bacterium is used as a bio-refinery to generate the terpene products.

Micro-organisms can be used as an adjunct therapy for diseases like protein energy malnutrition (PEM), anemia, diarrhea, cancer, obesity, ulcerative colitis, Crohn's disease, irritable bowel syndrome, and gluten therapy resistant celiac.

Microbes as Source of Antioxidants

Free radicals play an important role in the origin of numerous diseases, including lifestyle diseases and physiological diseases like high blood pressure, cancer, diabetes, cardiovascular, neurodegenerative, etc. Free radicals act by catalyzing the various toxic oxidative reactions that lead to the formation of toxic lipid peroxides. They also inhibit the enzymes of respiratory chain of mitochondria and damage DNA and proteins which is lethal for the cell. It is therefore important to search for newer alternatives to allopathic medicines with reduced or no side effects.

Antioxidants are biological molecules with the ability to protect vital metabolites from harmful oxidation. Due to this fascinating role, their beneficial effects on human health are of paramount importance. Traditional approaches using solvent-based extraction from food/non-food sources and chemical synthesis are often expensive, time-consuming, and detrimental to the environment. There are numerous medicinal plants that have been reported to possess strong antioxidant activity along with their free radical scavenging activity. Besides plants, various microbes including bacteria and fungi also possess powerful antioxidant activity. Some of the microbes belong to the probiotic group that has the potential to protect the body from dangerous free radicals. Thus various micro-organisms can be used as natural sources of antioxidants to develop unique functional foods for scavenging free radicals, thereby preventing many diseases (Gupta et al. 2013a). With the advent of metabolic engineering tools, the successful reconstitution of heterologous pathways in *Escherichia coli* and other micro-organisms provides a more exciting and amenable alternative to meet the increasing demand for natural antioxidants (Lin et al. 2014).

The antioxidant activity of organic extracts of eight fungal species, *Ganoderma lucidum, Ganoderma applanatum, Meripilus giganteus, Laetiporus sulphureus, Flammulina velutipes, Coriolus versicolor, Pleurotus ostreatus*, and *Panus tigrinus*, was evaluated for free radical (DPPH• and OH•) scavenging capacity. The highest DPPH scavenging activity was found in the methanol extract of *G. applanatum* (12.5 µg/mL; 82.80%) and the chloroform extract of *G. lucidum* (510.2 g/mL, 69.12%). The same extract also showed the highest LP inhibition (91.83%, 85.09%) at 500 µg/mL, while the methanol extracts of *G. applanatum* and *L. sulphureus* showed the highest scavenging effect on OH• radicals (68.47%, 57.06%, respectively) at 400 µg/mL. The antioxidative potencies correlated generally with the total phenol content (0.19–9.98 mg/g) (Karaman et al. 2010).

In another study, probiotic strain *Lactobacillus plantarum* that was used as a starter culture in the production of dried fermented meat products was assessed for its antioxidant activity against peroxide radicals by Nedelcheva et al. (2010). It was found that the inclusion of this culture provided the desired fermentation process in the raw sausage mass and reduced pathogenic flora. The application of starter cultures with antioxidant activity preserved the color of the meat products and delivered substances with antioxidant activity as well (Nedelcheva et al. 2010). Similarly, Miang, a kind of traditional fermented tea leaves containing several kinds of *Lactobacilli* spp., was investigated for its antioxidant activity. The study suggested that both *L. fermentum* FTL2311 and *L. fermentum* FTL10BR strains could liberate certain substances that possessed antioxidant activity expressed as trolox equivalent antioxidant capacity (TEAC) and equivalent concentration (EC) values for free radical scavenging and reducing mechanisms, respectively (Klayraung and Okonogi 2009).

Microbes as Source of Colors

There has been a growing interest in the use of natural ingredients, particularly in the food industry. Ingredients like colors can also be derived from natural sources such as microbes that produce various pigments like carotenoids, melanins, flavins, quinones, prodigiosins and, more specifically, monascins, violacein or indigo (Gupta et al. 2011). There are various examples of bacteria, fungi, and yeast that are colored and can be used to produce natural colors. The major advantage of using microbes for color production over plant sources is that they can be mass multiplied because of their high growth rate. The other advantages of producing pigments from micro-organisms include independence from weather conditions and geographic variability, easy and fast growth and colors of different shades can be obtained by growing on cheap substrates.

Before extracting colors from microbes, they are checked for safety and efficacy. Microbial colors should be non-toxic as they play a significant role as food colorants. Fungal cell production offers reliable scalable technology.

Bacteria such as *Serratia marcescens, S. rubidaea, Pseudomonas magneslorubra, Vibrio psychroerythrous, V. gazogenes, Bacillus* sp., *Alteromonas rubra, Rugamonas rubra,* and gram-positive actinomycetes, such as *Streptoverticillium rubrireticuli* and *Streptomyces longisporus,* produce colored secondary metabolites. The actinomycete *Streptomyces coelicolor* A3 produces a closely related linear tripyrrole, undecyl-prodigiosin, and a cyclic derivative, butylmeta- cycloheptyl-prodiginine, in a 2:1 ratio (Harris et al. 2004).

The red pigment of *S. marcescens* is known as prodigiosine, a non-diffusible pigment attached to the inner membrane (Khanafari et al. 2006). Prodigiosin is a multifaceted secondary metabolite produced by *S. marcescens, S. rubidaea, V. psychroerythrous, V. gazogenes, A. rubra, Lugomonas rubra* and gram-positive actinomycetes such as *Streptoverticillium rubrireticuli* and *Streptomyces longisporus* (Khanafari et al. 2006). This pigment possesses antifungal, immunosuppressive, antiproliferative, and anticancer activity (Pandey et al. 2007).

Color-producing fungi include *Penicillium, Blakeslea trispora, Xanthophyllomyces dendrorhous, Phycomyces blakesleeanus car S* mutant, *Ashbye gossypii, Eremothecium ashbyii, Epicoccum* sp., and genetically modified fungus *Fusarium sporotrichioides*. Similarly, some examples of color-producing yeast are *Xanthophyllomyces dendrorhous* and *Rhodotorula*. They produce astaxanthin pigment that imparts an orange-red color and *Rhodotorula* produces torulene, torularhodin, and carotene pigment (Dufossé 2009). Color-producing algae include *Dunaliella salina, D. bardwil* (produces β-carotene), *Haematococcus lacustris,* and *H. pluvialis* compactin resistant mutant (produces astaxanthin and canthaxanthin) (Chattopadhyay et al. 2008; Dufossé 2009).

Microbes as Source of Amino Acids

Studies have shown that microbes have the potential to produce some essential amino acids. Metabolic engineering constantly improves the productivity of amino acid-producing strains, mainly *Corynebacterium glutamicum* and *Escherichia coli* strains. Classic mutagenesis and screening have been accelerated by combination with intracellular metabolite sensing. Synthetic biology approaches have allowed access to new carbon sources to realize a flexible feedstock concept. Moreover, new pathways for amino acid production as well as fermentative production of non-native compounds derived from amino acids or their metabolic precursors have been developed. These include dipeptides, α,ω-diamines, α,ω-diacids, keto acids, acetylated amino acids, and ω-amino acids (Wendisch 2014).

There are numerous examples of bacteria that can produce different types of amino acids. Mycosporines and mycosporine-like amino acids (MAAs), including shinorine (mycosporine-glycine-serine) and porphyra-334 (mycosporine-glycine-threonine), are UV-absorbing compounds produced by cyanobacteria, fungi, and marine micro- and macroalgae. MAAs have the ability to protect these organisms from damage by environmental UV radiation. Its structure elucidation revealed that the novel MAA is mycosporine-glycine-alanine, which substitutes l-alanine for the l-serine of shinorine (Miyamoto et al. 2014).

In another recent study, it was found that a microbial-like pathway contributes to phenylalanine biosynthesis in plants via a cytosolic tyrosine: phenyl-pyruvate aminotransferase (Yoo et al. 2013). Phenylalanine is a vital component of proteins in all living organisms and in plants is a precursor for thousands of additional metabolites. Animals are incapable of synthesizing phenylalanine and must obtain it directly or indirectly from plants. While plants can synthesize phenylalanine in plastids through arogenate, the contribution of an alternative pathway via phenylpyruvate, as it occurs in most microbes, has not been demonstrated. This study established that plants also utilize the microbial-like phenylpyruvate pathway to produce phenylalanine, and flux through this route is increased when the entry point to the arogenate pathway is limiting. It was found that the plant phenylpyruvate pathway utilizes a cytosolic aminotransferase that links the co-ordinated catabolism of tyrosine to serve as the amino donor, thus interconnecting the extraplastidial metabolism of these amino acids. This discovery unlocks another level of complexity in the plant aromatic amino acid regulatory network, unveiling new targets for metabolic engineering.

The bacteria produce amino acids but they also help in reducing D-amino acid toxicity with the help of their enzyme "racemases." It is known that D-amino acids are toxic for life on earth, yet they are formed constantly due to geochemical racemization and bacterial growth (the cell walls of which contain D-amino acids). Studies have shown that D-amino acids are recycled by some bacteria, which use D-amino acids as a source of nitrogen by running enzymatic racemization in reverse. Consequently, when soils are inundated with racemic amino acids, resident bacteria consume D- as well as L-enantiomers, either simultaneously or sequentially depending on the level of their racemase activity. Bacteria thus protect life on earth by keeping environments free of D-amino acid (Zhang and Sun 2014).

All bacteria are capable of reverse racemization. Specifically, when soils are inundated with LD-alanine, LD-aspartic acid, LD-glutamic acid, and LD-leucine, resident bacteria absorb L- and D-enantiomers in equal or nearly equal rates. In the case of alanine, the conversion of D-enantiomers appears to be enabled by constitutive alanine racemases that are ordinarily anabolic in function (i.e., cell wall synthesis). In the case of the other three amino acids, exposure to L-enantiomers appears to induce catabolic racemases. As a result, as soon as L-enantiomers are exhausted, the organisms could consume D-enantiomers (Zhang and Sun 2014).

Microbes as Source of Vitamins

Humans and other organisms have essential requirements for a range of different vitamins to assist in metabolic pathways, and as such, a plentiful dietary intake of vitamins is linked with improved health. Vitamins are essential micronutrients. Their key role is to enter the biochemical processes as co-enzymes. The B-group or B-complex vitamins include thiamin (B1), riboflavin (B2), niacin (B3), pyridoxine (B6), pantothenic acid (B5), biotin (B7 or H), folate (B11−B9 or M) and cobalamin (B12). These molecules are water soluble and play an important role in metabolism, particularly the cellular metabolism of carbohydrates (thiamin), proteins and fats (riboflavin and pyridoxine). B-group vitamins, normally present in many foods, can be easily removed or destroyed during cooking and food processing, so their deficiency is rather common in the human population. For this reason, several countries have adopted laws requiring the fortification of certain foods with specific vitamins and minerals (Burgess et al. 2009; LeBlanc et al. 2010).

Where humans have vitamin deficiencies, supplementation with non-food vitamins is a standard remedy. Addition of vitamins to animal feeds also promotes healthy livestock. Microbes possess the simplicity and flexibility needed for economic vitamin production. Many micro-organisms can be genetically modified to overexpress enzymes in the synthesis pathway of various vitamins. With the ability to grow on cheap feedstocks and produce an improved yield compared to chemical synthesis, microbial production of vitamins also boasts a much lower cost in terms of the energy input necessary. Maintaining microbial fermentation conditions is significantly less expensive than the different conditions needed for a variety of reaction steps in the chemical process.

The adaptability of lactic acid bacteria (LAB) to the fermentation process, their biosynthetic capacity and metabolic versatility are some of the principal features that facilitate the application of LAB in foods for producing, releasing and/or increasing specific beneficial compounds.

Among these, vitamin production by LAB has recently gained attention. The proper selection and exploitation of nutraceutical-producing LAB is an interesting strategy to produce novel fermented foods with increased nutritional and/or health-promoting properties. Fermented milks or bread with high levels of B-group vitamins (such as folate and riboflavin) can be produced by LAB-promoted biosynthesis (Arena et al. 2014).

Vitamin production by lactobacilli is becoming a focus point to obtain novel fermented foods through nutraceutical-producing LAB (Capozzi et al. 2012; LeBlanc et al. 2011). Several lactobacilli have the genes involved in riboflavin biosynthesis, although the genetic capability to biosynthesize riboflavin is species and/or strain specific. In fact, the rib operon has been shown to have interruptions in some bacteria, for example *L. plantarum* WCFS1, resulting in the inability to produce riboflavin (Capozzi et al. 2011).

Conversely, *L. plantarum* NCDO 1752, *L. plantarum* JDMI and others strains of *L. plantarum* recently isolated from cereals-derived products (Burgess et al. 2006; Capozzi et al. 2011) can grow in the absence of riboflavin since they are capable of synthesizing this protein. The riboflavin operon in all lactobacilli studied displays the same gene organization coding for a riboflavin-specific deaminase and reductase (ribG), a riboflavin synthase α subunit (ribB), a bifunctional enzyme which also catalyzes the formation of 3,4-dihydroxy-2-butanone 4-phosphate from ribulose 5-phosphate (ribA), and a riboflavin synthase β subunit (ribH). There are still challenges that face microbial production of vitamins that are difficult to address. For example, in engineered microbes that can express every enzyme necessary to produce the desired vitamin, allosteric inhibition from the vitamin or intermediates in the synthesis pathway can hamper the yield. Also, particularly in the case of riboflavin production, very little of the vitamin is secreted by the microbes, meaning that the cultured cells must be lyzed in order to obtain the vitamin and replaced with a new culture (i.e., difficulties in feasibility of continuous culture).

Microbes as Source of Proteins

Global population increase in the past few decades has intensified protein malnutrition especially in the developing world where agricultural industry is not reliable. Promising biotechnological methods have been established to alleviate the world's protein deficit.

Single cell protein (SCP) production from lignocellulose biomass is part of the upcoming technology aimed at providing protein supplement for both human food and animal feeds. Micro-organisms such as algae, fungi, yeast, and bacteria are involved in the bioconversion of low-cost feed stocks such as lignocelluloses to produce biomass rich in proteins and amino acids. Even the cladodes of *Opuntia ficus indica* (cactus pear) have potential for SCP production in arid and semi-arid regions (Gabriel et al. 2014). Various bacteria, molds, yeasts, and algae have been employed for the production of SCP. The bacteria include *Brevibacterium, Methylophilus methylitrophus, Achromobacter delvaevate, Acinetobacter calcoacenticus, Aeromonas hydrophilla, Bacillus megaterium, B. subtilis, Lactobacillus* sp., *Cellulomonas* sp., *Methylomonas methylotrophus, Pseudomonas fluorescens, Rhodopseudomonas capsulata, Flavobacterium* sp., *Thermomonospora fusca* and others. Some of the algae used are *Chlorella pyrenoidosa, C. sorokiana, Chondrus crispus, Scenedesmus acutus, Porphyrium* sp. and *Spirulina maxima* (Mahasneh 1997). The filamentous fungi that have been used include *Chaetomium celluloliticum, Fusarium graminearum, Aspergillus fumigatus, A. niger, A. oryzae, Cephalosporium cichhorniae, Penicillium cyclopium, Rhizopus chinensis, Scytalidium acidophilum, Tricoderma viridae, T. alba*, and *Paecilomyces varioti*. Yeasts such as *Candida utilis* (Torula yeast), *C. lipolytica, C. tropicalis, C. novellas, C. intermedia*, and *Saccharomyces cerevisiae* are among the various organisms that have been used for the production of SCP (Becker 2007). The micro-organisms to be cultured should be non-pathogenic, must have good nutritional values, must be usable as food and feed, should not contain toxic compounds, and production costs should be low. For example, Quorn, a leading brand of mycoprotein food product in the UK and Ireland, is extracted from a fungus, *Fusarium venenatum*. It is high in protein and dietary fiber and low in saturated fat and salt.

In a recent study, researchers demonstrated the production of SCP from different varieties of orange peel using *Aspergillus niger* and *Saccharomyces cerevisiae*. The bioconversion of fruit wastes into valuable products like SCP has the ability to solve the worldwide food protein deficiency by obtaining an economical product for food and feed. In addition, using wastes as a substrate for the production of high nutritious product may also alleviate environmental pollution to some extent (Azam et al. 2014).

In another study, the conversion of the enzymatic hydrolyzate of shell-fish chitin to SCP was investigated. The end product of chitin hydrolysis by *Penicillium ochrochloron* chitinase was mainly N-acetyl-D-glucosamine; its further utilization as a substrate for SCP production using *Yarrowia lipolytica* NCIM 3450 was studied. The 2% chitin hydrolyzate was found to be optimal for SCP production. Fish diets were formulated to replace fishmeal partially by SCP from *Yarrowia lipolytica* using chitin hydrolyzate in diets of *Lepidocephalus thermalis*. The result indicated that a 50% yeast SCP diet gave a better growth response in *Lepidocephalus thermalis* than other formulations (Patil and Jadhav 2014).

Microbes as Source of Natural Biopreservatives

The biopreservation of foods using bacteriocinogenic LAB from foods is an innovative approach. Biopreservation or biocontrol refers to the use of natural or controlled microbiota or its antibacterial products to extend the shelf-life and enhance the safety of foods (Stiles 1996). Since LAB occur naturally in many food systems and have a long history of safe use in fermented foods, and are thus classed as Generally Regarded As Safe (GRAS), they have great potential for extended use in biopreservation.

One such class is defined as bacteriocins. Bacteriocins are heat-stable ribosomally synthesized antimicrobial peptides produced by various bacteria, including food-grade LAB. These antimicrobial peptides have huge potential as both food preservatives and next-generation antibiotics targeting multiple drug-resistant pathogens (Perez et al. 2014). LAB bacteriocins are inherently tolerant to high thermal stress and are known for their activity over a wide pH range. These antimicrobial peptides are also colorless, odorless, and tasteless, which further enhances their potential usefulness. Bacteriocins have a fast-acting mechanism, which forms pores in the target membrane of bacteria, even at extremely low concentrations. They are also easily degraded by proteolytic enzymes due to their proteinaceous nature. Therefore, bacteriocin fragments do not live long in the human body or in the environment, which minimizes the opportunity of target strains to interact with the degraded antibiotic fragments; this is the common starting point in the development of antibiotic resistance. The most significant advantage of bacteriocins over conventional antibiotics is their primary metabolite nature since they have relatively simple biosynthetic mechanisms compared with conventional antibiotics, which are secondary metabolites. This fact makes them easily amenable through bioengineering to increase either their activity or specificity towards target micro-organisms (Perez et al. 2014).

Besides bacteriocins, LAB also produce a wide variety of other active antagonistic metabolites such as organic acids (lactic, acetic, formic, propionic, butyric, hydroxyl-phenyl-lactic, and phenyllactic), diverse antagonistic compounds (carbon dioxide, ethanol, hydrogen peroxide, fatty acids, acetoin, diacetyl, reuterin, reutericyclin), antifungal compounds (propionate, phenyl-lactate, hydroxyphenyl-lactate, cyclic dipeptides, phenyllactic acid, and 3-hydroxy fatty acids), and bacteriocins (such as nisin, pediocins, lacticins, enterocins and many others) (Oliveira et al. 2014). Other bacterial groups (especially those from the genus *Bacillus*) are also attracting attention because of the diversity of antimicrobial peptides they produce, some of which could also be exploited as biopreservatives.

Another compound produced by the group of LAB is reutericyclin. This unique tetramic acid is a negatively charged, highly hydrophobic antagonist. Reutericyclin acts as a proton ionophore, resulting in dissipation of the proton motive force (Gänzle 2004). It lacks activity towards yeasts and fungi, but is active on gram-positive bacteria including *Lactobacillus* spp., *Bacillus subtilis*, *B. cereus*, *Enterococcus faecalis*, *S. aureus*, and *Listeria innocua*. Spore germination of *Bacillus* species was inhibited by this antimicrobial compound, but the spores remained unaffected under conditions that do not permit germination. As in many other antagonists, inhibition of gram-negative bacteria (*E. coli* and *Salmonella*) is observed under conditions that disrupt the outer membrane, including truncated lipopolysaccharides (LPS), low pH and high salt concentrations. Reutericyclin was shown to be produced in concentrations active against competitors during growth of *Lactobacillus reuteri* in sourdough. It was proposed that reutericyclin-producing strains may have applications in the biopreservation of foods (Gänzle 2004).

Microbes as Source of Anticancer Foods

Cancer is a group of diseases characterized by unregulated growth and spread of abnormal cells, which can result in death if not controlled (Bicas et al. 2008). Sauerkraut can be an important part of diets designed for healing cancer. Sauerkraut is fermented white cabbage. Lacto-fermented cabbage has a long history of providing benefits for many different health conditions, and now it is proving to be beneficial for cancer. Cabbage, by itself, offers a number of health benefits, but the fermentation process increases the bioavailability of nutrients, rendering sauerkraut even more nutritious than the original cabbage (Lipski 2013).

Epidemiological studies have shown that consumption of cabbage and sauerkraut is connected with significant reduction of breast cancer incidence. Estrogens are considered a major breast cancer risk factor and their metabolism by P450 enzymes substantially contributes to carcinogenic activity. The effect of cabbage and sauerkraut juices of different origin on the expression profile of the estrogen metabolism key enzymes (CYP1A1, CYP1A2, CYP1B1) was investigated in breast cell lines MCF7, MDA-MB-231, and MCF10A. The effects of cabbage juices were compared with that exerted by indole-3-carbinol (I3C) and 3,3'-diindolylmethane (DIM). Treatment with cabbage juices or indoles for 72 hours affected the expression of CYP1 family genes in a cell type-dependent manner. The results supported epidemiological observations and partly explained the mechanism of the chemopreventive activity of white cabbage products (Szaefer et al. 2012a).

In another study, it was shown that the breakdown products of glucosinolates in cabbage may affect both the initiation phase of carcinogenesis, by decreasing the amount of DNA damage and cell mutation, and the promotion phase, by blocking the processes that inhibit programmed cell death and stimulate unregulated cell growth. A study published in 2012 in the journal *Nutrition Cancer* showed that consumption of cabbage and sauerkraut is connected with significant reduction of breast cancer incidence. This research supported the observation that the consumption of sauerkraut was beneficial for the prevention of breast cancer in women (Licznerska et al. 2013; Szaefer et al. 2012b).

The anticancer enzyme glutaminase is derived from a microbial source. Glutaminase is widely distributed in micro-organisms including bacteria, yeasts, and fungi. It is a potent antileukemic agent. L-Glutaminase, in combination with or as an alternative to asparaginase, could be of significance in enzyme therapy for cancer, especially acute lymphocytic leukemia. The enzyme mainly catalyzes the hydrolysis of amido bond of L-glutamine. In addition, some enzymes also catalyze glutamyl transfer reaction. A highly savory amino acid, L-glutamic acid and a taste-enhancing amino acid of infused green tea, theanine can be synthesized by employing a hydrolytic or transfer reaction catalyzed by glutaminase. Therefore, glutaminase is one of the most important flavor-enhancing enzymes. The detection of glutamine was accomplished by a coupled enzyme system composed of glutaminase plus glutamate oxidase, while the detection of glutamic acid was carried out by a single enzyme: glutamate oxidase.

However, microbial sources are preferred for large-scale production. Major genera of micro-organisms reported to produce glutaminase are bacteria species including *Acetobacter liquefaciens, Micrococcus luteus, Achromobacter* sp., *Nocardia* sp., *Acinetobacter glutaminasificans, Proteus morganii, Aerobacter aerogenes, Proteus vulgaris, Streptomyces netropsis, Erwinia carotovora, Streptomyces olivochromogenes, Escherichia coli, Vibrio cholerae, Flavobacterium flavescens, Vibrio costicola, Micrococcus glutamicus, Xanthomonas juglandis,* and *Micrococcus lysodeikticus.* Yeasts include *Cryptococcus albidus, Candida scottii, Cryptococcus nodaensis, Cryptococcus* sp. and *Debaryomyces* sp. and fungi like *Aspergillus oryzae, Tilachlidium humicola,* and *Verticillium malthousei* (Nandakumar et al. 2003).

Anticancer Compounds from Endophytic Sources

Drugs for cancer treatment show non-specific toxicity to proliferating normal cells, possess severe side effects, and are not effective against many forms of cancer (Gangadevi and Muthumary 2008). Thus, the treatment of cancer has been enhanced mainly by improvements in diagnosis which allow earlier and more precise therapy (Pasut and Veronesi 2009). The anticancer properties of several secondary metabolites from endophytes have been investigated. Taxol, a diterpenoid, is a well-known anticancer compound isolated from *Taxus brevifolia*. These trees are rare, slow growing, and produce small amounts of taxol, and hence expensive when obtained from their natural source (Gangadevi and Muthumary 2008). However, an endophyte called *Taxomyces andreanae* has provided an alternative approach to obtain cheaper and more available taxol through micro-organism fermentation. Another important drug for cancer is the alkaloid camptothecin, a potent antineoplastic agent. Camptothecin and 10-hydroxycamptothecin are two important precursors for the synthesis of the clinically useful anticancer drugs topotecan and irinotecan (Uma et al. 2008). The products were obtained from the endophytic fungi *Fusarium solani* isolated from *Camptotheca acuminata* (Kusari et al. 2009b; Shukla et al. 2014).

Phenylpropanoids have attracted much interest for use as anticancer, antioxidant, antimicrobial, anti-inflammatory, and immunosuppressive agents (Korkina 2007). The endophytic *Penicillium brasilianum*, found in root bark of *Melia azedarach*, promotes the biosynthesis of phenylpropanoid amides (Fill et al. 2010). Likewise, two monolignol glucosides, coniferin and syringin, are not only produced by the host plant but were also recognized by the endophytic Xylariaceae species as chemical signals during the establishment of fungus plant interactions. Koshino and his co-workers in 1988 characterized two phenylpropanoids and lignin from stromata of *Epichloe typhina*.

Lignans are another kind of anticancer agent originated as secondary metabolites through the shikimic acid pathway and display different biological activities that make them interesting in several lines of research (Gordaliza et al. 2004). Although their molecular backbone consists only of two phenylpropane units (C6-C3), lignans show enormous structural and biological diversity, especially in cancer chemotherapy (Korkina 2007). Some well-known alkaloids were first reported to be present in an endophytic fungus, *Alternaria* sp., isolated from the phloem of *Catharanthus roseus* that had the ability to produce vinblastine. Later, Zhang et al. (2000) discovered an endophytic *Fusarium oxysporum* from the phloem of *C. roseus* that could produce vincristine. Yang et al. (2004) also found an unidentified vincristine-producing endophytic fungus from the leaves of *C. roseus*. The aryl tetralin lignans, such as podophyllotoxin, are naturally synthesized by *Podophyllum* sp., are clinically relevant mainly due to their cytotoxicity and antiviral activities and are also valued as the precursor to useful anticancer drugs like etoposide, teniposide, and etopophos phosphate (Kour et al. 2008; Kusari et al. 2009a).

Another study reported on a novel fungal endophyte, *Trametes hirsuta*, that produces podophyllotoxin and other related aryl tetralin lignans with potent anticancer properties (Puri et al. 2006). Novel microbial sources of podophyllotoxin were reported from the endophytic fungi *Aspergillus fumigatus*. A new compound, ergoflavin, of the ergochrome class, was isolated from endophytic fungi growing on the leaves of an Indian medicinal plant, *Mimusops elengi*, family Sapotaceae (Deshmukh et al. 2009). Secalonic acid D, a mycotoxin also belonging to the ergochrome class, has shown potent anticancer activities. It was isolated from the mangrove endophytic fungus and demonstrated high cytotoxicity on HL60 and K562 cells by inducing leukemia cell apoptosis (Zhang et al. 2009). Wagenaar and co-workers reported identification of three novel cytochalasins, bearing antitumor activity, from the endophyte *Rhinocladiella* sp.

Finally, many compounds with anticancer properties isolated from endophytic microbes have been reported such as cytoskyrins, rubrofusarin B, phomoxanthones A and B, photinides A–F (Ding et al. 2009), and (+)- epiepoxydon (Shukla et al. 2014).

Microbes as Source of Antidiabetic Foods

Diabetes is a common and sometimes fatal disease that occurs when the supply of insulin is insufficient for the body to break down sugar properly. The majority of insulin used to manage diabetes is produced using biotechnology. Bacterial cells are genetically modified to produce large quantities of human insulin, which is then purified for therapeutic use. Millions of people worldwide now use Humuline, which is a major brand name for "human" insulin produced using genetically modified (GM) bacteria (Bajzer and Seeley 2006).

Researchers have created a strain of non-pathogenic *E. coli* bacteria that produce a protein called GLP-1. This protein triggers cells in the pancreas to make insulin. In one study, scientists fed the engineered bacteria to diabetic mice. After 80 days, the mice went from being diabetic to having normal glucose blood levels. Diabetic mice that were not fed the engineered bacteria still had high blood sugar levels. The promise is that a diabetic could eat yoghurt or drink a smoothie as glucose-responsive insulin therapy rather than relying on insulin injections. Creating bacteria that produce the protein has a number of advantages over using the protein itself as the treatment. The bacteria can secrete just the right amount of the protein in response to conditions in the host that could ultimately minimize the need for self-monitoring and allow the patient's own cells (or the cells of the commensal *E. coli*) to provide the appropriate amount of insulin when needed (Bronzwaer 2008; Cani et al. 2009).

Gut-friendly microbes have been engineered to make a specific protein that can help to regulate blood sugar in diabetic mice. Although this research is still in the very early stages, the microbes can be grown in yoghurt and may provide an alternative treatment for people with diabetes (Hossain et al. 2007). Probiotics are cheap, less than a dollar per dose. This is especially helpful in Third World countries where cheaper delivery methods of treatment could be easily distributed.

Interestingly, results from several genomic, metagenomic and metabolomic studies have provided substantial information to target gut microbiota by dietary interventions for the management of type 2 diabetes (T2D). Recent studies have proposed multifactorial interventions including dietary manipulation in the management of T2D. The same interventions have also been recommended by many national and international diabetes associations. These studies are aimed at deciphering the gut microbial influence on health and disease (Turnbaugh et al. 2009).

Probiotics, particularly lactobacilli and bifidobacteria, have recently emerged as prospective biotherapeutics with proven efficacy demonstrated in various *in vitro* and *in vivo* animal models adequately supported with their established multifunctional roles and mechanism of action for the prevention and treatment of disease. Dietary intervention in conjunction with probiotics is a novel multifactorial strategy to abrogate progression and development of diabetes that holds considerable promise through improving the altered gut microbial composition and targeting all the possible risk factors (Panwar et al. 2013). However, protein-based drugs can be very expensive to make and often strike problems as they degrade during digestion (Gupta et al. 2013b, 2014).

Microbes as Source of Antianemia Agents

Anemia is more prevalent in developing and Third World countries because of malnutrition, poor availability of iron, and chronic blood loss. The other factors of anemia are deficiency in folic acid, vitamin B12, and iron. Vitamin B12, along with folate, is involved in making the heme molecule, an integral part of hemoglobin. Vitamin B12 is important in living beings and it is used to treat pernicious anemia and peripheral neuritis. Folates perform important roles as co-factors in 1-carbon transfer reactions occurring in purine and pyrimidine biosynthesis. They are also required for efficient DNA replication, repair, and methylation. For these key roles in the cellular cycle, tissues with a high cell growth rate or turnover, such as hemopoietic cells and intestinal mucosa, have a high requirement for folate. Deficiency leads to megaloblastic anemia, resulting in inhibition of DNA synthesis in red blood cell production (Bronzwaer 2008).

There are some folic acid-producing bacteria and yeasts that can be cultured in whey or milk plasma, thereby accumulating high concentrations of folic acid in the medium. The various identified bacteria that produce and enhance the uptake of folic acid are *Lactococcus lactis* subsp. *cremoris*, *L. lactis* subsp. *lactis*, *Bifidobacterium adolescentis*, *B. pseudocatenulatum*, and yeasts like *Candida famata*, *C. guilliermondii*, *C. glabrata*, *Yarrowia lipolytica*, *Saccharomyces cerevisiae*, *Pichia glucozyma*, and *Yarrowia lipolytica*. The vitamin B12-producing bacteria are *Pseudomonas denitrificans* and *Propionibacterium shermanii* (Kassinen et al. 2007). Application of probiotic bacteria for the prevention of megaloblastic anemia is a novel scientific approach that involves lactic acid-fermented foods, increases iron absorption by optimization of pH in the digestive tract, activates enzyme phytases, and produces organic acids and other digestive enzymes (Scarpignato 2008).

A strain of probiotic bacteria developed by Swedish firm Probi doubled the absorption of iron from food in women. *Lactobacillus plantarum* 299v (Lp299v) not only helps our digestive system but also helps our immune system. It is also good for the heart and helpful in reducing gas and bloating conditions. It improves bowel movement by making it more normal and regular. It also reduces the negative effects of antibiotic drugs on colonic fermentation (Liu et al. 2001).

Several medicines are produced using genetically engineered bacteria or fungi that synthesize the medicine in giant bioreactors. Erythropoietin can be man-made in bioreactors by bacteria and is used to treat anemia but the treatment requires frequent and sometimes daily injections. Researchers have purified specialized cells from human blood that normally repair the lining of blood vessels. These cells were also genetically engineered to express erythropoietin. They were then mixed with mesenchymal stem cells, which are able to form blood vessels. This mixture was then injected underneath the skin of anemic mice. The cell mixture spontaneously formed networks of blood vessels underneath the skin. The vessel lining secreted erythropoietin and cured anemia in both types of mice (Gupta et al. 2014; Kassinen et al. 2007).

Microbes as Source of Antiobesity Agents

Obesity is a major public health concern, caused by a combination of increased consumption of energy-dense foods and reduced physical activity, with contributions from host genetics, environment, and adipose tissue inflammation. In recent years, the gut microbiome has also been found to be implicated and research in mice and humans has attributed to it both the manifestation and/or exacerbation of this major epidemic and vice versa. At the experimental level, analysis of fecal samples revealed a potential link between obesity and alterations in the gut flora (drop in Bacteroidetes and increase in Firmicutes), the specific gut microbiome being associated with the obese phenotype. Efforts to identify new therapeutic strategies to modulate gut microbiota would be of high priority for public health, and to date, probiotics and/or prebiotics seem to be the most effective tools (Katerina et al. 2014).

Lactobacillus (*L. sporogenes* and *L. acidophilus* NCFB 1748) and *Bifidobacterium* genus representatives have been reported to play a critical role in weight regulation as an antiobesity effect in experimental models and humans (Mercenier et al. 2002). *Lactobacillus sporogenes* has the ability to lower cholesterol levels. It produces a significant reduction in low-density lipoprotein (LDL) levels and a small but significant increase in high-density lipoprotein (HDL) cholesterol. *L. sporogenes* can be used as a side effect-free alternative to drug therapy in the treatment of high cholesterol and heart disease. Clinical studies have revealed that *L. sporogenes* can be successfully implanted in the intestine. *L. sporogenes* satisfies the essential requirement of an efficient probiotic. The spores of *L. sporogenes* are resistant to heat and other adverse environmental conditions. This property of spore formation by *L. sporogenes* is the main characteristic that makes it the probiotic of choice in clinical applications. Being sporulated, they germinate under favorable conditions and produce sufficient viable cells which proliferate and perform vital healthful functions. In addition, *L. sporogenes* spores are semi-resident and are slowly excreted out of the body. *L. sporogenes* is effective in the form of dietary supplements as well as when added to food products (Gupta et al. 2014; Montrose and Floch 2005).

A literature study was conducted focusing on clinical trials that examined the effect of specific micro-organisms on body weight control. Analysis of the eligible articles suggested that *Lactobacillus gasseri* SBT 2055, *L. rhamnosus* ATCC 53103, and the combination of *L. rhamnosus* ATCC 53102 and *Bifidobacterium lactis* Bb12 may reduce adiposity, body weight, and weight gain. This suggests that these microbial strains can be applied in the treatment of obesity. Furthermore, short chain fatty acid production and low-grade inflammation were found as the underlying mechanisms of action that influence metabolism and affect body weight. These findings might contribute to the development of probiotic treatments for obesity. Further research should be directed to the most effective combination and dosage rate of probiotic micro-organisms (Mekkes et al. 2014).

Another potential microbiome-altering strategy is the incorporation of modified bacteria that express therapeutic factors into the gut microbiota. For example, N-acetylphosphatidylethanolamines (NAPEs) are precursors to the N-acylethanolamide (NAE) family of lipids, which are synthesized in the small intestine in response to feeding and reduce food intake and obesity. A study demonstrated that administration of engineered NAPE-expressing *E. coli* Nissle 1917 bacteria in drinking water for 8 weeks reduced the level of obesity in mice fed a high-fat diet. Mice that received modified bacteria had dramatically lowered food intake, adiposity, insulin resistance, and hepatosteatosis compared with mice receiving standard water or control bacteria (Zhongyi et al. 2014).

Plants also represent the most studied natural source of antiobesity bioactives. *Camellia sinensis* is the most representative species exerting several antiobesity effects. Well-known probiotics (bacteria which bestow health benefit), such as strains of *Bifidobacterium* and *Lactobacillus* families, and certain prebiotics (non-viable food components that confer a health benefit on the host associated with modulation of microbiota effects), such as insulin-type fructans, have also shown capability to combat obesity. Finally, compounds from animal sources, in particular bioactive peptides from milk-derived whey and casein protein digestion, high dietary calcium and ω-3s polyunsaturated fatty acids (ω-3 PUFA) present in fish oils, have also shown potential antiobesity effects (Cristina et al. 2013).

Another study reported that there is a virus that causes obesity. Adenovirus-36 (Adv 36), commonly known as a human "common cold virus," is easily caught from an infected person who is coughing or sneezing. It is believed that at least one-third of people affected by obesity have been infected. Researchers at the University of Wisconsin began experimenting with Adv36 around 1995 and found that when they experimentally infected chickens and mice, the animals increased their body fat by 50–150%. Compared to uninfected animals, about 60–70% of infected animals became obese. The mode of action of this virus is that the DNA (genetic material) of the virus gets into the fat cells of the person or animal and causes them to bring in more fat and glucose from the blood and to make fat out of the glucose. The viral DNA also causes adult stem cells in the fat tissue to turn into fat cells, so the total fat cell number increases. Thus, an infected person will have bigger fat cells and more of them. Scientists have figured out the sequence of the DNA in Adv36, which gene in the virus causes the effect and how this gene changes the chemistry inside fat cells to cause obesity. Research is ongoing to identify antiviral agents that appear to work against Adv36 infection. A vaccine has been developed that is still in very early studies, to prevent infection with Adv36. More research is needed, but it appears that eventually we will be able to prevent Adv36-induced obesity (Cristina et al. 2013)).

Microbes as Source of Antiallergens

An allergy is a hypersensitivity disorder of the immune system. Symptoms include redness, itchiness, and runny nose, eczema, hives, or an asthma attack. Allergies can play a major role in conditions such as asthma. In some people, severe allergies to environmental or dietary allergens or to medication may result in life-threatening reactions called anaphylaxis (Kay 2000).

The rapid increase in immune-mediated disorders such as allergic disease is strongly linked to reduced early microbial exposure (Bach 2002). The gut microbiota represents the greatest microbial exposure by far and is central to the development of immune regulation. The specific composition of the gut microbiota may affect the risk of developing allergic disease (Penders et al. 2007). This finding provided the foundation for intervention studies designed to modify gut microbial composition for the treatment of allergic disease. The effects of beneficial bacteria (probiotics) or resistant starches or fiber (prebiotics) that selectively stimulate a limited number of beneficial bacteria have been evaluated in allergy treatment studies.

Studies have shown that probiotics and prebiotics can be used for the treatment of allergic diseases (Johannsen and Prescott 2009; Ozdemir 2010; Tang et al. 2010; Thomas and Greer 2010; Yao et al. 2010). In an experimental mouse model, researchers discovered a highly interesting effect that mice who were very susceptible to neurodermatitis developed the disease less frequently when a lyzate of certain pathogens was applied to the skin. The susceptibility of the lyzate-treated mice to developing neurodermatitis was considerably lower than that of control mice. More detailed analyses showed that the animals had a higher concentration of immune-modulating interleukin-10 and a lower concentration of the pro-inflammatory mediator interferon-γ. Interferon-γ is secreted when the immune system recognizes foreign invaders and fine-tunes the immune system to effectively get rid of foreign intruders. The secretion of interferon-γ causes inflammatory reactions that create a hostile environment for the intruders, thereby initiating their elimination (Thomsen et al. 2007). Non-pathogenic and probiotic bacteria do not cause inflammation, but are nevertheless recognized by the human immune system. The elevated production of interleukin-10 in certain immune system cells leads to anti-inflammatory reactions, which in turn create an active immunological tolerance. This means that the human body is able to learn and tolerate self-peptides and harmless allergens when it is exposed to specific non-pathogenic micro-organisms (Thomsen et al. 2007; Yan et al. 2007).

It has also been observed that use of probiotic bacteria or lyzates of non-pathogenic micro-organisms leads to a significant and permanent improvement in neurodermatitis. The application of non-pathogenic or probiotic bacteria to the skin is more effective in reducing the recurrence of diseases following the successful primary treatment of diseases such as neurodermatitis. The bacterium leads to permanent stabilization of the immune system once glucocorticoid treatment has alleviated the acute reaction (Sicherer and Sampson 1999).

Probiotics are perceived to exert beneficial effects in the prevention and treatment of allergic diseases by modifying the gut ecosystem. The effect of ingestion of fermented milk containing *Lactobacillus paracasei*-33 (LP-33) was observed in patients with perennial allergic rhinitis and it was found that ingestion of LP-33-fortified fermented milk can effectively and safely improve the quality of life of patients with allergic rhinitis, and may serve as an alternative treatment (Wang et al. 2004).

Conclusion

Microbes have great significance as food ingredients. They can be used as a source of proteins, vitamins, enzymes, preservatives, antiallergens, anticancer, antidiabetes, anemia, obesity, and cholesterol agents. Thus it can be safely concluded that microbes have tremendous potential as nutraceuticals and in designing various disease-targeting preventive functional foods and dietary supplements.

Acknowledgments

Thhe authors are grateful to Dr Ashok K. Chauhan, Founder President, Amity Group of Institutions, and Mr. Atul Chauhan, Chancellor, Amity University UP, Noida, for encouragement, research facilities, and financial support.

References

Arena M.P., D. Fiocco, S. Massa, V. Capozzi, P. Russo and G. Spano (2014) *Lactobacillus plantarum* as a strategy for an *in situ* production of vitamin B$_2$. *J. Food Nutr. Disord.*, S1–004.

Azam S., Z. Khan, B. Ahmad, I. Khan and J. Ali (2014) Production of single cell protein from orange peels using *Aspergillus niger* and *Saccharomyces cerevisiae*. *Global J. Biotechn. Biochem.*, 9: 14–18.

Bach J.F. (2002) The effect of infections on susceptibility to autoimmune and allergic diseases. *N. Engl. J Med.*, 347: 911.

Bajzer M. and R. Seeley (2006) Physiology: obesity and gut flora. *Nature*, 444: 1009–1010.

Becker E.W. (2007) Micro-algae as a source of protein. *Biotechnol. Adv.*, 25: 207–210.

Bicas J.L., F.C.C. Barros, R. Wagner, H.T. Godoy and G.M. Pastore (2008) Optimization of R-(+)-α-terpineol production by the biotransformation of R-(+)-limonene. *J. Indust. Microbiol. Biotechnol.*, 35: 1061–1070.

Bronzwaer S. (2008) EFSA scientific forum "from safe food to healthy diets". EU risk assessment –past, present and future. *Trends Food Sci. Technol.*, 19: S2–S8.

Burgess C.M., E.J. Smid, G. Rutten and D. van Sinderen (2006) A general method for selection of riboflavin-overproducing food grade micro-organisms. *Microb. Cell Fact.*, 5: 24.

Burgess C.M., E.J. Smid and D. van Sinderen (2009) Bacterial vitamin B_2, B_{11} and B_{12} overproduction: an overview. *Int. J. Food Microbiol.*, 133: 1–7.

Cani P.D., S. Possemiers, T. van de Wiele et al. (2009) Changes in gut microbiota control inflammation in obese mice through a mechanism involving GLP-2-driven improvement of gut permeability. *Gut*, 58: 1091–1103.

Capozzi V., V. Menga, A.M. Digesu et al. (2011) Biotechnological production of vitamin B_2-enriched bread and pasta. *J. Agric. Food Chem.*, 59: 8013–8020.

Capozzi V., P. Russo, M.T. Dueñas, P. López and G. Spano (2012) Lactic acid bacteria producing B-group vitamins: a great potential for functional cereals products. *Appl. Microbiol. Biotechnol.*, 96: 1383–1394.

Chen Z., L. Guo, Y. Zhang et al. (2014) Incorporation of therapeutically modified bacteria into gut microbiota inhibits obesity. *J. Clin. Invest.*, 124: 3391–3406.

Chattopadhyay P, S. Chatterjee and S.K. Sukanta (2008) Biotechnological potential of natural food grade biocolourants. *Afr. J. Biotechnol.*, 7: 2972–2985.

Cristina T.F., H. Schellekens, T.G. Dinan et al. (2013) A natural solution for obesity: bioactives for the prevention and treatment of weight gain. A review. *Nutr. Neurosci.*, 10:1–18.

Deshmukh S.K., P.D. Mishra, A.A. Kulkarni et al. (2009) Anti-inflammatory and anticancer activity of ergoflavin isolated from an endophytic fungus. *Chem. Biodivers.*, 6: 784–789.

Ding G., Z. Zheng, S. Liu, H. Zhang, L. Guo and Y.Y. Che (2009) Photinides AF, cytotoxic benzofuranone-derived γ-lactones from the plant endophytic fungus *Pestalotiopsis photiniae*. *J. Natl Prod.*, 72: 942–945.

Doria A., M.E. Patti and C.R. Kahn (2008) The emerging genetic architecture of type 2 diabetes. *Cell Metab.*, 8: 186–200.

Dufossé L. (2009) Pigments. *Microb. Encyc. Microbiol.*, 4: 457–471.

Eyberger A.L., R. Dondapati and J.R. Porter (2006) Endophyte fungal isolates from *Podophyllum peltatum* produce podophyllotoxin. *J. Natl Prod.*, 69: 1121–1124.

Fill T.P., B.F. da Silva and E. Rodrigues-Fo (2010) Biosynthesis of phenylpropanoid amides by an endophytic *Penicillium brasilianum* found in root bark of *Melia azedarach. J. Microbiol. Biotechnol.*, 20: 622–629.

Gabriel A., N. Victor and D.P. Jams (2014) Cactus pear biomass, a potential lignocelluloses raw material for single cell protein production (SCP): a review. *Int. J. Curr. Microbiol. Appl. Sci.*, 3: 171–197.

Gangadevi V. and J. Muthumary (2008) Taxol, an anticancer drug produced by an endophytic fungus *Bartalinia robillardoides* Tassi, isolated from a medicinal plant, *Aegle marmelos* Correaex Roxb. *World J. Microbiol. Biotechnol.*, 24: 717–724.

Gänzle M.G. (2004) Reutericyclin: biological activity, mode of action, and potential applications. *Appl. Microbiol. Biotechnol.*, 64: 326–332.

Gordaliza M., P.A. Garcıa, J.M. Miguel, Corral, M.A. Castro, M.A. Gomez Zurita (2004) Podophyllotoxin: distribution, sources, applications and new cytotoxic derivatives. *Toxicon*, 44: 441–459.

Gupta C., A.P. Garg, D. Prakash, S. Goyal and S. Gupta (2011) Microbes as potential source of biocolours. *Pharmacol. Online*, 2: 1309–1318.

Gupta C., D. Prakash and S. Gupta (2013) Functional foods enhanced with microbial antioxidants. *Acad. J. Nutr.*, 2: 10–18.

Gupta C., D. Prakash and S. Gupta (2013) Role of microbes in functional foods for the treatment of diabetes. *World J. Pharm. Pharmaceut. Sci.*, 2: 4626–4638.

Gupta C., D. Prakash, A.P. Garg and S. Gupta (2014) Nutraceuticals from microbes. In: Prakash D. and G. Sharma (eds) *Phytochemicals of Nutraceutical Importance*. Wallingford: CABI International Publishers, pp 79–102.

Harris K.P., R. Williamson, H. Slater et al. (2004) The *Serratia* gene cluster encoding biosynthesis of the red antibiotic, prodigiosin, shows species and strain dependent genome context variation. *Microbiology*, 150: 3547–3560.

Hossain P., B. Kawar and M. El Nahas (2007) Obesity and diabetes in the developing world – a growing challenge. *N. Engl. J. Med.*, 356: 213–215.

Hugenholtz J. and E.J. Smid. (2002) Nutraceutical production with food-grade microorganisms. *Curr. Opin. Biotechnol.*,13: 497–507.

Johannsen H. and S.L. Prescott (2009) Practical prebiotics, probiotics and synbiotics for allergists: how useful are they? *Clin. Exp. Allergy*, 39: 1801.

Karaman M.E., R. Jovin, M. Malbasa, Matavuly and M. Popovi (2010) Medicinal and edible lignicolous fungi as natural sources of anti oxidative and antibacterial agents. *Phytother. Res.*, 24: 1473–1481.

Kassinen A., L. Krogius-Kurikka, H. Makivuokko et al. (2007) The fecal microbiota of irritable bowel syndrome patients differs significantly from that of healthy subjects. *Gastroenterology*, 133: 24–33.

Katerina K., J. Evangelos, Giamarellos-Bourboulis and G. Stavrou (2014) Obesity as a consequence of gut bacteria and diet interactions. *ISRN Obesity*, Article ID 65189.

Kay A.B. (2000) Overview of "Allergy and allergic diseases: with a view to the future". *Br. Med. Bull.*, 56: 843–64.

Khanafari A., M.M. Assadi and F.A. Fakhr (2006) Review of prodigiosin, pigmentation in *Serratia marcescens*. *Online J. Biol. Sci.*, 6: 1–13.

Klayraung S. and S. Okonogi (2009) Antibacterial and antioxidant activities of acid and bile resistant strains of *Lactobacillus fermentum* isolated from miang. *Brazil. J. Microbiol.*, 40: 757–766.

Korkina L.G. (2007) Phenylpropanoids as naturally occurring antioxidants: from plant defense to human health. *Cell. Mol. Biol.*, 53: 15–25.

Kotzampassi K., E.J. Giamarellos-Bourboulis and G. Stavrou (2014) Obesity as a consequence of gut bacteria and diet interactions. *ISRN Obesity*, Article ID 651895. doi: 10.1155/2014/651895.

Kour A., A.S. Shawl, S. Rehman et al. (2008) Isolation and identification of an endophytic strain of *Fusarium oxysporum* producing podophyllotoxin from *Juniperus recurva*. *World J. Microbiol. Biotechnol.*, 24: 1115–1121.

Kusari S., M. Lamshoft and M. Spiteller (2009a) *Aspergillus fumigatus* Fresenius, an endophytic fungus from *Juniperus communis* L.Horstmann as a novel source of the anticancer pro-drug deoxypodophyllotoxin. *J. Appl. Microbiol.*, 107: 1019–1030.

Kusari S., S. Zuhlke and M. Spiteller (2009b) An endophytic fungus from *Camptotheca acuminata* that produces camptothecin and analogues. *J. Natl Prod.*, 72: 2–7.

LeBlanc J.G., M.P. Taranto, V. Molina and F. Sesma (2010) B-group vitamins production by probiotic lactic acid bacteria. In: Mozzi F., R. Raya and G. Vignolo (eds) *Biotechnology of Lactic Acid Bacteria: Novel Applications.* Ames: Wiley-Blackwell, pp. 211–232.

LeBlanc J.G., J.E. Laiño, M. Juarez del Valle et al. (2011) B-Group vitamin production by lactic acid bacteria – current knowledge and potential applications. *J. Appl. Microbiol.*, 111: 1297–1309.

Licznerska B.E., H. Szaefer, M. Murias, A. Bartoszek and W. Baer-Dubowska (2013) Modulation of CYP19 expression by cabbage juices and their active components: indole-3-carbinol and 3,3'-diindolylmethene in human breast epithelial cell lines. *Eur. J. Nutr.*, 52: 1483–1492.

Lin Y., R. Jain and Y. Yan (2014) Microbial production of antioxidant food ingredients via metabolic engineering. *Curr. Opin. Biotechnol.*, 26: 71–78.

Lipski E. (2013) *Digestion Connection: The Simple, Natural Plan to Combat Diabetes, Heart Disease, Osteoporosis, Arthritis, Acid Reflux – And More!* Emmaus: Rodale.

Liu Q., S. Nobaek, D. Adawi et al. (2001) Administration of *Lactobacillus plantarum* 299v reduces side-effects of external radiation on colon anastomotic healing in an experimental model. *Colorect. Dis.*, 3: 245–252.

Mahasneh I.A. (1997) Production of single cell protein from five strains of the micro-alga *Chlorella* sp. (Chlorophyta). *Cytobiosciences*, 90: 153–161.

Mekkes M.C., T.C. Weenen, R.J. Brummer and E. Claassen (2014) The development of probiotic treatment in obesity: a review. *Benef. Microbes*, 5: 19–28.

Mercenier A., S. Pavan and B. Pot (2002) Probiotics as biotherapeutic agents: present knowledge and future prospects. *Curr. Pharmaceut. Design*, 8: 99–110.

Miyamoto K.T., M. Komatsu and H. Ikeda (2014) Discovery of gene cluster for mycosporine-like amino acid biosynthesis from Actinomycetales microorganisms and production of a novel mycosporine-like amino acid by heterologous expression. *Appl. Environ. Microbiol.*, 80: 5028–5036.

Montrose D.C. and M.H. Floch (2005) Probiotics used in human studies. *J. Clin. Gastroenterol.*, 39: 469–484.

Nandakumar R., K. Yoshimune, M. Wakayama and M. Moriguchi (2003) Microbial glutaminase: biochemistry, molecular approaches and applications in the food industry. *J. Mol. Catalysis B: Enzymatic*, 23: 87–100.

Nedelcheva P., Z. Denkova, P. Denev, A. Slavchev and A. Krastanov (2010) Probiotic strain *Lactobacillus plantarum* NBIMCC 2415 with antioxidant activity as a starter culture in the production of dried fermented meat products. *Biotechnol. Biotechnol. Equipment*, 24: 1624–1630.

Oliveira P.M.L., E. Zannini and E.K. Arendt (2014) Cereal fungal infection, mycotoxins, and lactic acid bacteria mediated bioprotection: from crop farming to cereal products. *Food Microbiol.*, 37: 78–95.

Ozdemir O. (2010) Various effects of different probiotic strains in allergic disorders: an update from laboratory and clinical data. *Clin. Exp. Immunol.*, 160: 295–304.

Pandey R, R. Chander and K.B. Sainis (2007) Prodigiosins: a novel family of immuno-suppressants with anticancer activity. *Indian J. Biochem. Biophys.*, 44: 295–302.

Panwar H., H.M. Rashmi, V.K. Batish and S. Grover (2013) Probiotics as potential biotherapeutics in the management of type 2 diabetes – prospects and perspectives. *Diabet. Metab. Res. Rev.*, 29: 103–112.

Pasut G. and F.M. Veronese (2009) PEG conjugates in clinical development or use as anticancer agents: an overview. *Adv. Drug Deliv. Rev.*, 61: 1177–1188.

Patil N.S. and J.P. Jadhav (2014) Single cell protein production using *Penicillium ochrochloron* chitinase and its evaluation in fish meal formulations. *J. Microb. Biochem. Technol.*, S4: 005.

Penders J., E.E. Stobberingh, P.A. van den Brandt and C. Thijs (2007) The role of the intestinal microbiota in the development of atopic disorders. *Allergy*, 62: 1223–1236.

Perez R.H., T. Zendo and K. Sonomoto (2014) Novel bacteriocins from lactic acid bacteria (LAB): various structures and applications. *Microb. Cell Fact.*, 13 (Suppl 1): S3.

Puri S.C., A. Nazir, R. Chawla et al. (2006) The endophytic fungus *Trametes hirsuta* as a novel alternative source of podophyllotoxin and related aryl tetralin lignans. *J. Biotechnol.*, 122: 494–510.

Scarpignato, C. (2008) NSAID-induced intestinal damage: are luminal bacteria the therapeutic target? *Gut*, 57: 145–148.

Shukla S.T., P.V. Habbu, V.H. Kulkarni, K.S. Jagadish, A.R. Pandey and V.N. Sutariya (2014) Endophytic microbes: a novel source for biologically/pharmacologically active secondary metabolites. *Asian J. Pharmacol. Toxicol.*, 2: 1–16.

Sicherer, S.H. and H.A. Sampson (1999) Food hypersensitivity and atopic dermatitis: pathophysiology, epidemiology, diagnosis and management. *J. Allergy Clin. Immunol.*, 104: S114 –S122.

Stiles M. (1996) Biopreservation by lactic acid bacteria. *Antonie van Leeuwenhoek*, 70: 331–345.

Szaefer H., B. Licznerska, V. Krajka-Kuźniak, A. Bartoszek and W. Baer-Dubowska (2012a) Modulation of CYP1A1, CYP1A2 and CYP1B1 expression by cabbage juices and indoles in human breast cell lines. *Nutr. Cancer*, 64: 879–888.

Szaefer H., V. Krajka-Kuźniak, A. Bartoszek and W. Baer-Dubowska (2012b) Modulation of carcinogen metabolizing cytochromes P450 in rat liver and kidney by cabbage and sauerkraut juices: comparison with the effects of indole-3-carbinol and phenethyl isothiocyanate. *Phytother. Res.*, 26: 1148–1155.

Tang M.L., S.J. Lahtinen and R.J. Boyle (2010) Probiotics and prebiotics: clinical effects in allergic disease. *Curr. Opin. Pediatr.*, 22: 626–634.

Thomas D.W. and F.R. Greer (2010) Probiotics and prebiotics in pediatrics. *Pediatrics*, 126: 1217–1231.

Thomsen S.F., C.S. Ulrik, K.O. Kyvik et al. (2007) Importance of genetic factors in the etiology of atopic dermatitis: a twin study. *Allergy Asthma Proc.*, 28: 535–539.

Torres-Fuentes C., H. Schellekens, T.G. Dinan, and J.F. Cryan (2015) A natural solution for obesity: bioactives for the prevention and treatment of weight gain. *A review. Nutr. Neurosci.*, 18: 49–65.

Turnbaugh P.J., V.K. Ridaura, J.J. Faith, F.E. Rey, R. Knight and J.I. Gordon (2009) The effect of diet on the human gut microbiome: a metagenomic analysis in humanized gnotobiotic mice. *Sci. Translat. Med.*, 1: 1–10.

Uma S.R., B.T. Ramesha, G. Ravikanth, P.G. Rajesh, R. Vasudeva and K.N. Ganeshaiah (2008) Chemical profiling of *N. nimmoniana* form camptothecin, an important anticancer alkaloid: towards the development of a sustainable production system. In: Ramawat K.G. and J. Merillion (eds) *Bioactive Molecules and Medicinal Plants*. Berlin: Springer, pp. 198–210.

Wang M.F., H.C. Lin, Y.Y. Wang and C.H. Hsu (2004) Treatment of perennial allergic rhinitis with lactic acid bacteria. *Pediatr. Allergy Immunol.*, 15: 152–158.

Wendisch V.F. (2014) Microbial production of amino acids and derived chemicals: synthetic biology approaches to strain development. *Curr. Opin. Biotechnol.*, 30C: 51–58.

Yan F., H. Cao, T.L. Cover, R. Whitehead, M.K. Washington and D.B. Polk (2007) Soluble proteins produced by probiotic bacteria regulate intestinal epithelial cell survival and growth. *Gastroenterology*, 132: 562–575.

Yang X., L. Zhang, B. Guo and S. Guo (2004) Preliminary study of a vincristine producing endophytic fungus isolated from leaves of *Catharanthus roseus. Chinese Trad. Herb. Drugs*, 35: 79–81.

Yao T.C., C.J. Chang, Y.H. Hsu and J.L. Huang (2010) Probiotics for allergic diseases: realities and myths. *Pediatr. Allergy Immunol.*, 21: 900–919.

Yoo H., J.R. Widhalm, Y. Qian et al. *(*2013*)* An alternative pathway contributes to phenylalanine biosynthesis in plants via a cytosolic tyrosine: phenylpyruvate aminotransferase. *Nat. Commun.*, 4: 2833.

Zhang G. and H.J. Sun. (2014) Racemization in reverse: evidence that D-amino acid toxicity on earth is controlled by bacteria with racemases. *PLoS ONE*, 9: e92101.

Zhang L., B. Guo, H. Li et al. (2000) Preliminary study on the isolation of endophytic fungus of *Catharanthus roseus* and its fermentation to produce products of therapeutic value. *Chinese Trad. Herb. Drugs*, 31: 805–807.

Zhang Y.Z., L.Y. Tao, Y.J. Liang et al. (2009) Secalonic acid D induced leukemia cell apoptosis and cell cycle arrest of G1 with involvement of GSK-3β/β-catenin/c-Myc pathway. *Cell Cycle*, 8: 2444–2450.

Zhongyi C., L. Guo, Y. Zhang et al. (2014) Incorporation of therapeutically modified bacteria into gut microbiota inhibits obesity. *J. Clin. Invest.*, 124(8): 3391–3406.

Useful Resources

- www.biotecnika.org/content/december-2014/bacteria-found-be-rich-source-terpenes-important-ingredient-drugs-food-additiv?page=0,1
- Obesity Action Coalition: www.obesityaction.org/educational-resources/resource-articles-2/general-articles/obesity-due-to-a-virus-how-this-changes-the-game
- http://www.pharmainfo.net/venkatpasupuleti/glutaminase-anticancer-enzyme-microbe
- Patton D. (2005) Sauerkraut consumption may fight off breast cancer. www.nutraingredients.com/Research/Sauerkraut-consumption-may-fight-off-breast-cancer

7

Microbial Xanthan, Levan, Gellan, and Curdlan as Food Additives

Ozlem Ateş[1] and Ebru Toksoy Oner[2]*

[1] Department of Genetics and Bioengineering, Nisantasi University, Istanbul, Turkey
[2] IBSB, Department of Bioengineering, Marmara University, Istanbul, Turkey

*Corresponding author e-mail: ebru.toksoy@marmara.edu.tr

Introduction

Biopolymers have received considerable research attention over recent decades since they have potential applications in various fields of the economy such as food, packaging and chemical industries, agriculture, and medicine due to their compatibility with human lifestyle as well as their biocompatibility, biodegradability, and environmental credentials. These naturally occurring polymers have many uses as adhesives, absorbents, lubricants, soil conditioners, cosmetics, drug delivery vehicles, textiles, high-strength materials, implantable biomaterials, controlled drug carriers or scaffolds for tissue engineering (Guo and Murphy 2012; Ige et al. 2012; Kazak et al. 2010). Based on the production processes and origin of raw materials, these environmentally friendly biopolymers can be categorized as microbial biopolymers (polymers synthesized by micro-organisms like bacterial exopolysaccharides), natural biopolymers (obtained from biosources like plant polysaccharides), and synthetic polymers (Poli et al. 2011; Spizzirri et al. 2015).

Bacterial polysaccharides, one of the main groups of microbially produced biopolymers and hydrocolloids, show great diversity and functions with unique and commercially relevant material properties. Therefore, there is increased interest in bacterial polysaccharides in industrial, biomedical, chemical engineering, and food science areas (Fahnestock et al. 2011; Nwodo et al. 2012; Rehm 2010).

Based on their biological functions, microbial polysaccharides can be subdivided into the exopolysaccharides (EPSs) (xanthan, levan, alginate,

Microbial Functional Foods and Nutraceuticals, First Edition. Edited by Vijai Kumar Gupta, Helen Treichel, Volha (Olga) Shapaval, Luiz Antonio de Oliveira, and Maria G. Tuohy.

cellulose, etc.) and the intracellular storage polysaccharides (glycogen, starch) (Schmid and Sieber 2015). Bacterial EPSs such as dextran, xanthan, gellan, curdlan, pullulan, acetan, and levan are used commercially and are well-known industrial microbial polysaccharides with numerous applications and a considerable market (Freitas et al. 2011; Kazak et al. 2010; Llamas et al. 2012; Ptaszek et al. 2015).

Most polysaccharides are hydrocolloids that are widely used in the food industry for their extensive range of functionalities and applications. They can be utilized as stabilizers, emulsifiers, thickeners, and gelling agents primarily in food products such as bread, sauces, syrup, ice cream, instant food, beverages and ketchup, and most of these hydrocolloids are classified as food additives (Ahmad et al. 2015; Pegg 2012). In 2008, the global hydrocolloid market, currently dominated by the algal and plant polysaccharides such as starch, galactomannans, pectin, carrageenan, and alginate, had a market value of US$4 million and it is expected to reach US$3.9 billion by 2012 and US$7 billion by 2019 (Patel and Prajapati 2013; Williams et al. 2002). However, production of these algal and plant polysaccharides is highly affected by climatological and geological conditions whereas microbial polysaccharides are produced by fermentation under controlled conditions with high and stable yields, which make them economically competitive with polysaccharides of algal and plant origin (Kaur et al. 2014).

In this chapter, after a brief description of microbial exopolysaccaharides and their use in the food industry, the bacterial EPSs xanthan, levan, gellan, and curdlan are discussed in more detail with a special focus on their uses as food additives.

Microbial Exopolysaccharides (EPS)

In recent years, considerable attention has been given to the environmental and human compatible biopolymers which have been used as biotechnology products in numerous sectors such as food, chemical, agriculture, and health (Guo and Murphy 2012; Vroman and Tighzert 2009).

Polysaccharides are natural, non-toxic, biodegradable polymers that cover the surface of most cells and play important roles in various biological mechanisms such as immune response, adhesion, infection, and signal transduction (Yasar Yildiz and Toksoy Oner 2014). Besides the interest in their applications in the health and bionanotechnology sectors, polysaccharides are also used as thickeners, bioadhesives, stabilizers, probiotics, and gelling agents in the food and cosmetic industries and as emulsifiers, biosorbents, and bioflocculants in the environmental sector (Toksoy Oner 2013).

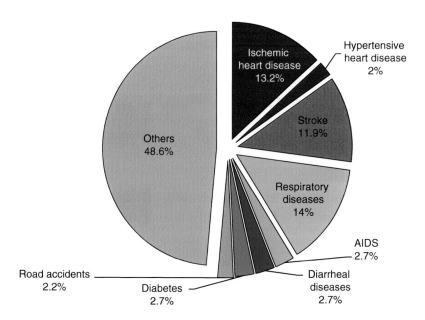

Figure 8.1 The proportion of deaths by cause in WHO regions in 2012.

Microbial Functional Foods and Nutraceuticals, First Edition. Edited by Vijai Kumar Gupta, Helen Treichel, Volha (Olga) Shapaval, Luiz Antonio de Oliveira, and Maria G. Tuohy. © 2018 John Wiley & Sons Ltd. Published 2018 by John Wiley & Sons Ltd.

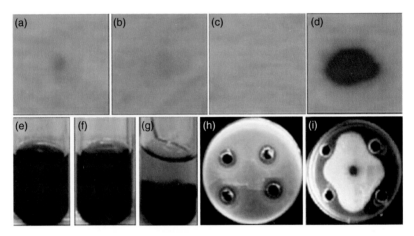

Figure 8.5 Efficiency of KSK-II in the hydrolysis of blood. Initially, blood stain washed from cotton fabrics after 30 min of incubation at room temperature with detergent only (a), enzyme only (b) and detergent with the aid of KSK-II (c) against the fabric stained with blood (d). Then natural blood clots were subjected to dissolution with KSK-II by incubation for 6 h at room temperature at pH 10.0 (e) and pH 7.4 (f) against control (g). The antimicrobial activity of KSK-II against the bacterium *S. aureus* and the fungus *R. solani* (b) is shown in panels (h) and (i), respectively.

Treated   Untreated

Figure 8.6 The antiviral effect of *E. foetida* protease on the CMV infection in cowpea plants. Local lesions were counted at the third day of virus inoculation.

(a)

(b)

Figure 12.1 *Auricularia auricula-judae* grown on decomposing logs in the Western Ghats (a) and *Termitomyces umkowaan* grown on termite mounds (b).

The plant, algal, and microbial polysaccharides are adequate for the market but there is a demand for eco-friendly technologies. Therefore, the use of bacterial polysaccharides for industrial applications is expected to increase sharply (Kumar et al. 2007). Since microbial sources allow high-yield, sustainable and economical production of polysaccharides under controlled conditions to the specifications required at industrial levels, they are preferred to plants and algae (Kazak et al. 2010; Toksoy Oner 2013). Production by microbial sources is achieved within days to weeks whereas 3–6 months are required for production from plants. Moreover, geographical or seasonal variations and sustainable use of agricultural land are ever increasing concerns for production processes relying on plant resources. Production from microalgae is dependent on solar energy but there is no such requirement for microbial production, which also utilizes different organic resources as fermentation substrates (Donot et al. 2012; Toksoy Oner 2013)

The term exopolysaccharide was first used by Sutherland (1972) to define high molecular weight carbohydrate polymers produced by marine bacteria. EPSs are synthesized and secreted by various microorganisms (gram-positive and gram-negative bacteria, fungi, and some algae) into the extracellular environment as either soluble or insoluble polymers. Therefore their composition, functions, and chemical and physical properties that establish their primary conformation vary from one bacterial species to another. EPSs are composed mainly of carbohydrates (a wide range of sugar residues) and some non-carbohydrate factors (such as acetate, pyruvate, succinate, and phosphate) (Llamas et al. 2012; Nicolaus et al. 2010; Staudt et al. 2012; Vu et al. 2009).

Exopolysaccharides can be classified based on monomeric composition as homopolysaccharides, composed of a single type of monosaccharide, like dextran, levan or mutan, and heteropolysaccharides which have complex structures and are made up of various types of monosaccharides. They also vary in size from disaccharides to heptasaccharides like xanthans or gellans (Donot et al. 2012; Llamas et al. 2012; Nicolaus et al. 2010; Nwodo et al. 2012; Schmid and Sieber 2015).

Bacterial EPSs have long been accepted as significantly important biomaterials in industrial areas. EPSs have found broad application in the food, pharmaceutical, cosmetics, and petroleum industries in which emulsifying, viscosifying, suspending, and chelating agents are required since they have a variety of unique and complex chemical structures that offer beneficial bioactive functions and potential applications. The first bacterial polymer, dextran, was discovered by Louis Pasteur (1861) in the mid nineteenth century as a microbial product in wine.Numerous articles are published on new microbial EPSs each year but few EPSs have continued to exist industrially (Donot et al.

2012; Kumar et al. 2007; Liang and Wang 2015; Llamas et al. 2012; Ordax et al. 2010).

In more recent years, several novel bacterial EPSs have been isolated and identified, but only a few of them have achieved significant commercial value due to high production costs and poor properties (Freitas et al. 2011; Kumar et al. 2007; Llamas et al. 2012; Mata et al. 2006; Nicolaus et al. 2010). Some of the bacterial EPS with superior physical properties (xanthan produced by *Xanthomonas campestris*, gellan produced by *Sphigomonas paucimobilis*, dextran produced by *Leuconostoc mesenteriodes* and curdlan produced by *Alcaligenes faecalis*) can be used instead of plant (guar gum or pectin) or algae (carrageenan or alginate) polysaccharides in traditional applications (Freitas et al. 2011; Liang and Wang 2015; Nicolaus et al. 2010). Other bacterial EPSs with unique properties such as levan provide various new commercial opportunities due to their biological activities (Freitas et al. 2011).

Although there is a wide range of microbial EPSs (Table 7.1) that have industrially promising physicochemical properties, the global market is still dominated by plant and algal polysaccharides. For commercialization of new polymers in large markets, it is crucial to overcome economic problems by lowering their production costs. The main costs are the substrate and infrastructures required for production, including bioreactors and maintaining asepsis (Donot et al. 2012). Use of cheaper sources, optimization of fermentation conditions to improve product yield, development of higher yielding strains (e.g., by mutagenesis or genetic manipulation) and improvement of downstream processing are involved in approaches to decrease production costs. Moreover, there is increased interest in studies of metabolic pathways of EPS synthesis and regulation to optimize microbial EPS production (Sousa et al. 2011).

Extremophilic micro-organisms isolated from extreme environments, such as deep-sea hydrothermal vents, Antarctic ecosystems, saline lakes, and geothermal springs, have also been studied as potential sources of highly valuable EPSs (Freitas et al. 2011). Since most commercial EPS-producing strains have been pathogenic, more research has been focused on discovering and developing novel and functional EPSs produced by extremophilic strains. Several bacterial strains of thermophiles (heat loving), psychrophiles (cold adapted), halophiles (hypersaline bodies), acidophiles (acid resistance), and alkaliphiles (alkaline environment survivors) have been reported as EPS producers (Kazak et al. 2010, 2015; Nicolaus et al. 2010; Yasar Yildiz et al. 2014).

Table 7.1 Commercially available significant microbial polysaccharides and their characteristics.

Microbial EPS	Monomeric units	Producer	Main properties	Industrial applications
Xanthan	Glucose, mannose and glucuronate	*Xanthomonas* spp.	Hydrocolloid, high viscosity yield at low shear rates even at low concentrations, stability over wide temperature, pH, and salt concentration ranges	Thickening, stabilizing agent, food additive
Levan	Fructose	*Halomonas smyrnensis* AAD6^T, *Zymomonas mobilis*, *Bacillus subtilis* (natto)	Low viscosity, high water solubility, biological activity, antitumor activity, anti-inflammatory, adhesive strength, film-forming capacity	Emulsifier, stabilizer and thickener, encapsulating agent, food and feed additive, osmoregulator, and cryoprotector
Curdlan	Glucose	*Agrobacterium* spp., *Rhizobium* spp. and *Cellulomonas* spp.	Gel-forming ability, water insolubility, edible and non-toxic, biological activity	Thickening, stabilizing agent, food additive
Gellan	Glucose, rhamnose, glucuronic acid	*Sphingomonas elodea, S. paucimobilis*	Hydrocolloid, stability, gelling capacity, thermo-reversible gels	Stabilizer, thickening agent, structuring and versatile gelling agent
Dextran	Glucose	*Leuconostoc* spp. and *Streptococcus* spp.	Non-ionic, good stability, Newtonian fluid behavior	Blood plasma extender and chromatography media
Pullulan	Glucose	*Aureobasidium pullulans*	Water soluble, adhesive ability, forms fibers, films: transparent, printable, heat sealable	Thickening, stabilizing, texturizing, gelling agent
Alginate	Mannuronic acid and glucuronic acid	*Pseudomonas* spp. and *Azotobacter* spp.	Hydrocolloid, gellling and film-forming capacity	Biomaterial, thickening, stabilizing, and gellifying agent
Cellulose	Glucose	Alphaproteobacteria, Betaproteobacteria, Gammaproteobacteria, and gram-positive bacteria	High crystallinity, insolubility in most solvents, high tensile strength, moldability	Food (nata de coco), diaphragms of acoustic transducers and wound dressings

EPS, exopolysaccharide.

Microbial Exopolysaccharides in Food Industry

Microbial EPSs have rheological properties that match industrial demands and can be produced in large amounts and with high purity (Patel and Prajapati 2013). They can be used as viscosifying agents, stabilizers, emulsifiers, gelling agents, or water-binding agents in foods. Xanthan and gellan, which are produced by *Xanthomonas campestris* and *Pseudomonas elodea* respectively, are the two main commercially available bacterial polysaccharides authorized for use as food additives in the United States and Europe (Morris 2006; Patel and Prajapati,2013). Dextran, curdlan, and cellulose, which are homopolysaccharides of glucose monomers, are neutral bacterial glucans (Nasab et al. 2010). Curdlan can be used as a gelling agent. The major market for pullulan is the food industry. Alginates produced by species of *Pseudomonas* and *Azotobacter* are widely used as thickening, stabilizing, and gelifying agents in the food, textile, paper, and pharmaceutical industries (Hay et al. 2014). Almost 30 species of lactic acid bacteria (LAB) are also known as polysaccharide producers but low production yields make it difficult to exploit them commercially. On the other hand, lactobacilli are GRAS (Generally Recognized as Safe) bacteria and their EPS could be utilized in foods (Badel et al. 2011). Besides the well-known levan-type and inulin-type fructans commonly used in the food sector and synthesized from sucrose by fructosyltransferases, alternan is an example of an α-glucan polymer often found in the food industry and produced by the alternan-sucrase enzyme of *Leuconostoc mesenteroides* (Ryan et al. 2015).

The market for hydrocolloids, which include many polysaccharides such as levan, pullulan, xanthan, alginate, dextran, cellulose, starch, agar, carrageenan, and pectin, was valued at more than US$4 million in 2008 and is estimated to reach US$7.911 billion by 2019. North America is the most important consumer and Asia-Pacific is the fastest growing market for hydrocolloids. Archer Daniels Midland Company, B&V SRL, Ceamsa, Danisco A/S, FMC Corporation, Gelnex, Kerry Group PLC, Taiyo Kagaku Co. Ltd, Ashland Inc., Cargill Inc., CP Kelco, and Fiberstar Inc. are the major companies in the global market(www.marketsandmarkets. com/Market-Reports/hydrocolloid-market231.html). Xanthan gum, which is the second EPS to be approved by the FDA as a safe food ingredient, is the only significant bacterial EPS in this market, with 6% of the total market value.

The principal reason for the extensive use of hydrocolloids in the food industry is their ability to bind with water and modify the properties of other ingredients. The modification of rheological characteristics is helpful in enhancing the sensory properties of foods. Hence hydrocolloids are employed as extremely significant food additives (Li and Nie 2016).

EPSs are alternatives to conventional food additives since there is a high consumer demand for products with low fat or sugar content and low levels of food additives, especially stabilizers and thickeners (Jolly 2002; Robitaille et al. 2009).

Since they improve the texture of fermented dairy products and the functional properties of foods (yoghurt, cheese, and bread) and also confer health benefits with their immunostimulatory, antitumoral or cholesterol lowering activity, EPSs are widely found in the dairy and food industries and in pharmaceuticals, where they are used directly as biological active agents, i.e., as a probiotic (Poli et al. 2011; Tieking and Gänzle 2005). EPS-producing starters are used in dairy products since with EPS, the amount of added milk solids is reduced, yoghurt has better water-binding capacity and its viscosity is improved, texture and mouth feel are enhanced and syneresis is avoided during fermentation or upon storage of fermented milk products such as dahi, yoghurt and cheeses, sour cream, and kefir. EPS cultures have also been shown to affect product rheology in European cultured dairy products (Patel and Prajapati 2013). Cheese fermented by EPS-producing *Streptococcus thermophillus* and *Lactobacillus delbrueckii* subsp. *bulgaricus* culture showed significantly lower hardness, consistency, adhesiveness, chewiness and relaxation in texture profile analysis (Ahmed et al. 2005). *Lactobacillus kefiri* and *L. kefiranofaciens* synthesize Kefiran heteropolysaccharide which is found in the fermented dairy beverage kefir (Maeda et al. 2004).

In bread production, EPSs produced by lactobacilli are used, since they favorably influence bread properties such as water absorption, softness, gluten content of dough and also improve the structure build-up, increase the specific volume of a loaf and prolong shelf-life (Tieking and Gänzle 2005).

EPSs like carrageenans, guar gum, and xanthan gum are used as ingredients and additives in flavored milks to provide sweetness, increase viscosity and enhance perception of texture and mouth feel. Some ice cream formulations also include fructans that influence texture and mouth feel and can allow the reduction of sugar and fat since they contain less energy (Early 2012).

In recent years, there has been a significant increase in the functional food market, including products formulated to maintain a "healthy" gut microbiota, sich as probiotics and prebiotics. EPSs that are related to the interaction of probiotics with the gut ecosystem are used in probiotics to improve bacterial persistence and colonization of a given niche (Salazar et al. 2016). Fructose-oligosaccharides (FOS) have interesting properties for food applications as they have a low sweetness compared to sucrose, are essentially calorie free, and non-carcinogenic. Food applications of inulin and FOS are based mainly on their prebiotic properties.

Remarkably, fructose-based EPSs can be fermented by gut microflora which leads to improvement of intestinal flora and increases mineral absorption (Patel and Prajapati 2013).

To encapsulate probiotic cells, various polysaccharides are employed to improve probiotic survival during processing and storage in food products or in gastrointestinal transit (Tripathi and Giri 2014). Gums and polysaccharides generally act as viscosifiers or stabilizers that are added to food products to avoid sedimentation of particles or creaming of emulsion droplets by increasing the viscosity of the product (Alting and van de Velde 2012).

Due to their flocculating ability, EPSs clarify liquids by precipitating colloidal suspensions. Therefore they can be employed as fining agents in the alcoholic beverage industry to remove hazes (haze stabilization) and to help filtration by increasing particle size (Buglass and Caven-Quantrill 2012).

Microbial polysaccharides such as gellan, pullulan, xanthan, bacterial cellulose, curdlan, and levan have been applied in the development of edible and/or biodegradable membranes and edible coatings, to serve as barriers in food packaging. (Espitia et al. 2014; Freitas et al. 2014).

Commercially available gum-based food thickeners have been widely used as additives in cold beverages and also they are used to obtain the optimal swallow response because of ease of preparation, convenience, reasonable cost, and the suspending ability of the thickened fluids (Cho and Yoo 2015).

Polysaccharides have been also used as fat replacers to improve technological characteristics or to enhance meat protein functionality (Villamonte et al. 2015).

Among several novel microbial EPSs, xanthan, levan, gellan, and curdlan are seen as the most promising polysaccharides for various industrial sectors (Donot et al. 2012).

Xanthan

Xanthan is the most commercially important microbial EPS with a rapidly growing demand (Frese et al. 2014; Vorhölter et al. 2008). Xanthan is a heteropolysaccharide composed of repetitive pentasaccharide units consisting of two glucose, two mannose, and one glucuronic acid residue with a backbone chain consisting of $(1 \rightarrow 4)$-β-D-glucan cellulose (Khan et al. 2007). Xanthan is a natural product produced by the plant pathogen proteobacterium *Xanthomonas campestris pv. campestris* (Xcc). Xanthan is a high molecular weight and anionic polysaccharide that can hinder the gel formation of myofibrillar proteins and enhance surface hydrophobicity under high pressure (Khan et al. 2007; Villamonte et al. 2015).

It can be recovered easily from culture broth by precipitation using ethanol or isopropanol (Schatschneider et al. 2013).

Xanthan shows high viscosity at low concentrations in solution and strong pseudoplasticity, and it is stable over a wide range of pH, temperature, and ionic strengths (Patel and Prajapati 2013). It shows weak gel-like properties at high polymer concentration (Milani and Maleki 2012). Since xanthan is resistant to pH variations, and stable in both alkaline and acidic conditions, it is used in a wide range of products. Due to its superior properties and rheological characteristics, xanthan is employed as a thickening or stabilizing agent in a wide range of industrial areas such as food, cosmetics, and oil drilling industries (Chivero et al. 2015; Schatschneider et al. 2013). It has been described as a "benchmark" product based on its significance in food and non-food applications which include dairy products, drinks, confectionery, dressing, bakery products, syrups and pet foods, as well as the oil, pharmaceutical, cosmetic, paper, paint, and textile industries (Cho and Yoo 2015; Patel and Prajapati 2013). Xanthan is also employed in non-food applications such as stabilizing cattle feed supplements, calf milk substitutes, agricultural herbicides, fungicides, pesticides, and fertilizers, and to impart thixotropy to toothpaste preparations (Morris 2006).

Xanthan was discovered in the 1950s by Allene Rosalind Jeanes at the United States Department of Agriculture, and it is the second microbial EPS approved as a food additive in 1969 by the US Food and Drug Administration. It was approved in Europe in 1982 under the E number E415. In 1988 the ADI value (acceptable daily intake) of xanthan was changed to "not specified" which confirmed it as a safe food additive (Khan et al. 2007; Petri 2015). Xanthan is commercially produced by several companies such as Monsanto/Kelco, SKW Biosystems, Cargill, Rhodia, and Archer Daniels Midland (Born et al. 2005). China has been one of the largest xanthan producers since 2005 (Petri 2015).

Aqueous dispersions of xanthan gum are thixotropic in either hot or cold water which means that products which are shaken and thinned during distribution will thicken when stored at the point of sale (Ahmad et al. 2015; Early 2012; Morris 2006). The thixotropy of xanthan samples allows manipulation and control of processes such as spreading, pumping, pouring, and spraying and has led to the development of a number of dry mix formulations such as sauces, gravies, and desserts. Moreover, xanthan can be heated or refrigerated without losing its desirable textural characteristics (Khan et al. 2007; Morris 2006).

The colloidal stability of emulsions and the shelf-life of edible emulsions of oil in water increase in the presence of xanthan. These superior properties and its shear-thinning behavior make xanthan gum an excellent emulsion stabilizer in salad dressings and sauces (Khan et al. 2007; Petri 2015).

In beverages, xanthan gum is effective for suspending fruit pulp and also enhances the taste of the drink. Therefore it can also be used in low-calorie drinks (Cho and Yoo 2015; Khan et al. 2007).

Xanthan gum in whipped cream to a level of 0.1% increases the consistency of the cream (Zhao et al. 2009). Simsek (2009) reported that the application of various xanthan concentrations influenced the syruping formation of refrigerated dough formulation. Texture and apparent viscosity of ice cream are very important since they determine the quality of the product (Dogan et al. 2013). Use of xanthan in ice cream has several advantages as it provides viscosity to the liquids during processing, reduces the freezing point and enhances both the taste and texture of the product (Dogan et al. 2013; Maletto 2005). Halim et al. (2014) observed that ice cream containing xanthan gum showed the best effect on over-run and the quality of ice cream was increased in terms of pH value and total plate (colony) count. Buttered syrups and chocolate toppings containing xanthan have excellent consistency and flow properties and because of their high viscosity at rest, appear thick and appetizing on products such as pancakes, ice cream, and cooked meats (Palaniraj and Jayaraman 2011).

Xanthan improves processing and storage and enhances elasticity of batters and doughs, and it can be used as a gluten replacement in the development of gluten-free breads (Khan et al. 2007; Morris 2006). For instance, the specific volume of wheat breads was significantly increased by xanthan at a dosage of 0.1% (w/w flour), whereas the crumb hardness of fresh breads could not be reduced by the addition of xanthan at all tested dosages (Guarda et al. 2004).

When xanthan is mixed with starch, the swelling of the starch effectively concentrates xanthan, leading to enhanced rheology of the bulk phase containing the enlarged swollen granules (Morris 2006; Petri 2015). The use of innovative packaging methods is possible due to the ability of xanthan to control pouring, pumping, spreading, and splashing. The mixture of xanthan and galactomannans, carob or tara forms transparent thermo-reversible gels (Morris 2006). Gellan-xanthan gum mixtures have been used for immobilization of bifidobacteria in beads to increase their tolerance of high acid environments (Sun and Griffiths 2000).

Levan

Levan is a homopolysaccharide of β(2-6)-linked fructose residues which is synthesized outside the cell from sucrose by the extracellular enzyme levansucrase (EC 2.4.1.10) by several bacteria, including *Acetobacter, Aerobacter, Azotobacter, Bacillus, Corynebacterium,*

Erwinia, *Gluconobacter*, *Mycobacterium*, *Pseudomonas*, *Streptococcus*, and *Zymomonas* (Donot et al. 2012; Freitas et al. 2011; Nakapong et al. 2013). In addition, halophilic bacterium *Halomonas smyrnensis* AAD6[T] has also been reported as the first high-level levan producer extremophile (Poli et al. 2009).

As a homopolysaccharide with many distinguished properties such as high solubility in oil and water, strong adhesivity, good biocompatibility and film-forming ability, levan has great potential as a novel functional biopolymer in foods, feeds, cosmetics, pharmaceutical and chemical industries (Kang et al. 2009; Kazak et al. 2010). In fact, a recent literature analysis on microbial EPSs named levan together with xanthan, curdlan, and pullulan as the most promising polysaccharides for various industrial sectors (Donot et al. 2012). Levan has many potential uses as an emulsifier, stabilizer and thickener, encapsulating agent, osmoregulator and cryoprotector in addition to its uses in medicine as plasma substitute, prolongator of drug activity, radioprotector, antitumor and antihyperlipidemic agent (Freitas et al. 2011; Kang et al. 2009; Sezer et al. 2011).

Levan is used as an industrial gum, a sweetener, a formulation aid, emulsifier, a carrier for flavor and fragrances, a surface-finishing agent, stabilizer and thickener, encapsulating agent, osmoregulator, and cryoprotectant (Esawy et al. 2013; Shih et al. 2010). It is also used as a food or feed additive with prebiotic and hypocholesterolemic effects (Esawy et al. 2013; Shih et al. 2010). In addition, it is used in edible food coatings (Shit and Shah 2014). A combination of fermented red ginseng and levan significantly inhibited body weight gain in HFD-induced obese mice and prevented insulin and leptin resistance (Oh et al. 2014).

Research has been reported on the potential use of levan by *Halomonas smyrnensis* as a bioflocculating agent (Sam et al. 2011), its suitability for peptide and protein-based drug nanocarrier systems (Sezer et al. 2011). With this microbial system, productivity levels were improved by use of cheap sucrose substitutes like molasses (Kucukasik et al. 2011) as well as other cheap biomass resources (Toksoy Oner 2013) as fermentation substrate. Moreover, levan and aldehyde-activated levan were successfully deposited by matrix-assisted pulsed laser evaporation (MAPLE) resulting in uniform, homogeneous, nanostructured, biocompatible, thin films (Sima et al. 2011, 2012) and furthermore, the feasibility of phosphanate-modified levan as adhesive mulitilayer film was demonstrated (Costa et al. 2013).

The antioxidant potential of this levan polysaccharide in high glucose conditions in pancreatic INS-1E cells by demonstrating a correlation between reduction in oxidative stress and apoptosis with its treatment has been reported (Kazak et al. 2014). *In vitro* anticancer activity of linear levan and its oxidized forms containing increasing amounts of aldehydes was also investigated and anticancer activity was found to depend on the dose as well as on the cell type and this effect became more apparent with increasing degree of oxidation (Kazak Sarilmiser and Toksoy Oner 2014). Ternary blend films of chitosan, PEO, and levan were also prepared to evaluate their morphological, thermomechanical, surface, and biological properties (Sennaroglu Bostan et al. 2014). Moreover, the generic metabolic model of *H. smyrnensis* AAD6^T was reconstructed to elucidate the relationship between levan biosynthesis and other metabolic processes (Ates et al. 2013). Recently, its whole genome sequence was announced (Sogutcu et al. 2012).

Fructan type polysaccharides that ferment in intestinal flora can enhance intestinal flora and mineral absorption. There are several reports of the prebiotic effect of levan-type fructans. Oligofructans obtained from hydrolization of levan from *Zymomonas mobilis* in a microwave oven beneficially affect the host by selective stimulation of probiotic bacteria in the colon (de Paula et al. 2008). The levan from *L. sanfranciscensis* LTH 2590 also exhibits prebiotic effects as demonstrated *in vitro* by different experimental approaches (Bello et al. 2001). The possibility of acting as prebiotic substrate has been demonstrated successfully for a fructan-type EPS produced by one strain of *L. sanfranciscensis* by Korakli et al. (2003). The bifidogenic effect was also reported for the levan-type EPS produced by another strain of the same species (Patel and Prajapati 2013). Recently, prebiotic activities of three different kinds of fructan, levan from *H. smyrnensis* AAD6T, inulin and *Agave tequilana* fructans, were compared and investigated for the effective formulation of symbiotics with antibiotic activity (Arrizon et al. 2014).

Levan can act as a prebiotic, cholesterol and triacylglycerol-lowering agent in the food industry (Nakapong et al. 2013). Commercial food-grade levan is produced by a number of companies, including RealBiotech Co. (www.realbio.com), Chungnam, Korea, and Advance Co. Ltd (www.advance.jp), Japan. Levan production and levan derivatives have also been developed by Montana Biotech and the Sugar Processing Research Institute (Kang et al. 2009).Various forms of levan like chewing gum and powder can be found as "functional food" at markets in Korea. Levan was initially industrialized as a food additive to support functional soluble fiber (Kang et al. 2009). Low molecular weight levan was found to show *in vitro* inhibitory effects against pathogenic bacteria and to have potential as a sweetener in food (Byun et al. 2014).

During 1-week storage of fresh wheat breads, an increased volume, clear softening effect and retarded staling were observed as a result of fructans of several acetic acid bacterial strains (Hermann et al. 2015). Dough rheology and bread texture were enhanced by levan from the sourdough isolate *Lactobacillus sanfranciscensis* (Brandt et al. 2003) and improved antistaling properties of bread were observed with levan addition (Kaditzky et al. 2008).

Levan is accepted as a non-toxic substance and has been verified as safe by many scientific studies (Kang et al. 2009). LD_{50} and NOEL (no observed effective level) values for levan are known to be 7.5 ± 0.5 g/kg bw and 1.5 g/kg/day, respectively (Kang et al. 2009). The Department of Food Chemistry, Ministry of Health, performed general toxicology tests and mutagenicity tests on levan from *B. subtilis* and verified its safety. Levan from *Z. mobilis* was approved as a food by the Korea Food and Drug Administration (KFDA) in 2001, and was listed in the Korea Health and Functional Food Code. A levan with the trade name FRUCTAN is approved as a food and food additive and marketed in many countries such as the USA, EU, Japan, Australia, and New Zealand (Kang et al. 2009).

Many applications for levan, including as a food additive, have been patented. Levan from *S. salivarius* was reported as a food additive with a hypocholesterolemic effect (Kazuoki et al. 1996). A GRAS-grade levan producer, *Lactobacillus reuteri*, can be used as a probiotic (van Hijum et al. 2004). The levan synthesized by *L. sanfranciscensis* can be used in human or pet food products (Vincent et al. 2005).

Gellan

Gellan is a high molecular weight, linear, anionic heteropolysaccharide that consists of tetrasaccharide chemical repeat units in which β-$(1 \rightarrow 4)$-linked glucose, glucuronic acid, glucose, and rhamnose in α-$(1 \rightarrow 3)$ linkage are bonded together (Ahmad et al. 2015; Wüstenberg 2015). Gellan, which is marketed as a broad-spectrum gelling agent, is highly viscous and has shear-thinning behavior in dilute aqueous solutions (Morris 2006). It is soluble in water, forming a viscous solution, and is insoluble in ethanol. The functionality of gellan and type of gellan gel are varied based on the degree of acylation and the ions present (Khan et al. 2007; Pegg 2012). An interesting feature of low acyl gels is their excellent flavor release properties (Morris 2006). Research on gellan gum is intensive since it is an appropriate model for the study of thermo-reversible sol–gel transition (Prajapati et al. 2013).

Gellan is extracellularly secreted by *Sphingmonas elodea* (originally called *Pseudomonas elodea*) by aerobic submerged fermentation in batch culture (Ryan et al. 2015). Kelco isolated the bacterium from water plant (elodea) tissue as part of a screening program intended to identify interesting new bacterial polysaccharides (Khan et al. 2007). Gellan was patented and produced commercially by Kelco. Gellan was first used in food in Japan in 1988 and the FDA has approved it for use in the United States and in Europe it is classified as E 148. It is sold commercially under the trade name Gelrit (Ahmad et al. 2015; Kang et al. 2014; Morris 2006). Specifications for gellan such as molecular mass, solubility, and functional uses were prepared at the 46th Joint Expert Committee on Food Additives (JECFA) in 1996 (Prajapati et al. 2013). It was also approved as a safe food ingredient and additive in various countries such as Australia, Canada, United States, Mexico, Chile, Japan, South Korea, and Philippines (Milani and Maleki 2012). Gellan gum products are available with the names Gelrite, Kelcogel, and Gel-Gro (Prajapati et al. 2013).

Toxicological analysis has been performed and the acute oral toxicity (LD_{50}) value for rats was found to be $>5\,g/kg$. Eye and skin irritation studies suggest that gellan is not an irritant. No significant effect on plasma biochemistry, hematological indices, urinalysis parameters, blood glucose, plasma insulin concentrations, or breath hydrogen concentration was observed for 23-day human dietary studies (Anderson et al. 1988).

Transparent and stable hydrogels of gellan gum have a wide range of applications in tissue engineering and drug delivery. Gellan, which is a non-toxic, biodegradable and biocompatible polymer, forms hydrogels with mechanical properties similar to normal human under definite conditions. Therefore it can be used as a versatile biomaterial (Kang et al. 2014).

Gellan can be employed as a thickener, binder, film former, and stabilizer in a wide range of food applications such as confectionery products, jams and jellies, fabricated foods, water-based gels, pie fillings and puddings, icings and frostings, and dairy products such as ice cream, yoghurt, milkshakes, and gelled milks. It can be used as a hydrocolloid coating for cheese. Gellan can replace agar in Asian foods: Japanese foods such as hard bean jelly, mitsumame jelly cubes, soft red bean jelly, and tokoroten noodles, in the preparation of other Japanese products including karasumi-like food and laver, and in the manufacture of tofu curd (Ahmad et al. 2015; Morris 2006; Prajapati et al. 2013; Wüstenberg 2015). Interactions with other polysaccharides and between the two gellan types are an advantage for using gellan gum for food product development since this allows the production of a wide range of textures and provides structure with superior properties (Danalache et al. 2015).

Gellan gum can be used in the pet food industry, often in combination with other hydrocolloids, to produce both meat chunks and gelled pet foods. The moisture barrier properties of edible gellan films have been investigated (Morris 2006).

Gellan gum is used as a stabilizer in water-based gels such as desserts, drinking jellies, and mango bars (Danalache et al. 2015; Wüstenberg 2015). In bakery products, it improves bake-stability of fillings with low to high soluble solids content. It is also used as a replacement for gelatin in cultured dairy products like yoghurt and sour cream in vegan, Kosher or Halal nutrition (Wüstenberg 2015). Fruit purees/juices along with gelling agents such as gellan can be used for development of novel food products with a variety of textures (Danalache et al. 2015). Gellan reduces clouding and pulp settling in beverages. It is employed in systems with low milk amounts, low-quality milk protein, or heat-damaged proteins that are found in spray-dried milk powders. It also provides a short texture to gelled confectionery, improves gelatin gummy candies, and reduces stickiness. Low-calorie (sugar-free) jams, fruit preparations for yoghurt, sauces, no-fat salad dressings with herbs, and films and adhesion systems (Wüstenberg 2015) and oil reduction in fried foods (Khan et al. 2007) are other applications.

Curdlan

Curdlan is a high molecular weight polysaccharide of glucose residues with the general formula $(C_6H_{10}O5)_n$. It is a a linear β-$(1 \rightarrow 3)$-glucan that is extracellularly produced under nitrogen-limited conditions by the non-pathogenic bacteria *Agrobacterium biobar* and mutants of *Alcaligenes faecalis* var. *myxogenes*, a group of micro-organisms occurring in the soil (Wu et al. 2015). Curdlan is insoluble in water, alcohols and the majority of organic solvents and is soluble in aqueous alkaline solutions and it shows gel strength between agar and gelatin (Khan et al. 2007; Morris 2006).

The term "curdlan" has been derived from "curdle" – describing its gelling behavior at high temperatures with different characteristics. A high-set, thermo-irreversible, strong elastic gel is formed upon heating its aqueous suspensions above 80 °C followed by cooling. Alternatively, a low-set, thermo-reversible gel can be obtained by heating it to 55 °C and subsequently cooling to ambient temperatures. The latter type of gel can be formed without heating, by the neutralization of its solutions with an alkali (Khan et al. 2007; Wu et al. 2015; Zhang and Edgar 2014).

Curdlan is widely used as a biothickening and gelling agent in foods such as tofu, noodles, and jellies since it has remarkable rheological properties among natural and synthetic polymers (Khan et al. 2007; Morris 2006). There are numerous applications of curdlan as a gelling agent in the food, construction, and pharmaceutical industries. Gelatin and agar are replaced with curdlan in the production of jellies, desserts, and confectionery. It can also be used as a thickener and binder in foods such as salad dressings, desserts, and pasta. Edible coatings of curdlan can be used to prolong shelf-life (Hu et al. 2015; Khan et al. 2007). Curdlan as a dietary fiber also provides a nutritional benefit and has the potential to be produced at a higher quality for related foods (Wu et al. 2015). Curdlan and microbial transglutaminase would improve the gel strength, water-holding capacity, and whiteness of heated fish meat gel (Hu et al. 2015).

Curdlan was first discovered in 1964 and research was started for production of curdlan for both food and general industrial applications by Takeda Chemical Industries Ltd in 1968. Although development work ceased in 1973, industrial-scale production was restarted in 1988 (Morris 2006). Its usage for food was approved in Korea, Taiwan, and Japan in 1989 (Hu et al. 2015).

After xanthan and gellan, curdlan was approved as the third microorganism-fermented hydrocolloid food additive by the US FDA in December 1996 (Wüstenberg 2015). Curdlan has been used in a wide range of food applications in Japan where bacterial polysaccharides are considered as natural products (Morris 2006). Toxic effects of curdlan have been extensively studied and it has been estimated as a safe product (Hu et al. 2015; Khan et al. 2007).

Curdlan is tasteless, odorless, colorless and has a high water-holding capacity that makes it an ideal textural agent in various food applications such as meat, dairy, sauces, baking, and nutraceutical products (Hu et al. 2015; Khan et al. 2007). Curdlan can be directly added to regular food systems at low levels (less than 1%) or used as a building block of the food structure at high levels to develop new food products. In Japan, it is employed mostly in the meat-processing industry for improvement of texture by enhancement of the water-binding capacity. Elasticity and strength of noodles are also enhanced by curdlan (Khan et al. 2007).

Rheological and thermal behaviors of curdlan have led to its application for the integrity of oil-moisture barriers in batter and coating systems, imitation seafood delicacies, vegetarian meat analogues, and tofu products having freeze/thaw and retort stability (Khan et al. 2007; Zhang and Edgar 2014). Curdlan gels have been used to develop calorie-reduced items, since there are no digestive enzymes for curdlan in the upper alimentary tract, and curdlan can be used for fat replacement (McIntosh et al. 2005; Zhang and Edgar 2014).

Non-food applications of curdlan have also been studied and antitumor, antioxidative, and immunomodulating activities have been discovered due to curdlan's unique helical structure, gel-forming capacity, and potent pharmacological properties. Several studies have been employed to improve the solubility of curdlan to increase its applications, develop nanostructures for encapsulation and release of biological active reagents, provide curdlan derivatives with antitumor, anti-HIV, and anticoagulation activity, improve wound healing activities and inhibit microbes, fungi, and viruses (Li et al. 2014; McIntosh et al. 2005; Zhang and Edgar 2014).

Conclusion

Exopolysaccharides produced by a diverse group of microbial systems are rapidly emerging as new and industrially important biomaterials. Microbially produced EPSs show great diversity and functions with unique and commercially relevant material properties. EPSs are mainly associated with high-value applications in various fields such as food, feed, packaging, chemicals, textiles, cosmetics and pharmaceutical industry, agriculture and medicine, and they have received considerable research attention over recent decades due to their biocompatibility, biodegradability, and both environmental and human compatibility.

Exopolysaccharides are widely used to enhance the quality, texture, mouth feel and flavor of food as thickeners, emulsifiers, stabilizers, texturizers, and gelling agents. Xanthan, levan, gellan, and curdlan, which are currently used as additives in the food industry, are well known industrial microbial polysaccharides with numerous applications and a considerable market. There is a growing body of research associated with these high-value microbial EPSs for development of low-cost production processes, improvement of properties, and optimization of performance as food additives. These natural polysaccharides will become increasingly essential since they are natural non-toxic alternatives to synthetic food additives with their potentially harmful characteristics.

References

Ahmad, N. H., Mustafa, S. and Che Man, Y. B. (2015) Microbial polysaccharides and their modification approaches: a review. *Int. J. Food Properties*, 18(2): 332–347.

Ahmed, N. H., El Soda, M., Hassan A. N. and Frank J. (2005) Improving the textural properties of an acidcoagulated (Karish) cheese using exopolysaccharide producing cultures. *Food Sci. Technol.*, 38: 343–347.

Alting, A. C. and van de Velde, F. (2012) Proteins as clean label ingredients in foods and beverages. In: Baines, D. and Seal, R. (eds) *Natural Food Additives, Ingredients and Flavourings*. Cambridge: Woodhead Publishing, pp. 197–211.

Anderson, D.M.V., Brydon, W.G. and Eastwood, M.A. (1988) The dietary effects of gellan gum in humans. *Food Addit. Contam.*, 5: 237.

Arrizon, J., Hernández-Moedano, A., Toksoy-Oner, E. and González-Avila, M. (2014) *In vitro* prebiotic activity of fructans with different fructosyl linkage for symbiotics elaboration. *Int. J. Probiot. Prebiot.*, 9(3): 69–76.

Ateş, Ö., Arga, K.Y. and Toksoy Öner, E. (2013) The stimulatory effect of mannitol on levan biosynthesis: lessons from metabolic systems analysis of *Halomonas smyrnensis* AAD6^T. *Biotechnol. Progr.*, 29(6): 1386–1397.

Badel, S., Bernardi, T. and Michaud, P. (2011) New perspectives for *Lactobacilli* exopolysaccharides. *Biotechnol. Adv.*, 29(1): 54–66.

Bello, F.D., Walter, J., Hertel, C. and Hammes, W.P. (2001) In vitro study of probiotic properties of levan-type exopolysaccharides from *Lactobacilli* and non-digestible carbohydrates using denaturing gradient gel electrophoresis. *Syst. Appl. Microbiol.*, 24: 232–237.

Born, K., Langendorff, V. and Boulenguer, P. (2005) Xanthan. In: Steinbuchel, A. and Rhee, S. K. (eds) *Polysaccharides and Polyamides in the Food Industry*, vol. 1. Weinheim: Wiley-VCH, pp. 481–519.

Brandt, M. J., Roth, K. and Hammes, W. P. (2003) Effect of an exopolysaccharide produced by *Lactobacillus sanfranciscensis* LTH1729 on dough and bread quality. In: *Sourdough from Fundamentals to Application*. Brussels: Vrije Universiseit Brussels (VUB), p. 80.

Buglass, A. J. and Caven-Quantrill, D.J. (2012) Applications of natural ingredients in alcoholic drinks. In: Baines, D. and Seal, R. (eds) *Natural Food Additives, Ingredients and Flavourings*. Cambridge: Woodhead Publishing, pp. 358–416.

Byun, B.Y., Lee, S.J. and Mah, J.H. (2014) Antipathogenic activity and preservative effect of levan (β-2, 6-fructan): a multifunctional polysaccharide. *Int. J. Food Sci. Technol.*, 49(1): 238–245.

Chivero, P., Gohtani, S., Yoshii, H. and Nakamura, A. (2015) Effect of xanthan and guar gums on the formation and stability of soy soluble polysaccharide oil-in-water emulsions. *Food Res. Int.*, 70: 7–14.

Cho, H.M. and Yoo, B. (2015) Rheological characteristics of cold thickened beverages containing xanthan gum-based food thickeners used for dysphagia diets. *J. Acad. Nutr. Dietet.*, 115(1): 106–111.

Costa, R.R., Neto, A.I., Calgeris, I. et al. (2013) Adhesive nanostructured multilayer films using a bacterial exopolysaccharide for biomedical applications. *J. Materials Chem. B*, 1: 2367–2374.

Danalache, F., Beirão-da-Costa, S., Mata, P., Alves, V. D. and Moldão-Martins, M. (2015) Texture, microstructure and consumer preference of mango bars jellified with gellan gum. *LWT–Food Sci. Technol.*, 62: 584–592.

de Paula, V. C., Pinheiro, I. O. and Lopes, C. E. (2008) Microwave-assisted hydrolysis of Zymomonas mobilis levan envisaging oligofructan production. *Bioresource Technol.*, 99(7): 2466–2470.

Dogan, M., Kayacier, A., Toker, Ö. S., Yilmaz, M. T. and Karaman, S. (2013) Steady, dynamic, creep, and recovery analysis of ice cream mixes added with different concentrations of xanthan gum. *Food Bioproc. Technol.*, 6(6): 1420–1433.

Donot, F., Fontana, A., Baccou, J.C. and Schorr-Galindo, S. (2012) Microbial exopolysaccharides: main examples of synthesis, excretion, genetics and extraction. *Carbohydr. Polym.*, 87(2): 951–962.

Early, R. (2012) The application of natural hydrocolloids to foods and beverages. In: Baines, D. and Seal, R. (eds) *Natural Food Additives, Ingredients and Flavourings*. Cambridge: Woodhead Publishing, pp. 175–196.

Esawy, M. A., Amer, H., Gamal-Eldeen, A. M. et al. (2013) Scaling up, characterization of levan and its inhibitory role in carcinogenesis initiation stage. *Carbohydr. Polym.*, 95(1): 578–587.

Espitia, P. J. P., Du, W. X., de Jesús Avena-Bustillos, R., Soares, N. D. F. F. and McHugh, T. H. (2014) Edible films from pectin: physical, mechanical and antimicrobial properties – a review. *Food Hydrocoll.*, 35: 287–296.

Fahnestock, K.J., Austero, M.S. and Schauer, C.L. (2011) Natural polysaccharides: from membranes to active food packaging. In: Kalia, S. and Averous, L. (eds) *Biopolymers: Biomedical and Environmental Applications*; Hoboken: John Wiley and Sons, pp. 59–78.

Freitas, F., Alves, V. D. and Reis, M. A. (2011) Advances in bacterial exopolysaccharides: from production to biotechnological applications. *Trends Biotechnol.*, 29: 388–398.

Freitas, F., Alves, V. D., Reis, M. A., Crespo, J. G. and Coelhoso, I. M. (2014) Microbial polysaccharide-based membranes: current and future applications. *J. Appl. Polym. Sci.*, 131(6), 40047.

Frese, M., Schatschneider, S., Voss, J., Vorhölter, F. J. and Niehaus, K. (2014) Characterization of the pyrophosphate-dependent 6-phosphofructokinase from *Xanthomonas campestris* pv. *campestris*. *Arch. Biochem. Biophy.*, 546: 53–63.

Guarda, A., Rosell, C. M., Benedito, C. and Galotto, M. J. (2004) Different hydrocolloids as bread improvers and antistaling agents. *Food Hydrocoll.*, 18(2): 241–247.

Guo, M. and Murphy, R. J. (2012) Is there a generic environmental advantage for starch-PVOH biopolymers over petrochemical polymers? *J. Polym. Environ.*, 20(4): 976–990.

Halim, N. R. A., Shukri, W. H. Z., Lani, M. N. and Sarbon, N. M. (2014) Effect of different hydrocolloids on the physicochemical properties, microbiological quality and sensory acceptance of fermented cassava (tapai ubi) ice cream. *Int. Food Res. J.*, 21(5): 1825–1836.

Hay, I.D., Wang, Y., Moradali, M.F., Rehman, Z.U. and Rehm, B.H.A, (2014) Genetics and regulation of bacterial alginate production. *Environ. Microbiol.*, 16(10): 2997–3011.

Hermann, M., Petermeier, H. and Vogel, R. F. (2015) Development of novel sourdoughs with *in situ* formed exopolysaccharides from acetic acid bacteria. *Eur. Food Res. Technol.*, 1–13.

Hu, Y., Liu, W., Yuan, C. et al. (2015) Enhancement of the gelation properties of hairtail (*Trichiurus haumela*) muscle protein with curdlan and transglutaminase. *Food Chemistry*, 176, 115–122.

Ige, O. O., Umoru, L. E. and Aribo, S. (2012) Natural products: a minefield of biomaterials. *ISRN Materials Sci.*, article no. 983062.

Jafar Milani and Gisoo Maleki (2012) Hydrocolloids in food industry. In: Valdez, B. (ed.) *Food Industrial Processes – Methods and Equipment*. Rijeka: InTech, pp. 17–38.

Jolly, L. (2002) Exploiting exopolysaccharides from lactic acid bacteria. *Antonie van Leeuwenhoek*, 82: 367–374.

Kaditzky, S., Seitter, M., Hertel, C. and Vogel, R.F. (2008) Performance of *Lactobacillus sanfranciscensis* TMW 1.392 and its levansucrase deletion mutant in wheat dough and comparison of their impact on bread quality. *Eur. Food Res. Technol.*, 227: 433–442.

Kang, D., Zhang, H. B., Nitta, Y., Fang, Y. P. and Nishinari, K. (2014) Gellan. In: Ramawat, K.G. and Merillion, J.M. (eds) *Polysaccharides*. New York: Springer, pp. 1–46.

Kang, S. A., Jang, K. H., Seo, J. W. et al. (2009) Levan: applications and perspectives. In: Bernd, H. A. and Rehm, B. H. (eds) *Microbial Production of Biopolymers and Polymer Precursors*. Norfolk: Caister Academic Press, pp. 145–161.

Kaur, V., Bera, M. B., Panesar, P. S., Kumar, H. and Kennedy, J. F. (2014) Welan gum: microbial production, characterization, and applications. *Int. J. Biol. Macromol.*, 65: 454–461.

Kazak, H., Toksoy Öner, E. and Dekker, R.F.H. (2010) Extremophiles as sources of exopolysaccharides. In: Columbus, E. (ed.) *Handbook of Carbohydrate Polymers: Development, Properties and Applications*. New York: Nova Science Publishers, pp. 605–619.

Kazak, H., Barbosa, A., Baregzav, B. et al. (2014) Biological activities of bacterial levan and three fungal β-glucans, botryosphaeran and lasiodiplodan under high glucose condition in the pancreatic β-cell line INS-1E. In: Popescu, L. M., Hargens, A. R. and Singal, P. K. (eds) *Adaptation Biology and Medicine Volume 7: New Challenges*. New Delhi: Narosa Publishing, pp.105–115.

Kazak, H., Ates, O., Ozdemir, G., Arga, K.Y. and Toksoy Oner, E. (2015) Effective stimulating factors for microbial levan production by *Halomonas smyrnensis* AAD6^T. *J. Biosci. Bioeng.*, 119(4): 455–463.

Kazak Sarilmiser, H. and Toksoy Oner, E. (2014) Investigation of anti-cancer activity of linear and aldehyde-activated levan from *Halomonas smyrnensis* AAD6^T. *Biochem. Eng. J.*, 92: 28–34.

Kazuoki, I. (1996) Antihyperlipidemic and antiobesity agent comprising levan or hydrolysis products thereof obtained from Streptococcus salivarius. United States Patent 5,527,784.

Khan, T., Park, J. K. and Kwon, J. H. (2007) Functional biopolymers produced by biochemical technology considering applications in food engineering. *Korean J. Chem. Eng.*, 24(5): 816–826.

Korakli, M., Pavlovic, M., Gänzle, M. G. and Vogel, R. F. (2003) Exopolysaccharide and kestose production by *Lactobacillus sanfranciscensis* LTH 2590. *Appl. Environ. Microbiol.*, 69: 2073–2079.

Kucukasik, F., Kazak, H., Guney, D. et al. (2011) Molasses as fermentation substrate for levan production by *Halomonas* sp. *Appl. Microbiol. Biotechnol.*, 89(6): 1729–1740.

Kumar, A.S., Mody, K. and Jha, B. (2007) Bacterial exopolysaccharides – a perception. *J. Basic Microb.*, 47: 103–117.

Li, J. M. and Nie, S. P. (2016) The functional and nutritional aspects of hydrocolloids in foods. *Food Hydrocoll.*, 53: 46–61.

Li, P., Zhang, X., Cheng, Y. et al. (2014) Preparation and in vitro immunomodulatory effect of curdlan sulfate. *Carbohydr. Polym.*, 102: 852–861.

Liang, T. W. and Wang, S. L. (2015) Recent advances in exopolysaccharides from Paenibacillus spp.: production, isolation, structure, and bioactivities. *Marine Drugs*, 13(4): 1847–1863.

Llamas, I., Amjres, H., Mata, J. A., Quesada, E. and Béjar, V. (2012) The potential biotechnological applications of the exopolysaccharide produced by the halophilic bacterium *Halomonas almeriensis*. *Molecules*, 17(6): 7103–7120.

Maeda, H., Zhu, X., Suzuki, S., Suzuki, K. and Kitamura, S. (2004) Structural characterization and biological activities of an exopolysaccharide kefiran produced by Lactobacillus kefiranofaciens WT-2bT. *J. Agric. Food Chem.*, 52: 5533–5538.

Maletto, P. (2005) Low carbohydrate ice cream. US Patent Application 10/701,254.

Mata, J. A., Béjar, V., Llamas, I. and (2006) Exopolysaccharides produced by the recently described halophilic bacteria *Halomonas ventosae* and *Halomonas anticariensis*. *Res. Microbiol.*, 157(9): 827–835.

McIntosh, M., Stone, B. A. and Stanisich, V. A. (2005) Curdlan and other bacterial $(1 \rightarrow 3)$-β-D-glucans. *Appl. Microbiol. Biotechnol.*, 68(2): 163–173.

Morris, V. J. (2006) Bacterial polysaccharides. In: Stephen, A.M., Philips, G.O. and Williams, P.A. (eds) Food Polysaccharides and Their Applications. Boca Raton: CRC Press, pp. 413–454.

Nakapong, S., Pichyangkuraa, R., Ito, K., Iizukab, M. and Pongsawasdia, P. (2013) High expression level of levansucrase from *Bacillus licheniformis* RN-01 and synthesis of levan nanoparticles. *Int. J. Biol. Macromol.*, 54: 30–36.

Nasab, M.M., Gavahian, M., Yousefi, A.R. and Askari, H. (2010) Fermentative production of dextran using food industry wastes. *World Acad. Sci. Eng. Technol.*, 44: 8–24.

Nicolaus, B., Kambourova, M. and Toksoy Oner, E. (2010) Exopolysaccharides from extremophiles: from fundamentals to biotechnology. *Environ. Technol.*, 31(10): 1145–1158.

Nwodo, U. U., Green, E. and Okoh, A. I. (2012) Bacterial exopolysaccharides: functionality and prospects. *Int. J. Mol. Sci.*, 13(11): 14002–14015.

Oh, J. S., Lee, S. R., Hwang, K. T. and Ji, G. E. (2014) The anti-obesity effects of the dietary combination of fermented red ginseng with levan in high fat diet mouse model. *Phytother. Res.*, 28(4): 617–622.

Ordax, M., Marco-Noales, E., López, M. M. and Biosca, E. G. (2010) Exopolysaccharides favor the survival of *Erwinia amylovora* under copper stress through different strategies. *Res. Microbiol.*, 161(7): 549–555.

Palaniraj, A. and Jayaraman, V. (2011) Production, recovery and applications of xanthan gum by *Xanthomonas campestris*. *J. Food Eng.*, 106(1): 1–12.

Pasteur, L. (1861) On the viscous fermentation and the butyrous fermentation. *Bulletin de la Société Chimique de France*, 11: 30–31 (in French).

Patel, A. and Prajapati, J. B. (2013) Food and health applications of exopolysaccharides produced by lactic acid bacteria. *Adv. Dairy Res.*, 1(107): 2.

Pegg, A.M. (2012) The application of natural hydrocolloids to foods and beverages. In: Baines, D. and Seal, R. (eds) *Natural Food Additives, Ingredients and Flavourings*. Cambridge: Woodhead Publishing, pp. 175–196.

Petri, D. F. (2015) Xanthan gum: a versatile biopolymer for biomedical and technological applications. *J. Appl. Polym. Sci.*, 132(23).

Poli, A., Kazak, H., Gurleyendag, B. et al. (2009) High level synthesis of levan by a novel *Halomonas* species growing on defined media. *Carbohydr. Polym.*, 78: 651–657.

Poli, A., Anzelmo, G., Fiorentino, G., Nicolaus, B., Tommonaro, G. and di Donato, P. (2011) Polysaccharides from wastes of vegetable industrial processing: new opportunities for their eco-friendly re-use. In: *Biotechnology of Biopolymers*. Rijeka: InTech, pp. 33–56.

Prajapati, V. D., Jani, G. K., Zala, B. S. and Khutliwala, T. A. (2013) An insight into the emerging exopolysaccharide gellan gum as a novel polymer. *Carbohydr. Polym.*, 93(2): 670–678.

Ptaszek, P., Kabziński, M., Kruk, J. et al. (2015) The effect of pectins and xanthan gum on physicochemical properties of egg white protein foams. *J. Food Eng.*, 144: 129–137.

Rehm, B. H. (2010) Bacterial polymers: biosynthesis, modifications and applications. *Nat. Rev. Microbiol.*, 8(8): 578–592.

Robitaille, G., Tremblay, A., Moineau, S., St Gelais, D., Vadeboncoeur, D. and Britten, M. (2009) Fat-free yoghurt made using a galactose-positive exopolysaccharide-producing recombinant strain of Streptococcus thermophilus. *J. Dairy Sci.*, 92: 477–482.

Ryan, P. M., Ross, R. P., Fitzgerald, G. F., Caplice, N. M. and Stanton, C. (2015) Sugar-coated: exopolysaccharide producing lactic acid bacteria for food and human health applications. *Food Function*, 679–693.

Salazar, N., Gueimonde, M., De Los Reyes-Gavilán, C. G. and Ruas-Madiedo, P. (2016) Exopolysaccharides produced by lactic acid bacteria and bifidobacteria as fermentable substrates by the intestinal microbiota. *Crit. Rev. Food Sci. Nutr.*, 56(9): 1440–1453.

Sam, S., Kucukasik, F., Yenigun, O., Nicolaus, B., Toksoy Oner, E. and Yukselen, M.A. (2011) Flocculating performances of exopolysaccharides produced by a halophilic bacterial strain cultivated on agro-industrial waste. *Bioresource Technol.*, 102: 1788–1794.

Sanchez, M. C. (2015) Food additives. In: *Food Law and Regulation for Non-Lawyers*. New York: Springer.

Schatschneider, S., Persicke, M., Watt, S. A. et al. (2013) Establishment in silico analysis, and experimental verification of a large-scale metabolic network of the xanthan producing *Xanthomonas campestris* pv. campestris strain B100. *J. Biotechnol.*, 167(2): 123–134.

Schmid, J. and Sieber, V. (2015) Enzymatic transformations involved in the biosynthesis of microbial exo-polysaccharides based on the assembly of repeat units. *ChemBioChem.*, 16(8): 1141–1147.

Sennaroglu Bostan, M., Cansever Mutlu, E., Kazak, H., Keskin, S.S., Toksoy Oner, E. and Eroglu, M.S. (2014) Comprehensive characterization of chitosan/PEO/levan ternary blend films. *Carbohydr. Polym.*, 102: 993–1000.

Sezer, A. D., Kazak, H., Toksoy Oner, E. and Akbuga, J. (2011) Levan-based nanocarrier system for peptide and protein drug delivery: optimization and influence of experimental parameters on the nanoparticle characteristic. *Carbohydr. Polym.*, 84: 358–363.

Shih, I. L., Chen, L. D. and Wu, J. Y. (2010) Levan production using *Bacillus subtilis natto* cells immobilized on alginate. *Carbohydr. Polym.*, 82(1): 111–117.

Shit, S. C. and Shah, P. M. (2014) Edible polymers: challenges and opportunities. *J. Polym.*, article ID 427259.

Sima, F., Mutlu Cansever, E., Eroglu, M. S. et al. (2011) Levan nanostructured thin films by MAPLE assembling. *Biomacromolecules*, 12: 2251–2256.

Sima, F., Axente, E., Sima, L. E. et al. (2012) Combinatorial matrix-assisted pulsed laser evaporation: single-step synthesis of biopolymer compositional gradient thin film assemblies. *Appl. Phys. Lett.*, 101: 233705.

Simsek, S. (2009) Application of xanthan gum for reducing syruping in refrigerated doughs. *Food Hydrocoll.*, 23(8): 2354–2358.

Sogutcu, E., Emrence, Z., Arikan, M. et al. (2012) Draft genome sequence of *Halomonas smyrnensis* AAD6T. *J. Bacteriol.*, 194(20): 5690–5691.

Sousa, S. A., Feliciano, J. R. and Leitão, J. H. (2011) Activated sugar precursors: biosynthetic pathways and biological roles of an important class of intermediate metabolites in bacteria In: *Biotechnology of Biopolymers*. Rijeka: InTech, pp. 257–274.

Spizzirri, U. G., Cirillo, G. and Iemma, F. (2015) Polymers and food packaging: a short overview. In: *Functional Polymers in Food Science: From Technology to Biology, Volume 1: Food Packaging*. Chichester: John Wiley and Sons.

Srivastava, A., Zhurina, D. and Ullrich, M. S. (2009) Levansucrase and levan formation in *Pseudomonas syringae* and related organisms. In: Ullrich, M. (ed.) *Bacterial Polysaccharides: Current Innovations and Future Trends*. Norfolk: Caister Academic Press, p. 213.

Staudt, A. K., Wolfe, L. G. and Shrout, J. D. (2012) Variations in exopolysaccharide production by *Rhizobium tropici*. *Arch. Microbiol.*, 194(3): 197–206.

Sun, W. and Griffiths, M. W. (2000) Survival of bifidobacteria in yogurt and simulated gastric juice following immobilization in gellan-xanthan beads. *Int. J. Food Microbiol.*, 61(1): 17–25.

Sutherland, I. W. (1972) Bacterial exopolysaccharides. *Adv. Microb. Physiol.*, 8: 143–213.

Tieking, M. and Ganzle, M. G. (2005) Exopolysaccharides from cereal associated *lactobacilli*. *Trends Food Sci. Technol.*, 16: 79–84.

Toksoy Oner, E. (2013) Microbial production of extracellular polysaccharides from biomass. In: Fang, Z. (ed.) *Pretreatment Techniques for Biofuels and Biorefineries*. Berlin: Springer, pp. 35–56.

Tripathi, M. K. and Giri, S. K. (2014) Probiotic functional foods: survival of probiotics during processing and storage. *J. Funct. Foods*, 9: 225–241.

Van Hijum, S., Van Geel-Schutten, G. H., Dijkhuizen, L. and Rahaoui, H. (2004) Novel fructosyltransferases. United States Patent 2004/0185537.

Villamonte, G., Jury, V., Jung, S. and de Lamballerie, M. (2015) Influence of xanthan gum on the structural characteristics of myofibrillar proteins treated by high pressure. *J. Food Sci.*, 80(3): 522–531.

Vincent, S., Brandt, M., Cavadini, C., Hammes, W. P., Neeser, J.R. and Waldbuesser, S. (2005) Levan-producing *Lactobacillus* strain and method of preparing human or pet food products using the same. United States Patent 6,932,991.

Vorhölter, F. J., Schneiker, S., Goesmann, A. et al. (2008) The genome of *Xanthomonas campestris* pv. *campestris* B100 and its use for the reconstruction of metabolic pathways involved in xanthan biosynthesis. *J. Biotechnol.*, 134(1): 33–45.

Vroman, I. and Tighzert, L. (2009) Biodegradable polymers. *Materials*, 2(2): 307– 344.

Vu, B., Chen, M., Crawford, R. J. and Ivanova, E. P. (2009) Bacterial extracellular polysaccharides involved in biofilm formation. *Molecules*, 14(7): 2535–2554.

Williams, P. A., Phillips, G. O. and Seisun, D. (2002) Overview of the hydrocolloid market. In: Williams, P. A. and Phillips, G. O. (eds) *Gums and Stabilizers for the Food Industry 11*. London: Royal Society of Chemistry.

Wu, C., Yuan, C., Chen, S., Liu, D., Ye, X. and Hu, Y. (2015) The effect of curdlan on the rheological properties of restructured ribbonfish (*Trichiurus* spp.) meat gel. *Food Chem.*, 179: 222–231.

Wüstenberg, T. (2015) General overview of food hydrocolloids. In: Cellulose and Cellulose Derivatives in the Food Industry: Fundamentals and Applications. Weinheim: Wiley.

Yamamoto, Y., Takahashi, Y., Kawano, M. et al. (1999) In vitro digestibility and fermentability of levan and its hypocholesterolemic effects in rats. *J. Nutr. Biochem.*, 10(1): 13–18.

Yasar Yildiz, S. and Toksoy Oner, E. T. (2014) Mannan as a promising bioactive material for drug nanocarrier systems. In: Sezer, A. D. (ed.) *Applications of Nanotechnology in Drug Delivery*. Rijeka: InTech, pp 312–342.

Yasar Yildiz, S., Anzelmo, G., Ozer, T. et al. (2014) *Brevibacillus themoruber*: a promising microbial cell factory for exopolysaccharide production. *J. Appl. Microbiol.*, 116(2): 314–324.

Yoo, S.H., Yoon, E.J., Cha, J. and Lee, H.G. (2004) Antitumor activity of levan polysaccharides from selected microorganisms. *Int. J. Biol. Macromol.*, 34(1): 37–41.

Zhang, R. and Edgar, K. J. (2014) Properties, chemistry, and applications of the bioactive polysaccharide curdlan. *Biomacromolecules*, 15(4): 1079–1096.

Zhao, Q., Zhao, M., Yang, B. and Cui, C. (2009) Effect of xanthan gum on the physical properties and textural characteristics of whipped cream. *Food Chem.*, 116(3): 624–628.

8

Microbial Fibrinolytic Enzyme Production and Applications

Essam Kotb

Research Laboratory of Bacteriology, Department of Microbiology, Faculty of Science, Zagazig University, Zagazig, Egypt

Corresponding author e-mail: ekotb@hotmail.com; essamkottb@yahoo.com

Introduction

Enzyme replacement therapy has become a common trend in medicine in the last 100 years. Fibrinolytic enzymes are a special type of proteases that can degrade the fibrin mesh of blood clots. Fibrin originates from the soluble protein precursor fibrinogen which then is laid down inside blood vessels that have been subjected to a disease or injury as minuscule strands that eventually harden, effectively capturing the blood components. There is no doubt that exogenous fibrin clots play a vital role in the healing of blood vessels; however, endogenous fibrin clots are a major risk for thrombotic or cardiovascular diseases (CVDs) (Kotb 2012).

Cardiovascular diseases, including ischemic heart disease, valvular heart disease, peripheral vascular disease, acute myocardial infarction, arrhythmias, high blood pressure and stroke, are the primary cause of death throughout the world, especially in Western countries. According to data provided by the World Health Organization (WHO) (Figure 8.1) in 2012, heart diseases are responsible for 27.1% (15.2 million deaths/year) of the total world mortality rate (Hassanein et al. 2011; Kotb 2013, 2014a).

The thrombolytic/fibrinolytic agents can be distinguished into two categories. The first is plasminogen activators (PAs), such as streptokinase, urokinase-type PA (uPA), and tissue-type PA (tPA), that activate plasminogen into active plasmin which eventually hydrolyzes fibrin. The second type of thrombolytic/fibrinolytic agent is plasmin-like enzymes, such as nattokinase (NK) and lumbrokinase (LK), that directly degrade fibrin, thereby dissolving blood thrombi rapidly and completely (Kotb 2014a, 2014b).

Microbial Functional Foods and Nutraceuticals, First Edition. Edited by Vijai Kumar Gupta, Helen Treichel, Volha (Olga) Shapaval, Luiz Antonio de Oliveira, and Maria G. Tuohy.
© 2018 John Wiley & Sons Ltd. Published 2018 by John Wiley & Sons Ltd.

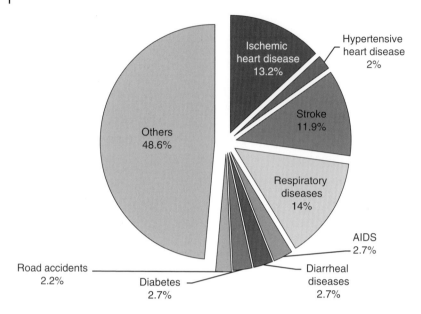

Figure 8.1 The proportion of deaths by cause in WHO regions in 2012. (*See color plate section for the color representation of this figure.*)

Although PAs are still widely used for the clinical thrombolytic/fibrinolytic therapy of CVDs, their side effects (non-specificity and short half-life) and expense have prompted researchers to search for cheaper and safer alternatives. Many thrombolytic/fibrinolytic enzymes have been isolated from various sources such as plants (Chung et al. 2010), animals (Zhang et al. 1995), and microbes (Kotb 2012, 2013, 2014a,b, 2015a,b). Among all these sources, microbial fibrinolytic enzymes are the best alternatives and are preferred due to feasibility of mass culture, low cost, broad biochemical diversity, and ease of genetic manipulation. Fibrinolytic enzymes were discovered in various microbes, the most important among which is *Bacillus* spp. which contributes to the traditional fermented food characteristics of various countries (Kotb 2015a, 2015b; Mine et al. 2005). The properties of these enzymes have been extensively studied, and their *in vitro* and *in vivo* action has been characterized. Therefore, microbial fibrinolytic enzymes, especially those from food-grade micro-organisms, have been used as functional food additives and drugs to cure CVDs (Hwang et al. 2002; Kotb 2013).

Fibrinolysis

To understand fibrinolytic activity, first we must understand the process of blood clotting or coagulation. The damaged blood vessel releases the compound thromboplastin. At the same time, platelets adhere to the edges of broken vessels and disintegrate to release platelet factor 3. Both compounds react with calcium ions and other protein factors to form the prothrombin activator. Once the prothrombin activator is formed, a clot follows. Blood thrombi can cause blockage of blood flow, preventing oxygen supply to organs that will eventually die. Deep vein thrombosis can result in pulmonary emboli while small thrombi can mobilize to the heart; this can result in myocardial infarction (heart attack) or move to the brain causing strokes or mini-strokes. All these events are life threatening. Of course, blood clotting is very important in tissue repair and saves the body from blood loss upon severe injury (Kotb 2012).

However, build-up of fibrin in the circulatory system is the main risk factor for CVD. Under normal healthy conditions, endogenous fibrin clots formed inside blood vessels rapidly dissolve and are effectively removed by the fibrinolysis system. The key enzyme in the fibrinolysis system is plasmin, which is a serine protease activated from the proenzyme plasminogen via a tPA (released from endothelial cells of vessel wall) or a uPA (released from kidney cells). The fibrinolysis system is well regulated by a number of sequential interactions between fibrin, plasminogen, and specific activators and inhibitors (Figure 8.2) with the aim

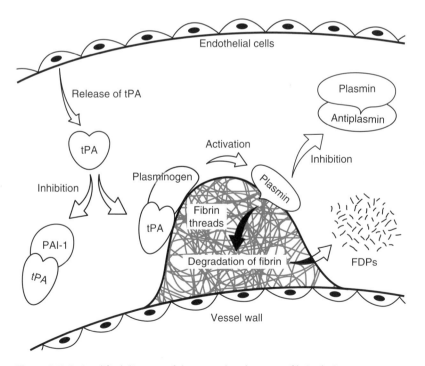

Figure 8.2 A simplified diagram of the natural endogenous fibrinolysis system.

of degrading fibrin into soluble fibrin degradation products (FDPs), followed by clot disintegration (Collen 1999; Kotb 2013, 2014a).

As the body ages, the ability of the kidneys and vessel walls to secrete uPA and tPA decreases, and thus the risk of thrombotic complications increases. There are several strategies for the treatment of such complications including the intake of anticoagulants (warfarin and heparin), antiplatelets (aspirin, clopidogrel, dipyridamole, and ticlopidine), surgical operations, or an external fibrinolytic enzyme. In fact, fibrinolytic enzymes are not only safer but also may prevent the incidence of endogenous thrombosis by intervention with some blood clotting factors (Hassanein et al. 2011; Kotb 2012, 2013, 2014a,b, 2015a,b).

Sources of Fibrinolytic Enzymes

Food Sources

Research on functional foods beyond their role as dietary essentials for growth is rapidly expanding. These foods could delay, manage, or prevent the premature onset of a chronic disease. Research indicates that diets high in fat may increase the risk of CVDs, whereas diets containing plant sterols, omega-3 fatty acids, soluble fibers, or antioxidants may help in preventing these diseases (Din et al. 2004).

Natto

Research has proved that fermented foods are the main source of fibrinolytic micro-organisms. The enzyme NK was first discovered in the traditional Japanese food natto. This food was originally introduced to human consumption by the historical warrior Minamoto in north-western Japan, during the Heian era (AD 794–1192). To this day, natto is very popular in that region as a remedy for beriberi, fatigue, dysentery, and thrombotic diseases. Natto has been introduced to other countries for the treatment of viral infection O-157, high blood pressure, osteoporosis, and atherosclerosis, in addition to fibrin dissolution (Sumi et al. 1987).

The recent use of NK goes back to Sumi and his co-workers in 1980 (Sumi et al. 1987). These workers were interested in testing the ability of foods to function as blood clot dissolvers so they conducted research on about 170 kinds of food from all over the world. One day, Sumi dropped a small portion of natto that he was eating into an artificial fibrin plate. The beans gradually dissolved the thrombus and had completely resolved it in 18 hours. Sumi and his co-workers presented the results for the first time to the Japan Agricultural Chemistry Society. Later they stated that the dietary intake of natto not only cures CVDs but could prevent the incidence of such diseases, thus suggesting that natto was a major contributor to the longevity of the Japanese people (Sumi et al. 1987, 1990, 1992).

The efficacy of 50 g of natto or 2000 fibrin units (FU) is equivalent to an amount of urokinase (UK) costing US$200; the efficacy of UK *in vivo* lasts only 4–20 minutes, whereas NK maintains activity for 8–12 hours (Fujita et al. 1995a, 1995b). NK produces its prolonged action in two ways: it dissolves existing thrombus and prevents coagulation of blood. Furthermore, it enhances the body's natural ability to fight blood clots due to its ability to convert prourokinase to active UK and its ability to increase the amount of endogenous tPA. It has many other benefits, including oral administration instead of injection, and can be used preventively. Subsequently, other microbial fibrinolytic enzymes have been discovered in other fermented foods (Kotb 2015a, 2015b).

Asian Fermented Shrimp Paste

Fermented shrimp paste is a popular traditional Asian food with a potent thrombolytic activity. It is traditionally made by salting fresh silver shrimp at a salt concentration of 10–15% (w/w). The mixture is then crushed and kept for 48 hours at room temperature (Wong and Mine 2004). The partially fermented paste is then blended into a smooth paste and sun-dried for 1 month. The final product is packed and stored for marketing.

The fermentation process of this food is not due to any commercial microbial culture but to at least four bacterial strains present naturally in air. Researchers discovered a potent fibrinolytic enzyme in this food. The fibrinolytic protein is a monomer of 18 kDa molecular weight and is a neutral metalloprotease showing broad pH stability. Best enzyme activity was found at 30–40 °C, and the enzyme was completely denatured over 60 °C. The enzyme is specific to fibrin and fibrinogen as substrates with no specificity towards other plasma proteins. The N-terminal amino acid sequencing is DPYEEPGPCENLQVA, thus it is considered as a novel enzyme with no homology to pre-existing fibrinolytic enzymes. It also had anticoagulant activity measured by prothrombin time (PT) and activated partial thromboplastin time (APTT) tests. In addition, it is resistant to pepsin and trypsin digestion so can be orally administered not only to cure but also to prevent the incidence of deep vein thrombosis.

Chungkook-jang

Chungkook-jang is a popular traditional Korean fermented soybean food which is similar to Japanese natto. The procedure of chungkook-jang preparation is shown in Figure 8.3. The chungkook-jang is inoculated with *B. subtilis* then left for fermentation to occur.

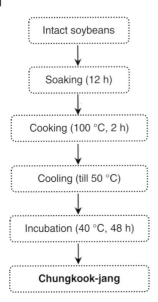

Figure 8.3 Flow diagram for the preparation of chungkook-jang.

Many researchers have studied the fibrinolytic activity of chungkook-jang and its functional mechanistics. However, a major problem with the fibrinolytic enzymes of chungkook-jang is their unstability under the acidic conditions of the stomach and heating conditions during cooking. Therefore, food microencapsulation techniques were applied to overcome this obstacle, improve the survival of micro-organisms, and protect sensitive food components (Desai and Park 2005).

Bacilli producing the fibrinolytic enzymes (CKs) in chungkook-jang have been isolated, the most potent among which is *Bacillus* sp. strain CK 11-4. The optimum temperature and pH for CK were 70°C and 10.5, respectively, and the molecular weight was 28.2 kDa. The N-terminal sequence of CK is AQTVPYGIPLIKAD which is identical to that of subtilisin Carlsberg and different from that of NK. But because the level of CK activity to fibrin was eight times higher than that of subtilisin Carlsberg, these enzymes are not considered to be identical. For the differentiation of CK and subtilisin Carlsberg, a restriction enzyme map of the CK gene was compared with that of subtilisin Carlsberg and its isoforms. The restriction enzyme sites of the CK gene were different from those of *B. licheniformis* NCIB6816 (subtilisin Carlsberg). It is interesting that when CK was added to a mixture of plasminogen and substrate, fibrinolytic activity was increased relative to the control level, although plasminogen did not show activity. Was CK activated by digest products produced from plasminogen as stimulators? In that case, because CK reacted with two substrates, chromogenic substrate and plasminogen, the activity must be less than that for one chromogenic substrate reaction.

Kim and his co-workers succeeded in the isolation of another fibrinolytic enzyme-producing bacterium, *B. vallismortis* Ace02, from chungkook-jang. The molecular mass of *B. vallismortis* fibrinolytic enzyme is 28 kDa and the N-terminal sequence is AQSVPYGVSQ with similarity with the bacteriolytic enzyme, L27, from *B. licheniformis*, which has strong lytic activity against *S. mutans*, a major causative strain of dental caries. This suggests the use of the fibrinolytic enzyme Ace02 for prevention of dental caries as well as being an effective thrombolytic agent (Kim et al. 2007).

Douchi

Douchi has been a popular traditional soybean-fermented food in China for at least the last 2000 years. The preparation and manufacturing process of douchi is more complicated than that of natto and chungkook-jang. Fibrinolytic activity has been found in douchi, from which some fibrinolytic bacterial strains were isolated such as *B. subtilis* and *B. amyloliquefaciens* DC-4. The fibrinolytic enzyme from *B. amyloliquefaciens* DC-4 is designated as subtilisin DFE and it displays thermophilic, hydrophilic, and thrombolytic activity. It has a molecular mass of 28 kDa with an optimal temperature, pH, and isoelectric point (pI) of 48 °C, 9.0, and 8.0, respectively. Subtilisin DFE is a serine fibrinase, similar to NK from *B. natto*. The N-terminal sequence is AQSVPYGVSQIKAPALHSQGFTGS, which is identical to that of subtilisin K-54, and different from that of NK and fibrinolytic enzymes of chungkook-jang (CKs). Gene sequence analysis showed that subtilisin DFE is unique among fibrinolytic enzymes (Peng et al. 2003).

The second fibrinolytic enzyme isolated from douchi was designated as subtilisin FS33 and *B. subtilis* DC33 was found to be the producing strain. This 30 kDa enzyme is more fibrin specific than subtilisin DFE fibrinase and has an optimum temperature, pH, and pI of 55 °C, 8.0, and 8.7, respectively. Its fibrinolytic activity was six times higher than that of subtilisin Carlsberg. The N-terminal sequence, AQSVPYGIPQIKAPA, of the enzyme is different from other fibrinolytic enzymes. Subtilisin FS33 was able to degrade fibrin clots in two ways: by converting the inactive plasminogen into active plasmin which in turn degrades fibrin, and by direct action upon fibrin (Wang et al. 2006).

Tempeh

Tempeh is a traditional fermented-soybean food popular in Indonesia. It has potent fibrinolytic/thrombolytic activity due to *B. subtilis* TP6 fibrinolytic enzyme. The enzyme is designated as TPase and has a molecular mass, optimal temperature, and pH of 27.5 kDa, 50 °C, and 7.0, respectively.

TPase is stable in the acidic range but not in the alkaline range and isa serine metalloprotease type. The N-terminal sequence, AESVPYGVSQIKAPALHSQGFTGS, is highly similar to other subtilisin-like serine proteases such as subtilisin K54 from *B. subtilis* K-54 (97%), subtilisin BPN´ from *B. amyloliquefaciens* (95%), and NK from *B. natto* (82%). The fibrinogen degradation pattern created by TPase was similar to that obtained with human plasmin. In addition, N-terminal sequence analysis of the fibrinogen degradation products demonstrated that TPase cleaves Glu or Asp near hydrophobic acids as a P1 site in the α- and β-chains of fibrinogen to generate D´ and E´ fragments similar to those generated by plasmin. The TPase enzyme had a direct action upon fibrin threads and on plasminogen-rich fibrin plates, and did not activate fibrin clot lysis. It also converted active PAI-1 to the latent form (Kim et al. 2006).

Doen-jang

Doen-jang is another traditional Korean fermented food. It contains a fibrinolytic enzyme termed subtilisin DJ-4 among five proteases from the *Bacillus* sp. strain DJ-4. The enzyme's molecular mass, optimal pH, and temperature are 29 kDa, 10 and 40 °C, respectively. The enzyme was stable in a pH range of 4–11 for 2 days and it retained 80% of fibrinolytic activity when incubated at 50 °C for 1 hour. The DJ-4 enzyme was classified as a subtilisin serine protease by the inhibition assay and amidolytic assay using chromogenic substrates. The N-terminal sequence, AQSVPYGVSQIKAP, is identical to that of subtilisin BPN´, while, the specific activity of subtilisin DJ-4 was 2.2 and 4.3 times higher than those of subtilisin BPN´ and subtilisin Carlsberg, respectively. Subtilisin DJ-4 was found to be plasmin-like in its action (direct action upon fibrin) (Kim and Choi 2000).

Jeot-gal

The bacterium *Bacillus* sp. KA38 produces a different fibrinolytic enzyme in jeot-gal, a salty fermented fish used primarily as a seasoning and appetizer in Korea. The optimal fibrinolytic activity was observed at pH 7.0 and 40 °C, and the pH stability was in the alkaline range of 7–9. This 41 kDa enzyme has a unique N-terminal sequence, VYPFPGPIPN. The fibrinolytic activity was not affected by the serine protease inhibitors, but was severely inhibited by the metalloprotease inhibitors; thus, KA38 is classified as a novel metalloprotease type with zinc ions as co-factors. The enzyme is highly specific for fibrin clots and directly degrades them. In preliminary *in vivo* experiments on mice, continued fibrinolytic activity was detected in blood after oral administration. Jeot-gal (KA38) enzyme is considered safe for long-term consumption, and is a potential candidate for oral enzyme therapy (Kim et al. 1997).

Skipjack

Katsuwokinase (KK) is a fibrinolytic enzyme found in skipjack (shiokara or *Katsuwonus pelamis*), a traditional salt-fermented food in Japan. Shiokara is prepared from the meat of various marine animals, especially shrimp, in a brown viscous paste of the animal's fermented viscera mixed with 10% salt and 30% malted rice, packed in a closed container, and fermented for up to 1 month. Although some researchers demonstrated protease activity in shiokara some time ago, fibrinolytic activity was first reported in 1995 (Sumi et al. 1995). Crude KK shows a very potent fibrinolytic activity above 45 CU/g in comparison with that of plasmin. Its fibrinolytic activity was not affected by 10% NaCl, and was stable in the pH range 1–10 at 37 °C for 30 minutes with PI of 5.0 and a molecular weight of 35 kDa. The N-terminal sequence of purified KK is IVGGYEQZAHSQPHQVSLNSG, with 80% homology with that of trypsin, and its fibrinolytic activity is 2.6 times greater than that of plasmin. Because of its potent activity, salt resistance and high stability, KK serves as an excellent oral fibrinolytic therapeutic enzyme (Sumi et al. 1995).

Kimchi

Kimchi or Kimchee is traditional fermented pickled vegetables prepared in different ways using radish, cabbage, cucumber, garlic, ginseng, hot peppers, Indian mustard leaves or sesame leaves. It is served as an appetizer or side dish in Korea. Various bacterial strains were found to secrete fibrinolytic enzymes in the food, three of which were identified as *B. amyloliquefaciens*, *B. brevis*, and *Micrococcus luteus*. They were found to produce 2.6, 1.5 and 2.0 plasmin unit/mL of fibrinolytic enzyme, respectively (Noh et al. 1999).

Kishk

Kishk is a traditional fermented food which originated during the Pharaonic period and is still made today. Egyptian farmers prepared kishk as a home-made product from parboiled wheat and milk. During the preparation of kishk, wheat grains are boiled until soft, dried, milled, and sieved. Milk is separately soured in clay containers, and mixed with the moistened wheat flour, resulting in a paste called hamma. The hamma is allowed to ferment for about 24 hours, following which it is kneaded and two volumes of soured salted milk are added prior to dilution with water. Alternatively, milk is added to the hamma and fermentation is allowed to proceed for a further 24 hour. The dough mass is thoroughly mixed, formed into balls and sun dried. It is a highly nutritious food, having a protein content of about 23.5%, and has a high digestibility and high biological values and therefore it has been promoted in Mexico and Europe (Benkerroum 2013; Kotb 2015a, 2015b).

Recently, a potent thrombolytic/fibrinolytic protease designated KSK-07 from *B. megaterium* KSK-07 was isolated from kishk for the first time with a molecular mass of 28.5 kDa and optimal reaction temperature, pH, and isoelectric values of 50 °C, 8.0, and 10.0, respectively (Kotb 2015a). Amidolytic and inhibition studies confirmed that KSK-07 is a chymotrypsin family serine protease. It demonstrated direct fibrinolytic action (plasmin-like protease) upon blood clots *in vitro* and prolonged the blood clotting time 1.9-fold due to proteolysis of some blood clotting factors leading to abolition of both thrombin generation and platelet aggregation. The enzyme could not degrade collagen which is abundant in blood vessels and gut, so it is concluded to be devoid of hemorrhagic activity and therefore may overcome the limitations of many of the currently available thrombolytic drugs, suggesting that this enzyme may be an effective thrombolytic agent with high specificity to fibrin and non-specificity to other plasma proteins (Kotb 2015a).

A second but different antimicrobial oxidant-stable fibrinolytic alkaline protease designated as KSK-II was produced by *L. plantarum* KSK-II isolated from kishk (Kotb 2015b), with molecular mass of 43.6 kDa. It is a subtilisin family metalloprotease with Fe^{2+} as co-factor, as indicated by inhibition and amidolytic studies. It is industrially important rather than physiologically significant from the perspective of its optimal activity at 50 °C (stable up to 70 °C), ability to function at alkaline pH (10.0), and stability at broad pH ranges (7.5–12.0), in addition to its stability towards SDS, H_2O_2, organic solvents, and detergents. KSK-II emphasized for the first time the potential of fibrinolytic activity for alkaline proteases used in detergents, especially in blood destaining from cloth (Kotb 2015b).

Table 8.1 summarizes the bacterial strains found in traditional fermented foods and some biochemical properties of their fibrinolytic enzymes.

Edible Mushrooms

There are approximately 14 000 species of mushrooms, among which edible and medicinal mushrooms account for only a small proportion. It is estimated that more than 10 million tons of edible and medicinal mushrooms are cultivated annually worldwide. Mushrooms have gained considerable attention from scientific and medicinal researchers investigating their biological functions. Several bioactive compounds from the submerged mycelia, fruiting bodies, and cultivation broth of these mushrooms have been found, such as cordycepin, ergosterol, mannitol, polysaccharides, polysaccharide-peptide complex, peptides, and fibrinolytic enzymes (Wong et al. 2011).

The fibrinolytic enzyme first studied as a therapeutic agent was from the fruiting bodies of the fungus *Flammulina velutipes* and subsequently, fibrinolytic enzymes were discovered in the fruiting bodies of *P. ostreatus*,

Table 8.1 Fibrinolytic bacteria present in traditional fermented foods.

Functional food	Fibrinolytic enzyme	Producing bacterium	Optimal pH and temperature	Molecular mass	Reference
Chungkook-jang	CK, a thermophilic alkaline serine protease	*Bacillus* sp. CK	pH 10, 70 °C	28.2 kDa	Kim et al. 1996a, 1996b
Doen-jang	bpDJ-2, serine protease	*Bacillus* sp. DJ-2	pH 9.0, 60 °C	42 kDa	Choi et al. 2005
	Subtilisin DJ-4, serine protease	*Bacillus* sp. DJ-4	pH 10.0, 40 °C	29 kDa	Kim and Choi 2000
Douchi	Subtilisin DFE, serine protease	*B. amyloliquefaciens* DC-4	pH 9.0, 48 °C	28 kDa	Peng et al. 2003
Fermented soybean	QK-1, serine protease	*B. subtilis* QK02	–	42 kDa	Ko et al. 2004
	QK-2, serine protease	*B. subtilis* QK02	pH 8.5, 55 °C	28 kDa	Ko et al. 2004
Jeot-gal	Jeot-gal enzyme, Zn^{2+} metalloprotease	*Bacillus* sp. KA38	pH 7.0, 40 °C	41 kDa	Kim et al. 1997
Natto	NK, serine protease	*B. natto*, NK	–	27.7 kDa	Fujita et al. 1993
	A bacillus protease	*B. firmus* NA-1	–	–	Seo and Lee 2004
	IMR-NK1, serine protease	*B. subtilis* IMR-NK1	pH 7.8, 55 °C	31.5 kDa	Chang et al. 2000
Tempeh	TPase, serine metalloprotease	*B. subtilis* TP6	pH 7.0, 50 °C	27.5 kDa	Kim et al. 2006
Kimchi	A bacillus protease	*B. amyloliquefaciens, B. brevis* and *Micrococcus luteus*	–	–	Noh et al. 1999
Kishk	KSK-07, serine protease	*B. megaterium* KSK-07	pH 8.0, 50 °C	28.5 kDa	Kotb 2015a
	KSK-II, Fe^{2+} metalloprotease	*L. plantarum* KSK-II	pH 10.0, 50 °C	43.6 kDa	Kotb 2015b

A. mellea, T. saponaceum, and *C. militaris* (Kotb 2012, 2013, 2014a,b, 2015a,b). In early times, fruiting bodies were cultured on growing media comprising sawdust, straw, woodchips, coffee grounds, logs, and similar organic matter. The fungus basidiospores or the underground mycelia were then seeded on the artificial solid or liquid culture medium under controlled temperature and moisture conditions. Improvement in mushroom cultivation has brought several benefits to the commercial production and laboratory applications of mushroom. In the last decade, an industrial-level technology for mushroom submerged path culture in the laboratory has been developed. When a fresh environment and nutrients are provided, the mycelium grows stably, creating a new generation without intervention of pollutants or contaminants.

Choi and Sa (2000) isolated a fibrinolytic metalloprotease from the cultured mycelia of *Ganoderma lucidum.* In the following decade, a number of fibrinolytic enzymes were isolated from mushroom mycelia, including metalloproteases from *A. mellea* (AMMP), *F. velutipes* FVP-1, and *Perenniporia fraxinea* (Park et al. 2007). Fibrinolytic metalloproteases are sensitive to the metalloprotease inhibitors ethylenediaminetetra-acetic acid (EDTA) and 1,10-phenanthroline in addition to metal ions. Changing the culture medium's physical or nutritional parameters, such as salt concentrations, osmotic pressure, nitrogen source, and carbon source, greatly alters the extracellular performance of fungal mycelia, thus contributing to a particular mushroom metabolite (Kadimaliev et al. 2008).

Culture broth of submerged cultivations is always rich in mushroom bioactive derivatives. Several extracellular fibrinolytic enzymes have been purified from the submerged broth of *Fomitella fraxinea* (FFP1 and FFP2), *C. sinensis* (CSP), and *S. commune* (MsK) (Lu et al. 2010a). The N-terminal sequences of most fibrinolytic enzymes have been determined. These fibrinolytic enzymes, except the metalloprotease derived from *A. mella* fruiting body and mycelia, have markedly different N-terminal sequences. The optimal pH for these mushroom fibrinolytic enzymes is 5–10 and the optimal temperature is 20–60 °C. Previous studies have shown that most fibrinolytic serine proteases such as NK, subtilisin DFE and subtilisin QK-1 from traditional fermented foods are subtilisins of *Bacillus* origin (Kotb 2014a). On study of N-terminal sequence alignment, a chymotrypsin-like serine protease from mushroom *C. militaris* had high sequence identity with subtilisin PR1J (Kim et al. 2006); FFP1 from *F. fraxinea* also has 20% identity with NK and subtilisin E (Lee et al. 2006). Serine fibrinolytic enzymes from *C. sinensis* (CSP) and *F. fraxinea* (FFP1) were found to have broad substrate specificity for synthetic substrates for subtilisin and plasmin in addition to the natural substrates like fibrin, fibrinogen, and casein.

Table 8.2 shows the most important fibrinolytic enzymes that have been successfully isolated from mushrooms.

Table 8.2 Fibrinolytic enzymes derived from mushrooms.

Mushroom	Fibrinolytic enzyme	Enzyme source	Optimal pH and temperature	Molecular mass	References
A. mellea	Metalloprotease	Fruiting body	pH 7, 55 °C	18.5 kDa	Kim and Kim 1999
	AMMP	Mycelium	pH 6, 33 °C	21 kDa	Lee et al. 2005
C. militari	Serine protease	Fruiting body	pH 7.4, 37 °C	52 kDa	Kim et al. 2006
	Serine metalloprotease	Fruiting body	pH 7, 40 °C	34 kDa	Choi et al. 2011
	CMase (metalloprotease)	Culture broth	pH 6, 25 °C	27.3 kDa	Cui et al. 2008
C. sinensis	CSP (serine protease)	Culture broth	pH 7, 40 °C	31 kDa	Li et al. 2007
Flammulina velutipes	Two metalloproteases	Fruiting body	pI 5.5 and 6.05	–	Morozova et al. 1982
	FVP-1 (metalloprotease)	Mycelium	pH 6, 20–30 °C	37 kDa	Park et al. 2007
Fomitella fraxinea	FFP1(serine protease)	Culture broth	pH 10, 40 °C	32 kDa	Lee et al. 2006
	FFP2 (metalloprotease)	Culture broth	pH 5, 40 °C	42 kDa	
Ganoderma lucidum	Zn^{2+} metalloprotease	Mycelium	pH 7–7.5, 45 °C	52.1 kDa	Choi and Sa 2000
Grifola frondosa	GFMEP (Zn^{2+} metalloprotease)	Fruiting body	pI 7.5	20 kDa	Nonaka et al. 1995
Paecilomyces tenuipes	PTEFP (serine protease)	Fruiting body	pH 5, 35 °C	14 kDa	Kim et al. 2011
Perenniporia fraxinea	Metalloprotease	Mycelium	pH 6, 35–40 °C	42 kDa	Kim et al. 2008a
P.eryngii	Serine protease	Fruiting body	pH 5, 40 °C	14 kDa	Cha et al. 2010
P. ostreatus	Metalloprotease	Fruiting body	pH 7.5–8, 50 °C	24 kDa	Choi and Shin 1998
	POMEP (metalloprotease)	Culture broth	pH 7.4, 45 °C	13.6, 18.2 kDa	Liu et al. 2014
	PoFE (metalloprotease)	Mycelium	pH 6.5, 35 °C	32 kDa	Shen et al. 2007
Schizophyllum commune	Metalloprotease	Culture broth	pH 5, 45 °C	21.3 kDa	Lu et al. 2010b
	MsK (metalloprotease)	Mycelium	pH 4–6, 45 °C	17 kDa	Park et al. 2010
Tricholoma saponaceum	TSMEP1, TSMEP2 (two metalloproteases)	Fruiting body	pH 7.5–9, 55 °C	18.1, 17.9 kDa	Kim and Kim 2001

Non-Food Sources

Earthworms

Earthworms have been used as a crude drug in East Asian traditional medicine for several thousand years. Five hundred years ago, Shizhen Li compiled the famous medical book *Compendium of Material*, in which earthworms were described as "earth's dragons" and it was suggested that they could be used as antipyretic, hypertension, sedation, detoxification, bronchodilation and thrombolytic agents in the form of dried powder. At the end of the nineteenth century, Fredericq (1878) discovered a fibrinolytic enzyme from secretions of the alimentary tract of earthworms. This was a preliminary study in which the precise biochemical and physiological mechanism of the fibrinolytic enzyme was not clear. More fibrinolytic proteases were then successively obtained from other species, such as the earthworm tissue PA, earthworm PA, component A of EFE (EFEa), and biologically active glycolipoprotein complex (G-90). In 1991, the fibrinolytic protease LK was first extracted from the earthworm *L. rubellus* (Mihara et al. 1991).

Earthworm fibrinolytic enzymes (EFEs) are mainly present inside the digestive organs and body cavity, with a molecular weight of 25–32 kDa. Recently, the mechanics of earthworm secretions have been intensively investigated. Studies have shown that a portion of orally administered EFEs could be absorbed from the intestine by villi into blood and thus can dissolve blood fibrin clots by inhibition of platelet activation and aggregation. For this reason, they could be used as a relatively inexpensive thrombolytic agent for prevention and treatment of CVDs. Absorption and efficacy of EFE-d were enhanced in rats when delivered in water/oil (w/o) microemulsions (Cheng et al. 2008). It is now established that the most common earthworm producers are *L. rubellus*, *L. bimastus*, *Eisenia andrei*, *E. fetida* and *Neanthes japonica* (Kotb 2014a).

Earthworm fibrinolytic enzymes such as EfP-III-1 have broad substrate specificities, such as trypsin (cleaving the carboxylic sites of Arg and Lys) and chymotrypsin (cleaving the carboxylic sites of Phe, Trp, Tyr, and Leu). Hydrolysis of fibrinogen α-chains by EfP is the fastest, and hydrolysis of β-chains is faster than that of γ-chains. EfP activates plasminogen cleavage at R_{557}-I_{558}; that is to say, EfP has the ability to hydrolyze fibrinogen and to activate plasminogen and prothrombin, playing a part not only in fibrinogenolysis but also in fibrogenesis (Figure 8.4).

EfPs are macromolecules and it is predicted that they would not to permeate easily through the intestinal epithelium upon oral adminstration. In addition, EfPs are subjected to degradation by alimentary tract pepsin, trypsin, and chymotrypsin. Furthermore, the acidity of gastric juice can denature EfPs. However, EfPs and some therapeutic proteins

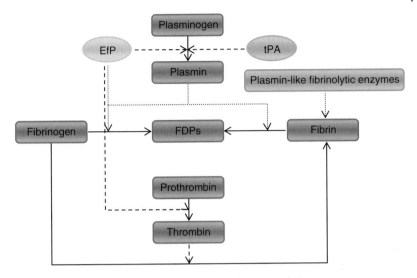

Figure 8.4 Role of EfP in the procoagulation and anticoagulation process. Conversion action is indicated by continuous arrows, activation is indicated by dashed arrows, degradation is indicated by dotted arrows.

can be transported into blood through the intestinal epithelium and perform their biological functions in the blood in the intact and active form before being degraded in the alimentary tract, such as bromelain, β-lactoglobulin, hepatitis-B surface antigen, and epoxy-β-carotenes. The N-terminal sequences of *L. rubellus* (LrP) such as LrP-III-1 and LrP-III-2 are similar to the protein transduction domain (Green and Loewenstein 1988). The sequences are rich in hydrophobic amino acid residues, which play an important role in membrane transportation of biological macromolecules.

Snakes

Fibrinolytic enzymes have been isolated from the toxins of snakes, especially the genera Viperidae, Elapidae, and Crotalidae. Snake venom consists of several biologically active peptides, which are lethal to humans by adversely altering hemostasis. The majority are Zn^{2+} metalloproteases and have been classified into serine proteases, metalloproteases, C-type lectins, disintegrins, and phospholipases.

Each venom has its own selective action on certain blood coagulation factors and/or blood cells (Clemetson et al. 2007). Venom fibrinolytic proteases may involve activation or inactivation of each factor related to fibrinolysis and/or coagulation, where the thrombin-like function involves stimulation of fibrinogen clotting processes and the formation of fibrin, while the fibrino(geno)lytic function involves the digestion of fibrin and fibrinogen.

Fibrino(geno)lytic venoms are classified in two groups, fibrin(ogen) olytic metalloproteases and serine proteases, with molecular masses of 25 and 60 kDa, respectively. The amino acid composition of these enzymes has high levels of Asx and Glx residues (Hahn et al. 1995). They have been found in *Agkistrodon rhodostoma*, *A. halys brevicaudus*, *A. contortrix*, *A. piscivorus piscivorus*, *A. piscivorus conami*, *A. acutus*, *A. caliginosus contortrix*, *A. halys pallas*, *A. halys ussuriensis*, *A. halys blomhoffii*, *A. saxatilis*, *Bothrops atrox*, *B. jararaca*, *B. alternates*, *B. asper*, *B. jararacussu*, *B. lanceolatus*, *B. neuwiedi*, *Cerastes cerastes*, *Crotalus atrox*, *Gloydius shedaoensis*, *G. ussuriensis*, *Trimeresurus elegans*, *T. flavoviridis*, *T. jerdonii*, *T. mucrosquamatus*, and *T. stejnegeri* (Kotb 2014a). The mechanism of action of the two fibrin(ogen)olytic enzymes differs, as they target different loci on the fibrin(ogen) peptide chain and thus can be classified into α-chain and β-chain fibrinogenases. Recently, several wild fibrin(ogen)olytic venoms and genetically recombined fibrin(ogen)olytic venoms have been examined in preclinical and clinical experiments for their thrombolytic efficacy and hemostatic safety in animal models and promising results were obtained.

Unconventional Sources

Fibrinolytic enzymes have been isolated from other sources such as the corn fly *Haematobia irritans irritans* (Dametto et al. 2000), the caterpillar *Lonomia oblique* (Pinto et al. 2004), the egg cases of the praying mantis *Tenodera sinensis* (Wang et al. 1989), the dung beetle *Catharsius molossus* (Ahn et al. 2003), the saliva of the vampire bat (Gardell et al. 1989), the plant *Allium tuberosum* (Chung et al. 2010), the herbal medicine *Spirodela polyrhiza* (Choi and Sa 2001), and the seaweed *Codiales codium* (Jeon et al. 1995).

Production of Fibrinolytic Enzymes from Metagenomes

Screenings for microbial fibrinolytic enzyme producers have mainly used the cultivation-dependent approach. However, the cultivable micro-organisms are less than 1% of the total microbial community. To overcome the problem of non-cultivability, metagenomic libraries are created. Metagenomes are the genomes of non-cultivable micro-organisms existing within a certain community or environment. Lee et al. (2007) discovered a novel fibrinolytic enzyme while constructing a metagenomic library using total genomic DNA extracted from deep sea sediments of the west coast of Korea. The genes were picked up, sequenced, and expressed in *E. coli*. One clone from recombinant *E. coli* clones showed potent fibrinolytic activity. The gene encoding enzyme activity was subcloned into the expression vector pUC19, and a database search for homologous sequences revealed it to be a zinc-dependent metalloprotease of 39 kDa molecular mass with optimal activity at 50 °C and pH 7.0.

Production of Fibrinolytic Enzymes from Transformed Plants and Animals

Transformed or recombinant plants have been introduced by nuclear transformation methods for the synthesis of edible vaccines and pharmaceuticals. The plastid or chloroplast genetic engineering technique has several benefits over the nuclear transformation method, such as high expression levels of the recombinant proteins, multigene engineering in a single transformation event, transgene containment by maternal inheritance expression, and prevention of positional and pleiotropic effects and incorporation of undesirable foreign DNA. Ko et al. (2009) assessed the possibility of producing recombinant serine fibrinolytic enzyme, fibrizyme, from the earthworm *E. andrei*, with plastid genetic engineering of the tobacco plant. After fibrizyme gene integration into the tobacco chloroplast genome, a plastid transformation vector was constructed by incorborating various regulatory elements into the protein cDNA. This vector was delivered by particle bombardment into tobacco leaf explants which subsequently regenerated into whole plants. The site-specific integration was stable as confirmed by PCR and Southern blot analysis. Northern and Western blot analysis revealed that mRNA and recombinant fibrizyme protein were highly expressed in the transformed tobacco plants.

The cDNAs of fibrinolytic enzymes from earthworms were previously successfully expressed in goat milk (Hu et al. 2004). Two copies of the fibrinolytic enzyme LK extracted from *L. rubellus* cDNA(w) obtained by RT-PCR and a synthesized LK cDNA(m) were cloned into a mammary gland-specific expression vector pIbCP. The pIbCP-LK-LK vector preparations were directly injected into lactating goat mammary glands. Results showed that both LK-w and LK-m were successfully expressed and translated in goat milk. The fibrinolytic activity of LK-w in milk was 225 000 tPA units/L, while that of LK-m was 550 000 tPA units/L. The molecular weight of recombinant LK is 31.8 kDa (Hu et al. 2004).

Production of Fibrinolytic Enzymes from Human Plasma

The human plasma fibrinolytic system is capable of degrading fibrin clots by the endogenous proteolytic enzyme plasmin. Between 1965 and 1970, a highly purified fibrinolytic zymogen was prepared from human and other mammal plasma, serum, and enriched plasma fractions for the dissolution of blood clots (Robbins 1978).

Production (Fermentation) Process

The cost of large-scale enzyme production is a major obstacle to the successful application of mirobial enzymes in industry. Many attempts have been made to improve the production and expression of fibrinolytic enzymes, including selection of the ideal strain, economical fermentation medium, optimization of nutritional (carbon, nitrogen, and trace elements) and environmental conditions, and overexpression by genetically engineered or mutated strains.

Strain Selection

The fermenting strain involved in a fermentation process plays a major role in the quality and amount of fibrinolytic enzyme. In the previous section, we discussed the wild microbial producers present in some functional foods. In addition, recombinant and mutagenic micro-organisms have been introduced to this field as new fibrinolytic enzyme producers which mainly increase production efficiency, with the overall aim of reducing production costs.

The traditional approach of strain improvement is mutagenesis of a suitable strain followed by random screening of mutants. This approach has been successful and has benefited in recent years from the application of important advances in science and technology. Random mutagenesis is used to induce mutation in organisms using UV, ethyl methyl sulfonate (EMS), and ethidium bromide (EtBr) for better distinctiveness.

In recent years, many attempts have been made to improve the strains using physical and chemical mutagenesis; for example, the production of thrombinase by *Streptomyces venezuelae* was greatly enhanced by strain improvement through mutagenesis by EMS (Naveena et al. 2012). The strain did not respond well to UV mutagenesis when compared to EMS. Since the strain DNA contains more GC base pairs than AT base pairs as it belongs to the genus *Actinomyces*, mutation by EMS caused GC transversions. The transversion mutation caused by EMS became permanent since the self DNA repair mechanism failed. The mutant strain was found to be more halo- and thermo-tolerant comparing to wild strains. The maximum specific growth rate, maximum specific consumption rate, and maximum production rate before and after strain improvement were also improved.

For recombinant fibrinolytic enzyme producers, it is well known that *B. subtilis* is an excellent host for the expression of foreign enzymes with pharmacological activity. For example, subtilisin DFE fibrinolytic enzyme was actively expressed in protease-deficient *B. subtilis* WB600 (Peng et al. 2004). In a different experiment, the native promoter of the subtilisin DFE gene was replaced by that of the α-amylase gene from *B. amyloliquefaciens* DC-4, resulting in a sharp increase in fibrinolytic activity from 80 to 200 U/mL (Xiao et al. 2004). Liu and Song (2002) succeeded in the functional expression of NK in *B. subtilis* as well.

In a failed trial, when subtilisin DFE and NK were expressed in an *E. coli* host system, insoluble aggregates without enzymatic activity were formed (Peng and Zhang 2002a). However, two studies reported the successful expression of both enzymes in *E. coli*. Both studies were based on the principle that extracellular subtilisins from *Bacillus* are synthesized as pre-proenzymes, and their propeptides may function as intramolecular chaperones to correct the folding of protease domain (Chiang et al. 2005; Zhang et al. 2005).

Zhang et al. (2005) succeded in the expression of subtilisin DFE via the expression vector pET32a in *E. coli* BL21 as a fusion protein (Trx-prosubtilisin DFE) and potent fibrinolytic activity was detected in both the soluble fraction and inclusion bodies fraction. In addition, the fusion protein was easily refolded and automatically cleaved to form the mature active subtilisin DFE. In the trial carried out by Chiang et al. (2005) on NK enzyme, NK or pro-NK was overexpressed in *E. coli* fused to the olesin C terminus by an intein polypeptide linker. After reconstitution, active NK was released through self-splicing of intein induced by temperature alteration and spontaneous cleavage of the propeptide. It may also be desirable to manipulate specific characteristics such as the ability to utilize cheaper substrates such as agroindustrial wastes.

Fermentation Parameters

Since micro-organisms possess diverse physiological characteristics, it is necessary to optimize nutritional and physical conditions as they are critical for the fermentative production of fibrinolytic enzymes. Temperature is the most crucial factor in fermentation processes for production of enzymes and other bioactive compounds. The optimal temperature of *Streptomyces megasporus* SD5 for enzyme synthesis is 55 °C because the strain was isolated from a hot spring (Chitte and Dey 2000). The pH of the fermentation medium plays a critical role in the electric charge of membrane and medium proteins, thus changing the H^+ pumbing ATPase and the transport of nutrient components across the cell membrane. The lowering of enzyme productivity at high pH or temperature may be due to enzyme aggregation, incorrect structure formation, hydrolysis of asp-x, destruction of S-S bonds, and deamidation.

Another major criterion for the production of enzymes is substrate selection. Various agroindustrial substrates are utilized in order to minimize production cost, reduce environmental pollution generated by those wastes, and enhance production yield as they are rich in carbon and nitrogen substrates. Since fibrinolytic enzymes play a key role in the treatment of blood clots, waste products like animal wastes are avoided since they contain harmful molecules that may affect the health of the individual. Therefore, components like soybean meal, glucose, magnesium sulfate, and calcium chloride have been used in the fermentation process.

In general, complex organic N-sources are preferable for enzyme production by bacteria than simple inorganic N-sources, because most bacteria possess the enzyme system including protease necessary for attacking these complex proteins. On studying C-sources, generally we found that simple C-sources are preferred by bacteria for protease production rather than complex C-sources. A plausible explanation may be the fact that bacteria are weak attackers for C-sources in comparison with fungi, so they prefer utilizing simple C-sources (Price and Stevens 1999). In some studies, fibrin was found to enhance enzyme production, suggesting that fibrin, as a substrate of fibrinolytic enzymes, could activate or induce enzyme production by bacteria during cultivation (Chitte and Dey 2000; Peng and Zhang 2002b). Above a certain level of nutrient or substrate concentration, enzyme productivity and/or activity became fixed or ceased. One possible reason for this repressive effect of high concentrations of sugars may be carbon catabolite repression (CCR). It has recently been established that a catabolite control protein (Ccp A) is responsible for the regulatory mechanisms of carbohydrate catabolism, and acts as a signal for the repression of enzyme synthesis.

Concerning the effect of minerals such as Cu^{2+}, Fe^{2+}, Ca^{2+}, and Mg^{2+}, enzyme production by micro-organisms becomes inhibited when some cations are added to the culture medium, while it is promoted by others. The inhibitory effect of metal ions may be due to their attachment to metabolic enzyme SH groups of cys-x breaking S-S bridges and thus changing the enzyme conformation or may be due to replacement with metals originally present in enzymes. The enhancement of enzyme production by some metals may be due to the fact that these metals are utilized by the micro-organisms in the biosynthesis of enzymes.

It is noticed, especially for bacteria, that in unagitated cultures, enzyme is produced at low levels and the cultures show an increase in enzyme production with an increase in agitation rate. This may be attributable to the aerobic growth of strains and the homogenous distribution of nutrients and cells within the culture. Fibrinolytic enzyme production by *S. commune* BL23 at high speeds (above 150 rpm) adversely affected both growth and enzyme production (Pandee et al. 2008). Immobilization of microbial cells may also increase enzyme production when compared with free cells. The promotion effect of immobilization may be because entrapment of cells provides more stabilized conditions than their free equivalents; moreover, cell autolysis is inferior to that observed in static or shaken free cultures. On the other hand, enzyme production is decreased at higher volumes and/or ages of inocula. It decreases at higher volumes of inocula due to the dilution of minerals below their optimal values. Wang et al. (2009) found that 100 mL of medium was more suitable for NK production by *Pseudomonas* sp. TKU015 than 50, 150, and 200 mL in a 250 mL flask.

Statistical Optimization

Microbial fermentation is inherently a complex process, so designing an appropriate medium for maximum production is of critical importance since the medium composition significantly affects product yield and cost. Nutritional and environmental requirements can be optimized by either conventional or statistical methods. The conventional method involves changing one independent variable at a time, while keeping others at a fixed level. Although this method is simple and easy, it is extremely time consuming and often the optimal conditions are misidentified because the comprehensive interaction between factors is not taken into consideration. These limitations can be eliminated by modeling or statistical methods, which are reliable, shortlist significant nutrients, help in understanding the interactions among different parameters and reduce the total number of experiments, resulting in saving time and resources. Therefore, developing mathematical models of fermentation kinetics has become the primary objective for fermentation behavior.

The most common statistical methods are the Taguchi orthogonal array design, response surface methodology (RSM), and fractional factorial design (FFD) (Mahajan et al. 2010). Taguchi's design uses a special set of arrays called orthogonal arrays, which provide full information on all factors affecting enzyme production, and gives the minimal number of experiments. RSM is a collection of statistical techniques which can be used for obtaining optimum conditions of independent factors, evaluating the factors' effects, designing experiments, building up models, and predicting alterations in production by analyzing the interactions between components. RSM is an excellent technique for process optimization because the interactions between variables can be identified and quantified, and it is suitable for testing multiple variables as fewer experimental trials are needed (Chen et al. 2009). FFD is especially suitable for studying the interactions and identifying the most significant components in the fermentation medium. A combination of factors generating a certain optimum response can be identified through the use of FFD and RSM (Liu et al. 2005).

Media optimization of fibrinolytic enzyme production by *B. subtilis* ICTF-1 using the L_{18}-orthogonal array method resulted in enhanced production of fibrinolytic enzyme (8814 U/mL) which was 2.6-fold higher than in unoptimized medium (3420 U/mL) (Mahajan et al. 2012). Both FFD and central composite design (CCD) were employed to optimize the fermentation media for production of NK from *B. natto* NLSSE and finally increased the fibrinolytic activity to 1300 U/mL, about 6.5-fold higher than the original (Liu et al. 2005). RSM was applied to optimize the URAK fibrinolytic enzyme production medium by *B. cereus* NK1. The production of URAK by the optimized medium reached 6326.98 FU/ml (Deepak et al. 2010). Process parameters for NK production using *B. natto* NRRL 3666 were successfully optimized by implementing conventional and various statistical methods (Mahajan et al. 2010). The optimized medium showed enhanced production of NK (1191 U/mL) which was 6.3-fold higher than the original medium (188 U/mL). In addition, a 1.6-fold increase in NK production (1932 U/mL) was achieved with a significant reduction in batch time (from 40 to 26 h) during the fermenter studies as compared to shake flask studies.

Solid-State Fermentation

Solid-state fermentation (SSF) is an alternative cultivation system for the production of valuable microbial products. It has numerous advantages over liquid-state fermentation (LSF), including higher product concentration, less effluent generation, simpler fermentation equipment an media, less fermentation space, absence of rigorous control of fermentation parameters, easier aeration and conditions closer to natural habitats of micro-organisms. Fungi have been widely utilized in SSF for production of enzymes and other biomaterials while bacteria are rarely utilized. The production of a fibrinolytic enzyme from *F. oxysporum* using different SSF methods on rice chaff largely increased with minimal cost (Tao et al. 1997, 1998). SSF of red bean (*Phaseolus radiatus* L.) by *B. subtilis* at 1/10 inoculum level (culture/red bean, v/w), and 30 °C for 96 hours for potent subtilisin-like serine fibrinolytic enzyme production increased (Chang et al. 2012). In fact, production of bacterial fibrinolytic enzymes like NK and CK in fermented foods is simply a natural SSF process.

Thrombolytic Activity of Fibrinolytic Enzymes *In Vivo*

Although PAs like UK are still in use for thrombolytic therapy, their price and side effects have prompted researchers to search for cheaper and safer resources. The microbial fibrinolytic enzymes, especially those from food-grade micro-organisms, were found to be the best alternatives for PAs and their effectiveness *in vivo* has been identified. Therefore, these functional foods have the potential to be used as drugs to prevent or cure thrombotic diseases (Hwang et al. 2002; Kotb 2015a, 2015b).

The bacterial fibrinolytic enzyme streptokinase produced by β-hemolytic streptococci was found to be immunogenic *in vivo*. Several antistreptokinase platelet-activating antibodies have been identified and extensive studies have been done for the identification of antigenic regions of streptokinases and production of recombinant forms with reduced immunogenicity (Regnault et al. 2003). A mutant streptokinase that lacked the C-terminal 42 amino acids was found to be less immunogenic than the native molecule (Torrens et al. 1999).

The sister bacterial enzyme, staphylokinase, produced by staphylococci also faced many problems upon its application *in vivo*. Its half-life in plasma is very short and high levels of antistaphylokinase antibodies were measured. However, the levels of antistaphylokinase antibodies in seum are lower than those of antistreptokinase antibodies. A clinical trial on more than 300 thrombotic patients reported that most of them developed high titers of neutralizing specific IgG antibodies which would predict therapeutic refractoriness upon repeated administration; however, rare allergic reactions to staphylokinase adminstration were found (Collen 1998). Wild-type staphylokinase contains three non-overlapping immunodominant epitopes, at least two of which (the variant Lys^{35}, Glu^{38}, Lys^{74}, Glu^{75}, Arg^{77} and the variant Lys^{74}, Glu^{75}, Arg^{77}, Glu^{80}, Arg^{82}) could be eliminated with partial inactivation of the molecule by site-directed substitution of clusters of two or three charged amino acids with alanine (Collen 1998). In patients with peripheral arterial occlusion given intra-arterial doses of 6.5–12 mg of transformed compound, these variants induced significantly less neutralizing antibodies and staphylokinase-specific IgG than wild-type staphylokinase.

On the other hand, the bacterial enzyme NK obtained from *B. subtilis* is particularly potent because it increases the body's natural ability to fight blood clots by enhancing production of plasmin and UK. NK has many other benefits, including oral administration due to its stability in the digestive system, confirmed efficacy, prolonged action (8–12 h), and cost-effectiveness, and it can prevent conditions leading to blood clots with an oral daily dose of as little as 2000 FU or 50 g of natto (Fujita et al. 1995a, 1995b). Furthermore, NK has the ability to regulate the total fibrinolytic activity in the body through cleavage of PAI-1, which is the primary inhibitor of fibrinolysis (Urano et al. 2001).

So far, NK is the only fibrinolytic enzyme from micro-organisms whose fibrinolytic/thrombolytic *in vivo* mechanisms have been explored. It was the subject of several *in vivo* trials, including human trials. In a dog trial, after induction of blood clotting in a major leg vein, dogs were orally administered four capsules of NK (250 mg/capsule). Within 5 hours of treatment, blood clots were completely dissolved and normal blood circulation was restored (Sumi et al. 1990). A similar experiment was done with rats and it was found that natto suppressed intimal thickening and modulated lysis of mural thrombi after endothelial injury in femoral arteries (Suzuki et al. 2003). Another rat trial was undertaken on thrombi induced in the common carotid, where the endothelial cells of the vessel wall had been injured by acetic acid. Animals treated with NK recovered 62% of the arterial blood flow, while those treated with plasmin regained just 15.8% recovery (Fujita et al. 1995b). The antithrombotic effect of NK is due to the absorption of enzyme into the bloodstream, prolonging both PT and APTT, with inhibition of the aggregation of rat platelets induced by thrombin (Lee et al. 2007).

When NK was given orally to human volunteers, lysis of blood thrombi was observed by angiography and, more importantly, the fibrinolytic activity, t-PA, and FDP concentrations in the plasma doubled (Kim et al. 1996a, 1996b). Another human trial involved 12 Japanese volunteers, each of whom was given 200 g of natto daily before breakfast, and their plasma fibrinolytic activity was tested. Administration of natto enhanced the ability of participants to dissolve blood clots and the volunteers retained such enhanced fibrinolytic activity for 2–8 hours after administration (Sumi et al. 1989).

Unconventional Applications of Fibrinolytic Enzymes

Some think that fibrinolytic enzymes are enzymes that are capable of degrading fibrin clots and hence their applications are restricted to the treatment of thrombotic diseases. In fact, fibrinolytic enzymes have some unconventional applications in several fields, which will be discussed here.

Fibrinolytic Enzymes as Detergent Additives and Antimicrobials

Some fibrinolytic enzymes work better on the alkaline side and are unspecific in their action, and so have a wide spectrum for the hydrolysis of several proteins in addition to fibrin. The best example for this type of fibrinolytic enzyme is KSK-II from the bacterium *L. plantarum* (the biochemical properties are mentioned above). The relative activity of KSK-II for the hydrolysis of plasma proteins in addition to fibrin and collagen is very high, and this has been attributed to the large number of possible cleavage sites. This property prohibits the use of KSK-II in the dissolution of blood clots *in vivo* because their unspecificity can lead to hemolysis of RBCs and can cause internal hemorrhage due to their action on hemoglobin and collagen (a major structural component of blood vessels and gut), respectively (Kotb 2015b). KSK-II has an optimal pH of 10.0 and is maximally stable at pH 7.5–12.0 and 45 °C. The optimum reaction temperature was 50 °C and it was catalytically stable up to 70 °C for 1 hour; thus, it was considered a suitable additive for commercial detergents.

Referring to detergent-compatible alkaline enzymes currently in use, they have pH optima in the range of 9.0–12.0 and thermostabilities at laundry temperatures (40–70 °C). During the washing experiment, the enzyme showed excellent stablility and compatibility with detergent formulations, Persil, X-tra°, and Ariel°, retaining about 112%, 98%, and 92% of its initial activity with superiority compared to alkaline proteases from other species of *Bacillus*. The potency of blood stain removal from cotton fabrics shown by a mixture of detergents and KSK-II was significantly higher (p < 0.05) compared with the detergent solutions alone or purified protease solution alone (Figure 8.5a–d). The removal of blood stain from fabrics was explained on the basis of degradation of blood fibrin (Figure 8.5e–g) and RBCs due to the fibrinolytic and hemolytic activity of the enzyme.

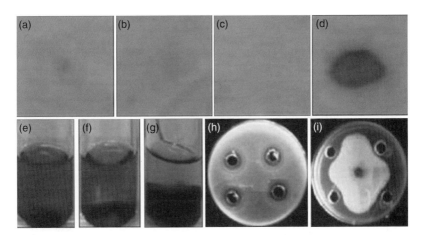

Figure 8.5 Efficiency of KSK-II in the hydrolysis of blood. Initially, blood stain washed from cotton fabrics after 30 min of incubation at room temperature with detergent only (a), enzyme only (b) and detergent with the aid of KSK-II (c) against the fabric stained with blood (d). Then natural blood clots were subjected to dissolution with KSK-II by incubation for 6 h at room temperature at pH 10.0 (e) and pH 7.4 (f) against control (g). The antimicrobial activity of KSK-II against the bacterium *S. aureus* and the fungus *R. solani* (b) is shown in panels (h) and (i), respectively. (*See color plate section for the color representation of this figure.*)

One of the unique characters of KSK-II is its ability to inhibit the growth of *S. aureus*, *B. cereus*, *P. aeruginosa*, *P. vulgaris*, and *E. coli* to 29%, 21%, 13%, 10%, and 7%, respectively post 24 hours of incubation at 37 °C. Figure 8.5 h,i shows the antibacterial and antifungal activity of KSK-II protease against the skin pathogen *S. aureus* and the soil-borne plant pathogenic fungus *Rhizoctonia solani*, respectively. This type of antimicrobial protease is particularly applied as an antimicrobial agent in medical and food applications (Masschalck and Michiels 2003) and for wall perforation to release endogenous metabolites (Zukaite and Biziulevicius 2000).

Fibrinolytic Enzymes for Prevention of Dental Caries

The full gene sequemce of the fibrinolytic enzyme Ace02 isolated by Kim et al. (2007) from chungkook-jang bacterium, *B. vallismortis* Ace02, exactly coincided with the bacteriolytic enzyme L27, from the soil bacterium *B. licheniformis*, which has a strong lytic activity against *S. mutans*, a major causative agent of dental caries. This suggested the use of Ace02 for lysis of *S. mutans* and hence the prevention of dental caries as well as being an effective thrombotic agent. It might be possible to manufacture confectionery covered with freeze-dried fibrinolytic enzyme for the prevention of dental caries.

Fibrinolytic Enzymes as Plant Virus Inhibitors

Phytoviruses represent significant threats to agriculture worldwide, and their biological control remains a challenge, underscoring the need for safer and more effective agents. The vast majority of plant viruses are non-enveloped positive-stranded RNA viruses, so their protein codes are naked.

Proteases are the most extensively studied class of enzymes and are associated with many biological functions including food digestion, lysosomal degradation, signaling cascades, fibrinolysis and thrombosis, cancer invasion and metastasis, tissue growth and remodeling, and others. However, several questions still remain about their mechanism of action and functions.

Amaravadi et al. (1990) found that cowpea mosaic virus and TMV lost their virulence when viral suspensions were incubated with the earthworm enzyme extract as indicated by local lesion assay. They could not isolate and purify the viricidal substance from *E. foetida* extract. Earthworms lack antibodies, and instead have efficient innate immune systems to defend themselves against invaders (Cooper 2002). Liu et al. (2004) reported that antimicrobial peptides isolated from the coelomic cells of earthworms act as immunity agents. Thus, it was suggested that the antiviral serine protease from earthworm extract is also acting as one of the immunity substances in earthworms.

Treated   Untreated

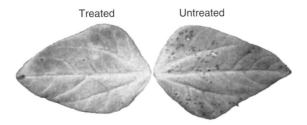

Figure 8.6 The antiviral effect of *E. foetida* protease on the CMV infection in cowpea plants. Local lesions were counted at the third day of virus inoculation. (*See color plate section for the color representation of this figure.*)

Ueda et al. (2008) have screened for antiviral substances in microorganisms, insects, and soil invertebrates. Interestingly, a protease active at a wide pH range (6–12) and also at low temperature (4 °C) with antiplant viral activity was found along with cold-adapted rawstarch-digesting amylases from an extract of the earthworm *E. foetida*. The molecular weight and the N-terminal amino acid sequence of the antiviral protease were 27.0 kDa and VIDGGXNAIIG, respectively, which are similar to those of LrP-F1 to LrP-F6, fibrinolytic serine proteases (24.6–33.0 kDa) from *L. rubellus* (Cho et al. 2004) and EFE-a to EFE-g, fibrinolytic serine proteases (23.0–29.5 kDa) from *E. foetida* (Wang et al. 2003). There is 60% inhibition of CMVand TMV lesions formed on enzyme-treated leaves (Figure 8.6). The antiviral activity was explained on the basis of the hydrolytic action of enzyme on the capsid proteins of both CMV and TMV but the physiological significance has not been explained. These results suggest that the enzyme could be applicable as a potential antiviral agent against CMV, TMV, and other plant viruses.

Fibrinolytic Enzymes as Anti-Inflammatory, Analgesic, Antibacterial, and Mucolytic Expectorant Agents

The fibrinolytic enzyme serrapeptase is secreted by the bacterium *Serratia* sp. E-15 living inside the intestine of the silkworm *Bombyx mori* to enable the emerging moth to dissolve its fibrous cocoon (Mazzone et al. 1990). Serrapeptase was found to have potent anti-inflammatory activity as well as several other beneficial properties in addition to its thrombolytic activity (Kotb 2012, 2013, 2014a,b, 2015a,b; Moriya et al. 1994).

Researchers in India conducted a study to assess the response to serrapeptase in patients with carpal tunnel syndrome (CTS) as an alternative non-surgical conservative approach. Twenty patients with CTS were evaluated clinically after 6 weeks of enzyme adminstration. Sixty-five percent of them showed significant improvement without measurable side effects (Panagariya and Sharma 1999). Another study was conducted to assess the efficacy of serrapeptase in the treatment of venous inflammatory disease.

The enzyme was also effective in 65% of patients (Bracale and Selvetella 1996). A clinical evaluation of serrapeptase was also conducted to determine its efficacy in reducing inflammation in patients with breast engorgement. Exactly 86% of patients receiving the enzyme had a moderate to marked improvement in breast pain, breast swelling and induration without adverse efffects. A prospective study was conducted on the effect of serrapeptase on postoperative swelling and pain of the ankle. The swelling decreased by 50% on the third day of enzyme adminstration, while in the control groups (treated with ice) there was no measurable reduction in the swelling (Esch et al. 1989). A multicenter, double-blind, placebo-controlled trial was carried out to investigate the clinical efficacy of serratiopeptidase in 174 patients who underwent Caldwell-Luc antrotomy for chronic empyema (puncturing and draining pus from the nasal maxillary sinus). The puncture was placed under the top lip and above the gum line. Changes in swelling at the puncture sites were observed and no adverse side effects were reported (Tachibana et al. 1984).

Moving to another part of the body, in a study conducted in men with amicrobial prostato-vesiculitis (APV, a non-infectious prostate inflammation), researchers found that treatment with non-steroidal anti-inflammatory drugs, including serrapeptase, could reduce inflammation and swelling of the prostate (Vicari et al. 2005).

The efficacy of serrapeptase was evaluated in 193 subjects suffering from acute or chronic ear, nose or throat disorders. After 3–4 days' treatment, significant symptom regression was observed. It was concluded that serrapeptase has anti-inflammatory and antiedemic activity and acts rapidly on localized inflammation (Mazzone et al. 1990). One research trial in Japan investigated the effect of serrapeptase on sputum properties and symptoms in patients with chronic airway diseases. After 4 weeks of serrapeptase treatment, sputum output, viscosity, and neutrophil count decreased significantly. In addition, the frequency of coughing and expectoration also decreased. Serrapeptase may exert a beneficial effect on mucus clearance by reducing neutrophil numbers and altering the viscoelasticity of sputum. Another clinical study evaluated the effect of serrapeptase on the elasticity and viscosity of nasal mucus in adult patients with chronic sinusitis. Serrapeptase was administered orally for 4 weeks. The dynamic viscosity of the mucus at week 4 was significantly lower than that at week 0 (Majima et al. 1988).

An unusual clinical trial evaluated the effectiveness of serrapeptidase in the eradication of a periprosthetic infection (infection at an implanted orthopedic device site) in an *in vivo* animal trial. Rats were inoculated with *S. epidermidis* at the prosthetic site. After 2 weeks, infection persisted in 63.2% of animals in the control group, 37.5% of animals in an antibiotic-only group, and only 5.6% of animals in the serrapeptidase and antibiotic group. Thus, serrapeptidase was concluded to be effective in infection eradication and enhancing antibiotic efficacy against staphylococcal infections (Mecikoglu et al. 2006).

It would seem that the potential applications of serrapeptase are many and varied. Whether the application is thrombolytic, anti-inflammatory, analgesic, antibacterial, a mucolytic expectorant, as well as others, serrapeptase is the cutting edge in systemic replacement enzyme therapy.

Fibrinolytic Enzymes as Blood Pressure Regulators

Hypertension is a significant public health problem worldwide. One important factor in the regulation of blood pressure is angiotensin I converting enzyme (ACE) which catalyzes the conversion of inactive decapeptide angiotensin I to the potent active vascular constricting octapeptide vasoconstrictor angiotensin II and inactivates the antihypertensive vasodilator bradykinin. ACE inhibition can be achieved with food-derived bioactive peptides. There is a traditional belief that increased natto intake plays an important role in preventing and treating hypertension (Kim et al. 2008b). This traditional belief has been confirmed by several clinical trials. In 1995, researchers from Miyazaki Medical College and Kurashiki University of Science and Arts in Japan studied the effects of NK on blood pressure in both animal and human trials. In a rat trial, after a single IP injection of NK, systolic blood pressure (SBP) significantly decreased from 166 to 145 mmHg in 2 hours, and decreased further to 144 mmHg in 3 hours after enzyme intake. NK was also given orally for 4 consecutive days to human volunteers with high blood pressure. The SBP decreased from 173.8 to 154.8 mmHg and diastolic blood pressure (DBP) decreased from 101.0 to 91.2 mmHg.The researchers confirmed that the lowering effect of NK on blood pressure was due to inhibition of ACE. ACE is known to cause blood vessels to narrow and blood pressure to rise (Kotb 2012).

Conclusion

Microbial fibrinolytic enzymes isolated from fermented foods like natto and Chungkook-jang were found to be more potent when compared with those isolated from non-microbial soures like earthworms and snakes. Some fibrinolytic enzymes have found application not only in blood clot removal but also as blood pressure regulators, plant virus inhibitors, detergent additives, antimicrobials, in the prevention of dental caries, as anti-inflammatory, analgesic, antibacterial, and mucolytic expectorant agents. However, the mechanics behind the unconventional applications of fibrinolytic enzymes are currently still unclear.

Abbreviations

ACE,	angiotensin converting enzyme
APTT,	activated partial thromboplastin time
APV,	amicrobial prostato-vesiculitis
CCD,	central composite design
Ccp A,	catabolite control protein A
CCR,	catabolite repression
cDNA,	complementary DNA
CMV,	cucumber mosaic virus
CTS,	carpal tunnel syndrome
CVD,	cardiovascular disease
DBP,	diastolic blood pressure
EDTA,	ethylenediaminetetra-acetic acid
EFE,	earthworm fibrinolytic enzyme
EMS,	ethyl methyl sulfonate
EtBr,	ethidium bromide
FDP,	fibrin degradation products
FFD,	fractional factorial design
FU,	fibrin unit
IP,	intraperitoneal
LK,	lumbrokinase
LrP,	*L. rubellus* protease
NK,	nattokinase
PA,	plasminogen activators
PAI-1,	plasminogen activator inhibitor-1
PI,	isoelectric point
PT,	prothrombin time
RSM,	response surface methodology
RT-PCR,	reverse transcriptase polymerase reaction
SBP,	systolic blood pressure
SDS,	sodium dodecyl sulfate
TMV,	tobacco mosaic virus
tPA,	tissue-type plasminogen activator
UK,	urokinase
uPA,	urokinase plasminogen activator
WHO,	World Health Organization

References

Ahn M.Y., B.S. Hahn and K.S. Ryu (2003) Purification and characterization of a serine protease with fibrinolytic activity from the dung beetles, *Catharsius molossus. Thromb. Res.,* 112: 339–347.

Amaravadi L., M.S. Bisesi and R.F. Bozarth (1990) Vermial virucidal activity: implication for management of pathogenic biological wastes on land. *Biol. Wastes*, 34: 349–358.

Benkerroum N (2013) Traditional fermented foods of North African countries: technology and food safety challenges with regard to microbiological risks. *Compr. Rev. Food. Sci. F.*, 12: 54–89.

Bracale G. and L. Selvetella (1996) Clinical study of the efficacy of and tolerance to seaprose S in inflammatory venous disease. Controlled study versus serratio-peptidase. *Minerva Cardioangiol.*, 44(10): 515–524.

Cha W.S., S.S. Park, Kim S.J. and D. Choi (2010) Biochemical and enzymatic properties of a fibrinolytic enzyme from *Pleorotus eryngii* cultivated under solid-state conditions using corn cob. *Bioresour. Technol.*, 101: 6475–6481.

Chang C.T., M.H. Fan, Kuo F.C. and H.Y. Sung (2000) Potent fibrinolytic enzyme from a mutant of *Bacillus subtilis* IMR0NK1. *J. Agric. Food Chem.*, 48(8): 3210–3216.

Chang C.T., P.M. Wang, Y.F. Hung, and Y.C. Chung (2012) Purification and biochemical properties of a fibrinolytic enzyme from *Bacillus subtilis*-fermented red bean. *Food Chem.*, 133: 1611–1617.

Chen X.C., J.X. Bai, J.M. Cao et al. (2009) Medium optimization for the production of cyclic adenosine 3',5'–monophosphate by *Microbacterium* sp. no. 205 using response surface methodology. *Bioresour. Technol.*, 100(2): 919–924.

Cheng M.B., J.C. Wang, Y.H. Li et al. (2008) Characterization of water-in-oil microemulsion for oral delivery of earthworm fibrinolytic enzyme. *J. Control Release*, 129(1): 41–48.

Chiang C.J., H.C. Chen, Y. Chao and J.T.C. Tzen (2005) Efficient system of artificial oil bodies for functional expression and purification of recombinant nattokinase in *Escherichia coli. J. Agric. Food Chem.*, 53(12): 4799–4804.

Chitte R.R. and S. Dey (2000) Potent fibrinolytic enzyme from a thermophilic *Streptomyces megasporus* strain SD5. *Lett. Appl. Microbiol.*, 31(6): 405–410.

Cho I.H., E.S. Choi, H.G. Lim and H.H. Lee (2004) Purification and characterization of six fibrinolytic serine-proteases from earthworm *Lumbricus rubellus. J. Biochem. Mol. Biol.*, 37: 199–205.

Choi D., W.S. Cha, N. Park et al. (2011) Purification and characterization of a novel fibrinolytic enzyme from fruiting bodies of Korean *Cordyceps militaris. Bioresour. Technol.*, 102(3): 3279–3285.

Choi H.S. and Y.S. Sa (2000) Fibrinolytic and antithrombotic protease from *Ganoderma lucidum. Mycologia*, 92: 545–552.

Choi H.S. and Y.S. Sa (2001) Fibrinolytic and antithrombotic protease from *Spirodela polyrhiza. Biosci. Biotechnol. Biochem.*, 65: 781–786.

Choi H.S. and P.H. Shin (1998) Purification and partial characterization of a fibrinolytic protease in *Pleurotus ostreatus. Mycologia*, 90(4): 674–679.

Choi N.S., K.H. Yoo, J.H. Hahm (2005) Purification and characterization of a new peptidase, bacillopeptidase DJ-2, having fibrinolytic activity: produced by *Bacillus* sp. DJ-2 from Doen-Jang. *J. Microbiol. Biotechnol.*, 15(1): 72–79.

Chung D.M., N.S. Choi, P.J. Maeng, H.K. Chun and S.H. Kim (2010) Purification and characterization of a novel fibrinolytic enzyme from Chive (*Allium tuberosum*). *Food Sci. Biotechnol.*, 19: 697–702.

Clemetson K.J., Q. Lu and J.M. Clemetson (2007) Snake venom proteins affecting platelets and their applications to anti-thrombotic research. *Curr. Pharma. Design*, 13(28): 2887–2892.

Collen D. (1998) Engineered staphylokinase variants with reduced immunogenicity. *Fibrinol. Proteol.*, 12(2): 59–65.

Collen D. (1999) The plasminogen (fibrinolytic) system. *Thromb. Haemost.*, 82(2): 259–270.

Cooper E.L. (2002) The earthworm: a new model with biomedical applications. In: Cooper E.L., A. Beschin and M. Bilej (eds) *A New Model for Analyzing Antimicrobial Peptides with Biomedical Applications.* Amsterdam: IOS Press, pp. 3–26.

Cui L., M.S. Dong, X.H. Chen et al. (2008) A novel fibrinolytic enzyme from *Cordyceps militaris*, a Chinese traditional medicinal mushroom. *World J. Microbiol. Biotechnol.*, 24: 483–489.

Dametto M., A.P. David, S.S. Azzolini et al. (2000) Purification and characterization of a trypsin-like enzyme with fibrinolytic activity present in the abdomen of horn fly, *Haematobia irritans irritans* (Diptera: Muscidae). *J. Protein Chem.*, 19: 515–521.

Deepak V., S. Ilangovan, M.V. Sampathkumar et al. (2010) Medium optimization and immobilization of purified fibrinolytic URAK from *Bacillus cereus* NK1 on PHB nanoparticles. *Enzyme Microb. Technol.*, 47(6): 297–304.

Desai K.G.H. and H.J. Park (2005) Recent developments in microencapsulation of food ingredients. *Drying Technol.*, 23: 1361–1394.

Din J.N., D.E. Newby and A.D. Flapan (2004) Omega 3 fatty acids and cardiovascular disease – fishing for a natural treatment. *Br. Med. J.*, 328: 30–35.

Esch P.M., H. Gerngross and A. Fabian (1989) Reduction of postoperative swelling. Objective measurement of swelling of the upper ankle joint in treatment with serrapeptase: a prospective study. *Fortschr. Med.*, 107(4): 67–68, 71–72.

Fredericq L. (1878) La digestion des matieres albuminoides chez quelques invertebres. *Arch. Zool. Exp. Gen.*, 7: 391.

Fujita M., K. Nomura, K. Hong, Y. Ito, A. Asada and S. Nishimuro (1993) Purification and characterization of a strong fibrinolytic enzyme (nattokinase) in the vegetable cheese natto, a popular soybean fermented food in Japan. *Biochem. Biophys. Res. Commun.*, 197(3): 1340–1347.

Fujita M., K. Hong and Y. Ito (1995a) Transport of nattokinase across the rat intestinal tract. *Biol. Pharm. Bull.*, 18(9): 1194–1196.

Fujita M., K. Hong, Y. Ito, R. Fujii, K. Kariya and S. Nishimuro (1995b) Thrombolytic effect of nattokinase on a chemically induced thrombosis model in rat. *Biol. Pharm. Bull.*, 18(10): 1387–1391.

Gardell S.J., L.T. Duong, R.E. Diehl et al. (1989) Isolation, characterization, and cDNA cloning of a vampire bat salivary plasminogen activator. *J. Biol. Chem.*, 264: 17947–17952.

Green M. and P.M. Loewenstein (1988) Autonomous functional domains of chemically synthesized human immunodeficiency virus tat trans-activator protein. *Cell*, 55(6): 1179–1188.

Hahn B.S., I.M. Chang and Y.S. Kim (1995) Purification and characterization of piscivorase I and II, the fibrinolytic enzymes from eastern cottonmouth moccasin venom (*Agkistrodon piscivorus piscivorus*). *Toxicon*, 33(7): 929–941.

Hassanein W.A., E. Kotb, N.M. Awny and Y.A. El-Zawahry (2011) Fibrinolysis and anticoagulant potential of a metallo protease produced by *Bacillus subtilis* K42. *J. Biosci.*, 36(5): 773–779.

Hu R., S. Zhang, H. Liang, N. Li and C. Tu (2004) Codon optimization, expression, and characterization of recombinant lumbrokinase in goat milk. *Protein Expr. Purif.*, 37: 83–88.

Hwang C.M. D.I., Kim, J.E. Kim et al.(2002) In vivo evaluation of lumbrokinase, a fibrinolytic enzyme extracted from *Lumbricus rubellus*, in a prosthetic vascular graft. *J. Cardiovasc. Surg.*, 43: 891–894.

Jeon O.H., W.J. Moon and D.S. Kim (1995) An anticoagulant fibrinolytic protease from *Lumbricus rubellus*. *J. Biochem. Mol. Biol.*, 28: 138–142.

Kadimaliev D.A., O.S. Nadezhina, N. A. Atykian et al. (2008) Increased secretion of lignolytic enzymes by the *Lentinus tigrinus* fungus after addition of butanol and toluene in submerged cultivation. *Prikl. Biokhim. Mikrobiol.*, 44(5): 582–588.

Kim H.C., B.S. Choi, K. Sapkota et al. (2011) Purification and characterization of a novel, highly potent fibrinolytic enzyme from *Paecilomyces tenuipes*. *Process Biochem.*, 46: 1545–1553.

Kim H.K., G.T. Kim, D.K. Kim et al. (1997) Purification and characterization of a novel fibrinolytic enzyme from *Bacillus* sp. KA38 originated from fermented fish. *J. Ferment. Bioeng.*, 84(4): 307–312.

Kim J.B., W.H. Jung, J.M. Ryu et al. (2007) Identification of a fibrinolytic enzyme by *Bacillus vallismortis* and its potential as a bacteriolytic enzyme against *Streptococcus mutans*. *Biotechnol. Lett.*, 29: 605–610.

Kim J.H. and Y.S. Kim (1999) A fibrinolytic metalloprotease from the fruiting bodies of an edible mushroom, *Armillariella mellea. Biosci. Biotechnol. Biochem.*, 63: 2130–2136.

Kim J.H. and Y.S. Kim (2001) Characterization of a metalloenzyme from a wild mushroom, *Tricholoma saponaceum. Biosci. Biotech. Biochem.*, 65(2): 356–362.

Kim J.S., J.E. Kim, B.S. Choi et al. (2008a) Purification and characterization of fibrinolytic metalloprotease from *Perenniporia fraxinea* mycelia. *Mycol. Res.*, 112: 990–998.

Kim J.Y., S.N. Gum, J.K. Paik et al. (2008b) Effects of nattokinase on blood pressure: a randomized, controlled trial. *Hypertens. Res.*, 31: 1583–1588.

Kim S.B., D.W. Lee, C.I. Cheigh et al. (2006) Purification and characterization of a fibrinolytic subtilisin-like protease of *Bacillus subtilis* TP-6 from an Indonesian fermented soybean, Tempeh. *J. Ind. Microbiol. Biotechnol.*, 33: 436–444.

Kim S.H. and N.S. Choi (2000) Purification and characterization of subtilisin DJ-4 secreted by *Bacillus* sp strain DJ-4 screened from Doen-Jang. *Biosci. Biotechnol. Biochem.*, 64: 1722–1725.

Kim W., K. Choi, Y. Kim et al. (1996a) Purification and characterization of a fibrinolytic enzyme produced from *Bacillus* sp. strain CK 11-4 screened from Chungkook-Jang. *Appl. Environ. Microbiol.*, 62(7): 2488–2482.

Kim W., K. Choi, Y. Kim et al. (1996b) Purification and characterization of a fibrinolytic enzyme produced from *Bacillus* sp. strain CK 11-4 screened from Chungkook-Jang. *Appl. Environ. Microbiol.*, 62(7): 2488–2482.

Ko J.H., J.P. Yan, L. Zhu and Y.P. Qi (2004) Identification of two novel fibrinolytic enzymes from *Bacillus subtilis* QK02. *Comp. Biochem. Physiol. C Toxicol. Pharmacol.*, 137: 65–74.

Ko S.M., B.H. Yoo, J.M. Lim et al. (2009) Production of fibrinolytic enzyme in plastid-transformed tobacco plants. *Plant Mol. Biol. Rep.*, 27(4): 448–453.

Kotb E (2012) *Fibrinolytic Bacterial Enzymes with Thrombolytic Activity.* New York: Springer.

Kotb E (2013) Activity assessment of microbial fibrinolytic enzymes. *Appl. Microbiol. Biotechnol.*, 97: 6647– 6665.

Kotb E (2014a) The biotechnological potential of fibrinolytic enzymes in the dissolution of endogenous blood thrombi. *Biotechnol. Progr.*, 30(3): 656–672.

Kotb E (2014b) Purification and partial characterization of a chymotrypsin-like serine fibrinolytic enzyme from *Bacillus amyloliquefaciens* FCF-11using corn husk as a novel substrate. *World J. Microbiol. Biotechnol.*, 30(7): 2071–2080.

Kotb E (2015a) Purification and partial characterization of serine fibrinolytic enzyme from *Bacillus megaterium* KSK-07 isolated from

Kishk, a traditional Egyptian fermented food. *Appl. Biochem. Microbiol.,* 51(1): 34–43.

Kotb E (2015b) The biotechnological potential of subtilisin-like fibrinolytic enzyme from a newly isolated *Lactobacillus plantarum* KSK-II in blood destaining and antimicrobials. *Biotechnol. Progr.,* 31(2): 316–324.

Lee C.K., J.S. Shin, B.S. Kim, I.H. Cho, Y.S. Kim and E.B. Lee (2007) Antithrombotic effects by oral administration of novel proteinase fraction from earthworm *Eisenia andrei* on venous thrombosis model in rats. *Arch. Pharm. Res.,* 30: 475–480.

Lee J.S., H.S. Bai and S.S. Park (2006) Purification and characterization of two novel fibrinolytic proteases from mushroom, *Fomitella raxinea.* *J. Microbiol. Biotechnol,* 16: 264–271.

Lee S.Y., J.S. Kim, J.E. Kim et al. (2005) Purification and characterization of fibrinolytic enzyme from cultured mycelia of *Armillaria mellea. Protein Expr. Purif.,* 43(1): 10–17.

Li H.P., Z. Hu, J.L. Yuan et al. (2007) A novel extracellular protease with fibrinolytic activity from the culture supernatant of *Cordyceps sinensis*: purification and characterization. *Phytother. Res.,* 21: 1234–1241.

Liu B.Y. and H.Y. Song (2002) Molecular cloning and expression of nattokinase gene in *Bacillus subtilis. Acta Biochim. Biophys. Sin. (Shanghai),* 34(3): 338–340.

Liu J.G., J.M. Xing, T.S. Chang, Z.Y. Ma and H.Z. Liu (2005) Optimization of nutritional conditions for nattokinase production by *Bacillus natto* NLSSE using statistical experimental methods. *Process Biochem.,* 40: 2757–2762.

Liu X., X. Zheng, P. Qian et al. (2014) Purification and characterization of a novel fibrinolytic enzyme from culture supernatant of *Pleurotus ostreatus. J. Microbiol. Biotechnol.,* 24(2): 245–253.

Liu Y.Q., Z.J. Sun, C. Wang, S.J. Li and Y.Z. Liu (2004) Purification of a novel antibacterial short peptide in earthworm *Eisenia foetida. Acta Biochim. Biophys. Sin.,* 36: 297–302.

Lu C.L., S. Chen and S.N. Chen (2010b) Purification and characterization of a novel fibrinolytic protease from *Schizophyllum commune. J. Food Drug Analysis,* 18: 69–76.

Lu F., Z. Lu, X. Bie et al. (2010a) Purification and characterization of a novel anticoagulant and fibrinolytic enzyme produced by endophytic bacterium *Paenibacillus polymyxa* EJS-3. *Thromb. Res.,* 126: 349–355.

Mahajan P.M., S.V. Gokhale and S.S. Lele (2010) Production of nattokinase using *B. natto* NRRL 3666: media optimization, scale up and kinetic modeling. *Food Sci. Biotechnol.,* 19: 1593–1603.

Mahajan P.M., S. Nayak and S.S. Lele (2012) Fibrinolytic enzyme from newly isolated marine bacterium *Bacillus subtilis* ICTF-1: media optimization, purification and characterization. *J. Biosci. Bioeng.,* 113(3): 307–314.

Majima Y., M. Inagaki, K. Hirata, K. Takeuchi, A. Morishita and Y. Sakakura (1988) The effect of an orally administered proteolytic enzyme on the elasticity and viscosity of nasal mucus. *Arch. Otorhinolaryngol.*, 44(6): 355–359.

Masschalck B. and C.W. Michiels (2003) Antimicrobial properties of lysozyme in relation to foodborne vegetative bacteria. *Crit. Rev. Microbiol.*, 29: 191–214.

Mazzone A., M. Catalani, M. Costanzo et al. (1990) Evaluation of *Serratia* peptidase in acute or chronic inflammation of otorhinolaryngology pathology: a multicentre, doubleblind, randomized trial versus placebo. *J. Int. Med. Res.*, 18(5): 379–388.

Mecikoglu M., B. Saygi, Y. Yildirim, E. Karadag-Saygi, S.S. Ramadan and T. Esemenli (2006) The effect of proteolytic enzyme serratiopeptidase in the treatment of experimental implant-related infection. *J. Bone Joint Surg. Am.*, 88(6): 1208–1214.

Mihara H., H. Sumi T., Yoneta et al. (1991) A novel fibrinolytic enzyme extracted from the earthworm *Lumbricus rubellus*. *Jpn J. Physiol.*, 41(3): 461–472.

Mine Y., A. Wong and B. Jiang (2005) Fibrinolytic enzymes in Asian traditional fermented foods. *Food Res. Int.*, 38: 243–250.

Moriya N., M. Nakata, M. Nakamura et al. (1994) Intestinal absorption of serrapeptase (TSP) in rats. *Biotechnol. Appl. Biochem.*, 20(Pt 1): 101–108.

Morozova E.A., N.N. Falina, N.P. Denisova et al. (1982) Heterogeneity and substrate specificity of the fibrinolytic preparation from the fungus *Flammulina velutipes*. *Biokhimiya*, 47: 1181–1185.

Naveena B., K.P. Gopinath, P. Sakthiselvan and N. Partha (2012) Enhanced production of thrombinase by *Streptomyces venezuelae*: kinetic studies on growth and enzyme production of mutant strain. *Bioresour. Technol.*, 111: 417–424.

Noh K.A., D.H. Kim, N.S. Choi and S.H. Kim (1999) Isolation of fibrinolytic enzyme producing strains from kimchi. *Kor. J. Food Sci. Technol.*, 31: 219–223.

Nonaka T., H. Ishikawa, Y. Tsumuraya, Y. Hashimoto and N. Dohmae (1995) Characterization of a thermostable lysine-specific metalloendopeptidase from the fruiting bodies of a basidiomycete, *Grifola frondosa*. *J. Biochem.*, 118(5): 1014–1020.

Panagariya A. and A.K. Sharma (1999) A preliminary trial of serratiopeptidase in patients with carpal tunnel syndrome. *J. Assoc. Physicians India*, 47(12): 1170–1172.

Pandee P., A. H. Kittikul, O. Masahiro and Y. Dissara (2008) Production and properties of a fibrinolytic enzyme by *Schizophyllum commune* BL23. *Songklanakarin J. Sci. Technol.*, 30(4): 447–453.

Park I.S., J.U. Park, M.J. Seo et al. (2010) Purification and biochemical characterization of a 17 kDa fibrinolytic enzyme from *Schizophyllum commune. J. Microbiol.*, 48(6): 836–841.

Park S.E., M.H. Li, J.S. Kim et al. (2007) Purification and characterization of a fibrinolytic protease from a culture supernatant of *Flammulina velutipes* mycelia. *Biosci. Biotechnol. Biochem.*, 71: 2214–2222.

Peng Y. and Y.Z. Zhang (2002a) Cloning and expression in *E. coli* of coding sequence of the fibrinolytic enzyme mature peptide from *Bacillus amyloliquefaciens* DC-4. *Chin. J. Appl. Environ. Biol.*, 8: 285–289.

Peng Y. and Y.Z. Zhang (2002b) Optimization of fermentation conditions of douchi fibrinolytic enzyme produced by *Bacillus amyloliquefaciens* DC-4. *Chin. Food Ferment. Ind.*, 28: 19–23.

Peng Y., Q. Huang, R.H. Zhang and Y.Z. Zhang (2003) Purification and characterization of a fibrinolytic enzyme produced by *Bacillus amyloliquefaciens* DC-4 screened from douchi, a traditional Chinese soybean food. *Comp. Biochem. Physiol. Biochem. Mol. Biol.*, 134: 45–52.

Peng Y., X.J. Yang, L. Xiao and Y.Z. Zhang (2004) Cloning and expression of a fibrinolytic enzyme (subtilisin DFE) gene from *Bacillus amyloliquefaciens* DC-4 in *Bacillus subtilis. Res. Microbiol.*, 155(3): 167–173.

Pinto A.F., R. Dobrovolski, A.B. Veiga and J.A. Guimaraes (2004) Lonofibrase, a novel alpha-fibrinogenase from *Lonomia obliqua* caterpillars. *Thromb. Res.*, 113(2): 147–154.

Price N. and L. Stevens (1999) *Fundamentals of Enzymology: Cell and Molecular Biology of Catalytic Proteins*. Oxford: Oxford University Press.

Regnault V., G. Helft, D. Wahl et al. (2003) Anti streptokinase platelet-activating antibodies are common and heterogeneous. *J. Thromb. Hemost.*, 1: 1055–1061.

Robbins K.C. (1978) The human plasma fibrinolytic system: regulation and control. *Mol. Cell. Biochem.*, 20(3): 149–157.

Seo J.H. and S.P. Lee (2004) Production of fibrinolytic enzyme from soybean grits fermented by *Bacillus firmus* NA-1. *J. Med. Food*, 7(4): 442–449.

Shen M.H., J.S. Kim, K. Sapkota et al. (2007) Purification, characterization, and cloning of fibrinolytic metalloprotease from *Pleurotus ostreatus* mycelia. *J. Microbiol. Biotechnol.*, 17(8): 1271–1283.

Sumi H., H. Hamada, H. Tsushima, H. Mihara and H. Muraki (1987) A novel fibrinolytic enzyme (nattokinase) in the vegetable cheese Natto, a typical and popular soybean food in the Japanese diet. *Experientia*, 43(10): 1110–1111.

Sumi H., H. Hamada, H. Mihara, K. Nakanishi and H. Hiratani (1989) Fibrinolytic effect of the Japanese traditional food natto (nattokinase). *Thromb. Haemost.*, 62(1): 549.

Sumi H., H. Hamada, K. Nakanishi and H. Hiratani (1990) Enhancement of the fibrinolytic activity in plasma by oral administration of nattokinase. *Acta Haematol.*, 84(3): 139–143.

Sumi H., N. Nakajima and H. Mihara (1992) In vitro and in vivo fibrinolytic properties of nattokinase. *Thromb. Haemost.*, 89: 1267.

Sumi H., N. Nakajima and C. Yatagai (1995) A unique strong fibrinolytic enzyme (datsuwokinase) in skipjack "Shiokara", a Japanese traditional fermented food. *Comp. Biochem. Physiol.*, 112: 543–547.

Suzuki Y., K. Kondo, Y. Matsumoto et al. (2003) Dietary supplementation of fermented soybean, natto, suppresses intimal thickening and modulates the lysis of mural thrombi after endothelial injury in rat femoral artery. *Life Sci.*, 73: 1289–1298.

Tachibana M., O. Mizukoshi, Y. Harada, K. Kawamoto and Y. Nakai (1984) A multi-centre, double-blind study of serrapeptase versus placebo in post-antrotomy buccal swelling. *Pharmatherapeutica*, 3(8): 526–530.

Tao S., L. Peng, L. Beihui, L. Deming and L. Zuohu (1997) Solid state fermentation of rice chaff for fibrinolytic enzyme production by *Fusarium oxysporum. Biotechnol. Lett.*, 19(5): 465–467.

Tao S., L. Peng, L. Beihui, L. Deming and L. Zuohu (1998) Successive cultivation of *Fusarium oxysporum* on rice chaff for economic production of fibrinolytic enzyme. *Bioprocess Eng.*, 18(5): 379–381.

Torrens I., A.G. Ojalvo, A. Seralena, O. Hayes and J. de la Fuente (1999) A mutant streptokinase lacking the C-terminal 42 amino acids is less immunogenic. *Immunol. Lett.*, 70: 213–218.

Ueda M., K. Noda, M. Nakazawa et al. (2008) A novel antiplant viral protein from coelomic fluid of the earthworm *Eisenia foetida*: purification, characterization and its identification as a serine protease. *Comp. Biochem. Phys. B*, 151(4): 381–385.

Urano T., H. Ihara, K. Umemura et al. (2001) The profibrinolytic enzyme subtilisin NAT purified from *Bacillus subtilis* cleaves and inactivates plasminogen activator inhibitor type 1. *J. Biol. Chem.*, 276: 24690–24696.

Vicari E., S. La Vignera, C. Battiato and A. Arancio (2005) Treatment with non-steroidal anti-inflammatory drugs in patients with amicrobial chronic prostato-vesiculitis: transrectal ultrasound and seminal findings. *Minerva Urol. Nefrol.*, 57(1): 53–59.

Wang C.T., B.P. Ji, Li B., R. Nout, P.L. Li, H. Ji and L.F. Chen (2006) Purification and characterization of a fibrinolytic enzyme of *Bacillus subtilis* DC33, isolated from Chinese traditional douchi. *J. Ind. Microbiol. Biotechnol.*, 33: 750–758.

Wang F., C. Wang, M. Li, L. Cui, J. Zhang and W. Chang (2003) Purification, characterization and crystallization of a group of earthworm fibrinolytic enzymes from *Eisenia fetida. Biotechnol. Lett.*, 25: 1105–1109.

Wang J.D., T. Narui, H. Kurata, K. Taeuchi, T. Hashimoto and T. Okuyama (1989) Hematological studies on naturally occurring substances II. Eject of animal crude drugs on blood coagulation and fibrinolysis systems. *Chem. Pharm. Bull.*, 37: 2236–2238.

Wang S., H. Chen, T. Liang and Y. Lin (2009) A novel nattokinase produced by *Pseudomonas* sp. TKU015 using shrimp shells as substrate. *Process Biochem.*, 44: 70–76.

Wong A.H.K. and Y. Mine (2004) A novel fibrinolytic enzyme in fermented shrimp paste. A traditional Asian fermented seasoning. *J. Agr. Food Chem.*, 52: 980–986.

Wong K.H., C.K.M. Lai and P.C.K. Cheung (2011) Immunomodulatory activities of mushroom sclerotial polysaccharides. *Food Hydrocoll.*, 25(2): 150–158.

Xiao L., R.H. Zhang, Y. Peng and Y.Z. Zhang (2004) Highly efficient gene expression of a fibrinolytic enzyme (subtilisin DFE) in *Bacillus subtilis* mediated by the promoter of α-amylase gene from *Bacillus amyloliquefaciens*. *Biotechnol. Lett.*, 26: 1365–1369.

Zhang R.H., L. Xiao, Y. Peng, H.Y. Wang, F. Bai and Y.Z. Zhang (2005) Expression and characteristics of a novel fibrinolytic enzyme (subtilisin DFE) in *Escherichia coli*. *Lett. Appl. Microbiol.*, 41: 190–195.

Zhang Y., A. Wisner, Y. Xiong and C. Bon (1995) A novel plasminogen activator from snake venom. Purification, characterization, and molecular cloning. *J. Biol. Chem.*, 270(17): 10246–10255.

Zukaite V. and G.A. Biziulevicius (2000) Acceleration of hyaluronidase production in the course of batch cultivation of *Clostridium perfringens* can be achieved with bacteriolytic enzymes. *Lett. Appl. Microbiol.*, 30: 203–206.

9

Microbial Products Maintain Female Homeostasis

Sarika Amdekar[1], Mayuri Khare[1], Vivek Kumar Shrivastava[2], Avnish Kumar[3], and Vinod Singh[1]*

[1] Department of Microbiology, Barkatullah University, Habibganj, Bhopal, Madhya Pradesh, India
[2] Department of Microbiology, College of Life Sciences, Cancer Hospital and Research Institute, Gwalior, Madhya Pradesh, India
[3] Department of Biotechnology, Dr. B. R. Ambedkar University, Agra, India

*Corresponding author e-mail: vsingh3@rediff.com

Probiotics as Therapeutics

The age-old saying of Hippocrates that food can be medicine and medicine can be food is still believed today. Functional food supplements our nutritional requirements as well as promoting good health. Such food products are enriched with active physiological substances which can improve nutritional values. The Food and Agriculture Organization (FAO) and the World Health Organization (WHO) define probiotics as "live micro-organisms that benefit the consumer when administered in adequate quantity/numbers" (WHO and FAO 2002). Use of probiotics and their products is recommended by physicians for children and others at high risk of infectious diseases (WHO and FAO 2001). Probiotics are sold in the form of dried powder and capsules and/or fermented milk products such as dahi, buttermilk, yoghurts, etc.

Probiotics work by various mechanisms including providing defensive barriers, enhancing immunity, and clearing harmful bacteria from the gastrointestinal tract (Elmer 2001; McFarland 2000) (Figure 9.1). They also help in improving the normal microflora (Alander et al. 1999), prevent infectious diseases (Neutra et al. 2001; Silva et al. 1999), alleviate food-related allergies (Cianferoni and Spergel 2009; Majamaa and Isolauri 1997; Nowak-Wegrzyn and Sampson 2006; Sicherer and Sampson 2009), reduce serum cholesterol levels (Fukushima and Nakano

Microbial Functional Foods and Nutraceuticals, First Edition. Edited by Vijai Kumar Gupta, Helen Treichel, Volha (Olga) Shapaval, Luiz Antonio de Oliveira, and Maria G. Tuohy.
© 2018 John Wiley & Sons Ltd. Published 2018 by John Wiley & Sons Ltd.

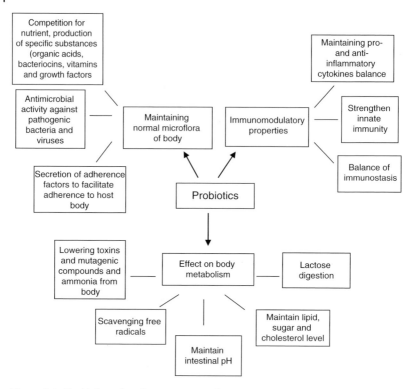

Figure 9.1 Health benefits of consuming probiotics. Probiotic effects on host body metabolism, maintaining normal microflora and ensuring the integrity of the immune system.

1996; Roos and Katan 2000), have anticancerous activity (de Moreno de LeBlanc and Perdigon 2004; Hirayama and Rafter 1999; Ishikawa et al. 2005), promote gut mucosal barrier function (Salminen et al. 1996), have immune adjuvant properties (Amdekar et al. 2012; Giacinto et al. 2005; Gill et al. 2000), alleviate bowel disease symptoms (Schultz and Sartor 2000), improve tolerance to lactose (de Vrese et al. 2001; Pelletier et al. 2001) and have antiarthritic activity (Amdekar et al. 2011a, 2013).

Probiotics play a wide role in both sexes in all age groups (de Vrese 2009; McCabe et al. 2013). They are equally advantageous for infants, adults, and older people (Gill et al. 2001; Peate 2010; Ritchie and Romanuk 2012 Weizman et al. 2005). Researchers have studied probiotics for their potential in a variety of conditions and published approximately 5000 articles regarding the importance of probiotics. However, very few studies have been carried out on the effect of probiotics on female health (Minelli et al. 1990; Uzar 2011). Various groups have studied the efficacy of probiotics for urinary tract infection in females

(Amdekar et al. 2011b; Anukam et al. 2009; Borchert et al. 2008; Boris et al. 1998; Bruce and Reid 1988, 2003; Homayouni et al. 2014; Parma et al. 2014; Reid et al. 2001; Strus et al. 2005). In both males and females, colonization of the gut by aerobes and anaerobes occurs in the same way, within 5 days of birth (Hoogkamp-Korstanje et al. 1979). Vaginal flora is responsible for maintaining homeostasis and hormone levels during various phases of female life. Probiotics can be used for the problems associated with women's health. Therefore, the purpose of this review is to explore the properties of probiotics with special reference to female health.

Concept of Probiotics

Probiotics (derived from Greek) means "for life" and denotes favorable micro-organisms which can treat pathogens in the gastrointestinal (GI) tract (Lilly and Stillwell 1965). The term was first used for substances which enhance the growth of other micro-organisms (Fuller 1989). The have also been defined as "live microbial feed supplement which benefits the host by maintaining the intestinal balance" (Havenaar and Huisin't Veld 1992) and "micro-organisms, when consumed in effective quantities, affect the overall health of the host" (Guarner and Schaafsma 1998). The importance of probiotics was first explored by Metchnikoff in the early 1900s (Salminen et al. 1996).

Bacterial species are of the utmost importance as probiotics. This mainly includes bacteria belonging to the genera *Lactobacillus* and *Bifidobacterium* (Botina et al. 2010; Kaushik et al. 2009; Singh et al. 2009). *Lactobacillus* is the predominant bacterium among the 1000 bacterial species which reside in the human intestinal gut (Lojo et al. 2004). These bacteria are the ingredients of many probiotic supplements (Frick et al. 2007; Gourbeyre et al. 2011; Macpherson and Harris 2004). In addition, *Lactococcus*, *Streptococcus*, *Propionibacterium*, *Saccharomyces*, *Enterococcus*, and *Bacillus* are also used as probiotics (Amdekar et al. 2010). Currently, to increase the efficiency of probiotics, prebiotics are used as well. They work together efficiently as synbiotics. Prebiotics selectively enhance the growth of probiotic bacteria in the colon and hence improve host health (Boehm et al. 2004).

Probiotics for Female Complications

Urinary Tract Infection (UTI)

Urinary tract infection is the most common problem in females of all age groups. Approximately 50–60% of females are affected with UTI during their life (Rahn 2008). *Escherichia coli*, *Proteus*, *Klebsiella*, and *Staphylococcus saprophyticus* are generally responsible for UTIs. Antibiotics are the drugs of choice; they cure disease but are not always effective and there is a chance of recurrence (Reid and Seidenfeld 1997). These drugs also have side effects and due to bacterial resistance are becoming ineffective for treating UTI.

Therefore alternative therapies are attracting patients and their physicians. *Lactobacillus* is an important microflora of the female genitourinary tract (Amdekar et al. 2011b; Kaminogawa and Nanno 2004). The mechanism behind the action of *Lactobacillus* in the urinary tract is still unexplained (Spiegel et al. 1980). Xia et al. (2006) proposed that *L. crispatus* blocks the adherence of pathogens. Hydrogen peroxide production (Czaja et al. 2000), inhibition of proinflammatory cytokines (Anukam et al. 2009) and release of 29 kDa biosurfactant protein (Osset et al. 2001) are responsible for their action. *Lactobacillus* has been shown to overcome UTIs in women (Grin et al. 2013; Gupta et al. 1998; Hemmerling et al. 2009; Redondo-Lopez et al. 1990). Probiotic strains of *Lactobacillus* are safe and effective in preventing UTI in adult women (Figure 9.2).

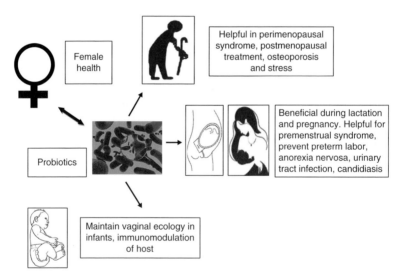

Figure 9.2 Effects of probiotics supplementation on female health. Probiotic supplementation in childhood maintains vaginal ecology and keeps the immune system strong. Probiotics are useful in middle-aged women to prevent urinary tract infections and candidiasis, but also help with premenstrual syndrome. Consumption of probiotics during pregnancy and lactation benefits the mother and fetus and lessens the chance of preterm labor. Probiotics are favorable in old age as they maintain bone mineral density (osteoporosis) and decrease perimenopausal and postmenopausal symptoms and associated stress.

Candidiasis

Candidiasis is the most important cause of vaginitis after vaginosis caused by bacteria. Candidiasis occurs in up to 40% of women with vagina-related problems (Vicariotto et al. 2012). White sticky discharge, persistent itching, and vaginal irritation are the classic symptoms of yeast candidiasis. Vulvovaginal candidiasis is commonly caused by *Candida albicans* while *C. glabrata, C. krusei*, and *C. tropicalis* may also cause candidiasis. At least one-third of women harbor *Candida* in the mouth, vagina, and intestine (Ott et al. 2008). Small numbers of *Candida* live as normal microflora but in immunosuppressed patients, can cause severe infection.

Studies suggest that normal flora has resistance against *Candida* in immunosuppressive states such as diabetes, pregnancy, and prolonged use of antibiotic therapy, and in such conditions *Candida* flourishes in the vaginal ecosystem (Payne et al. 2003). Incidence of candidiasis is due to use of broad-spectrum antibiotics that disturb the host's microflora (Fong 1994). Antifungal creams are usually given as treatment for candidiasis. These can relieve symptoms and local infection but are ineffective in episodes of relapse (Spinillo et al. 1992). Recurrence can be prevented by complete removal of *Candida* from the gut (Miles et al. 1977). Probiotics have been investigated as possible agents for the prevention of recurrences of candidiasis (Hilton et al. 1992; Lykova et al. 2000). Probiotics check the growth of *Candida* in the GI tract and vagina, but also they prevent *Candida* from adhering to epithelial surfaces. Probiotic bacteria also stimulate innate and acquired immune responses and increase opsonization in the host. These mechanisms collectively inhibit candidal growth in the vagina.

Anorexia Nervosa

Anorexia nervosa is a serious starvation disorder which manifests with physical, mental, and emotional factors. It is characterized by an intense fear of weight gain, distorted body image, and amenorrhea. It usually occurs among young women and adolescents. Over 7 million girls and women suffer from this disorder (American Psychiatric Association 2000). Anorexia nervosa is seen in 3.7% of females while 4.2% are affected by bulimia nervosa (overeating). Males can also be affected by these conditions.

Anorexia nervosa has one of the highest death rates of psychiatric-associated problems (Keel et al. 2003). Studies showed that such patients are generally lacking important nutrients like iron, vitamins A, C, and B6 (Abou-Saleh and Coppen 1986; Beaumont et al. 1981; Rock and Vasantharajan 1995), essential fatty acids (Langan and Farrell 1985), minerals like copper and zinc (Humphries et al. 1989), and omega-3 fatty acids (Katz et al. 1987). Thus, anorexic females are more prone to osteoporosis, pellagra, mood disorders, depression, and brain inco-ordination. Anorexic victims are mortally afraid of gaining weight so treatment strategies include positive thinking and adapting to eating healthy and nutritious food. Cognitive behavioral modification and natural or herbal supplements can be used, including wheat grass and sea kelp, fish oil supplements, ashwagandha, fenugreek, shatavari, chamomile, probiotics, psychotherapy, and acupressure.

Probiotic supplements can help in the treatment of anorexia nervosa by increasing the concentration of intestinal flora and providing nutrients. Anorexic females have symptoms of upset stomach including problems with stool passage, watery stool, excessive gas, and other digestive symptoms. Probiotics prevent lactose intolerance qs they convert lactose to lactic acid, which is better digested. They lower cholesterol in the body as bile in the gut. They can decrease inflammation, absorb mineral and ions and behave as an immunomodulator. Studies suggest that probiotics can be helpful in irritable bowel syndrome (IBS) leading to reduced abdominal pain, distension, flatulence, and borborygmi (Kajander et al. 2005). Treating subjects with probiotics has been shown to relieve general gastrointestinal problems (Ross 2007).

Preterm Delivery

Preterm delivery (PTD) is one of the most common causes of prenatal mortality and morbidity throughout the world (Myhre et al. 2011). Infant death and chances of diseases are increased with lower gestational age. Most cases of premature delivery are due to intrauterine infection (Goldenberg et al. 2002). Babies born preterm often have neurological disabilities like cerebral palsy and learning difficulties (Romero et al. 2000; Russell et al. 2007).

Risk factors for PTD include history of PTD, young age of mother, and intrauterine infection or bacterial vaginosis (Goldenberg et al. 2003; Mercer et al. 1996). Infectious bacterial vaginosis is responsible for 25–40% of the chance of PTD (Menon and Fortunato 2007). Organisms include anaerobes or facultative bacteria, mainly *Gardnerella vaginalis*, *Prevotella* sp., *Bacteroides* sp., *Mobiluncus* sp., some gram-positive round bacteria, and *Mycoplasma* (*M. hominis* and *Ureaplasma urealyticum*). Microbial invasion of the amniotic cavity and activation of the innate immune system are factors responsible for PTD. Microbes inhabiting the intrauterine portion release endotoxins and some proinflammatory cytokines that in turn increase prostaglandins and cause uterine contractility. These all leads to rupture of membranes and preterm labor (Goldenberg et al. 2000).

Being the normal flora of the vaginal ecosystem, *Lactobacillus* can be used to overcome this problem. Lactic acid is a major chemical factor which inhibits the growth of opportunistic vaginal pathogens. Lactobacilli reduce the growth of vaginal inhabitants including *Escherichia coli*, *C. albicans*, *Gardnerella vaginalis*, and *Mobiluncus* sp. by secreting hydrogen peroxide (H_2O_2) (Famularo et al. 2001). Studies have suggested that these probiotics also show a strong immunomodulatory effect and maintain vaginal immunity (Isolauri et al. 2004). A study conducted by Myhre et al. (2011) suggested that women who take probiotics during their pregnancy have a lower chance of preterm delivery. In another study, probiotics were given to pregnant women with a history of bacterial vaginosis. Probiotic treatment lowered the rate of preterm delivery (Krauss-Silva et al. 2010).

Pregnancy and Lactation

Probiotics are advantageous in pregnancy directly for the mother and indirectly for fetal health. Their use during pregnancy and lactation is safe and they do not circulate in the fetal body (Elias et al. 2011). During pregnancy, there is an increased requirement for nutrients to aid the growth of the fetus. Deficiency of nutrients during pregnancy leads to fetal abnormality. Moreover, in pregnancy, immune and metabolic functions of the fetus are dependent on the mother. Calcium (Ca) deficiency in the mother leads to abnormal fetal programming (Bhatia 2008), hypocalcemia (Camadoo et al. 2007), neonatal rickets, tetany (Sachan et al. 2005), and pre-eclampsia (Seely 2007). Iron (Fe) deficiency may cause anemia, premature birth and abnormal behavioral, emotional and cognitive development (Hernandez-Martinez et al. 2011; Rao and Georgieff 2007; Scholl 2005). Ca and Fe supplementation, and dietary Ca and Fe are usually given. Studies suggest that consumption of probiotics increases Ca serum levels in expectant mothers (Asemi et al. 2013). Several studies in animal models have suggested that probiotics increased absorption of Ca and Fe (Capcarova et al. 2010, 2011). In another study, iron fortified and probiotics fermented milk increased serum Ca and Fe levels in preschool children (Krauss-Silva et al. 2010). Limited data is available regarding the effect of probiotics in pregnant women with reference to Ca and Fe. Clinical trials in this direction may explore this fact in future.

There was an improvement in immunomodulatory properties in infants whose mothers received probiotics during pregnancy (Rautava et al. 2002). A study conducted by Dotterud et al. (2010) suggested that consumption of probiotics by pregnant women minimized the occurrence of atopic dermatitis. Some trials have studied the effect of probiotics taken during breast feeding. No adverse effects were observed in infants whose mothers took oral *L. reuteri* and *Bifidobacterium* during lactation, as probiotic bacteria do nottransfer from mother to infant via breast milk (Abrahamsson et al. 2009; Allen et al. 2010). However, more research is needed to validate these findings.

Osteoporosis

Osteoporosis is a silent disease manifesting as bone mineral density (BMD) or mass lost without any sign of disease or sickness. Women are more prone to osteoporosis but it also affects men (Naidu 2009). It is the major disease associated with aging and postmenopausal condition. Sufficient calcium intake has been reported to support bone growth and prevent bone loss during the aging process. Hormone replacement therapy is supportive but due to side effects is not generally preferred. Exercise and walking maintain bone density. In a study conducted on ovariectomized rats, probiotic bacteria increased BMD (McCabe et al. 2013). Studies with *L. paracasei*, *L. helveticus*, and *L. plantarum* showed an increase in BMD (Chiang et al. 2011; Narva et al. 2004). Probiotic bacteria synthesize vitamins which are required for calcium formation. Moreover, they produce fatty acids necessary for mineral absorption. They also decrease oxidative stress which increases osteoclast differentiation (Parvaneh et al. 2014). Also, probiotics decrease intestinal inflammation thereby increasing effective calcium absorption.

Hormone-Associated Imbalance

Hormonal imbalance is commonly observed in women. Doctors are prescribing natural medicines for such problems because of fewer side effects. Hormone replacement therapy is often not recommended.

Premenstrual Syndrome

Premenstrual syndrome (PMS) is a common cause of physical, psychological, and social problems in females who are having periods (Barnhart et al. 1995). Most women experience some mild emotional or physical premenstrual symptoms, but for 8–20% these symptoms can be so severe and disruptive that they seek medical treatment. This usually occurs about 5–11 days before a woman starts her monthly menstrual period, i.e., after ovulation. Typical symptoms include decreased energy, tension, irritability, depression, headache, altered sex drive, breast pain, backache, abdominal bloating, and edema of the fingers and ankles. PMS affects between 30% and 40% of menstruating women, with peak occurrence in the 30s and 40s. In most cases symptoms are relatively mild but in about 10% of women symptoms can be quite severe. Severe PMS with depression, irritability, and severe mood swings is referred to as premenstrual dysphoric disorder (PMDD) (Steiner 1996).

The cause of PMS is not known, but breakdown of progesterone and an increase in β-glucuronidase enzymes have been shown to have some relationship to PMS. β-glucuronidase is a bacterial enzyme that breaks the bond between glucuronic acid and a metabolite, a toxin or a hormone-like estrogen, which by binding to a glucuronic acid becomes soluble in water and can thus be excreted in the bile. The activity of this enzyme can be reduced by establishing a proper bacterial flora or by administration of certain probiotic bacteria (Goldin and Gorbach 1984). Probiotics, like *Lactobacillus acidophilus*, found in yoghurt, reduce β-glucuronidase activity. Probiotics and prebiotics (inulin and fructo-oligosaccharides) improve intestinal imbalance (dysbiosis) and facilitate the absorption of nutrients. No controlled trials have yet been done in this direction. Studies have shown that probiotics can help women with PMS (de Vrese 2009).

Perimenopausal/Postmenopausal Complications

The menopause usually occurs around the age of 45-55. In the perimenopause (the stage around the time before menopause) and postmenopause (end of menstruation) complications are very common in women. Symptoms includes hot flashes, mood swings, weight gain, trouble sleeping, fatigue, depression, irritability and anxiety, joint/muscle pain, and loss of libido. Complementary therapies such as probiotics, phytoestrogens, acupuncture, and herbal medicines are often used for the management of menopausal symptoms. These therapies have not been tested in clinical trials. Probiotic supplementation as part of a daily health regimen for women in their perimenopausal and postmenopause years is beneficial. *Lactobacillus* relieves chronic constipation, gas pain and other digestive discomforts as well as restoring beneficial bacteria in the digestive tract to healthy levels.

Breast Cancer

One in eight women will be diagnosed with breast cancer. It is the second most important cause of death in females (Jemal et al. 2010). Studies suggest that genetic and environmental factors are involved in the occurrence of breast cancer. Dietary factors, such as fermented dairy products, may decrease the chances of breast cancer development. Several epidemiology-based studies suggest the defensive role of probiotics in overcoming cancers (Commane et al. 2005). Their protective effect occurs by binding to mutagens (Kumar et al. 2010). Probiotics indirectly check the growth of cancer by inhibiting bacteria which convert procarcinogenic compounds (harmless) to carcinogenic compounds (harmful), thus reducing carcinogens in the intestinal tract. Reduction in β-glucuronidase and β-glucosidase enzymes is another mechansim responsible for their anticancerous activity. In addition, they prevent bile acid malabsorption. Enhancing host immunity is also a factor leading to the anticancer effect of probiotics (de Moreno de LeBlanc and Perdigon 2010). Administrating lactic acid bacteria (LAB) may lead to an increase in antioxidative enzymes, further decreasing oxidative stress and preventing damage caused by carcinogens (Kumar et al. 2011; Mohammadi and Mortazavian 2011). There is a need for more research into the anticancer effect of probiotics *in vivo* and *in vitro*.

Stress

Women are twice as likely to experience depression as men and three times more likely to suffer from anxiety disorders. Hypertension is usually associated with lipid abnormality and obesity. Stress may also lead to imbalance in hormones related to sexual behavior. The prevalence of hypertension increases in postmenopausal women due to hormonal changes in the ovaries (Dubey et al. 2002). Depression is common during the menopause.

Raised estrogen levels increase hypertension/stress so generally during the menopause hormone replacement therapy is prescribed to women. However, it has been observed that estrogen replacement therapy is of no use for women who have a surgical or unnatural menopause, and it does not affect the blood pressure of such patients (Reckelhoff and Fortepiani 2004). Phytoestrogens derived from food can be a natural alternative to cope with estrogen-based hypertension. Their efficacy can be enhanced by adding probiotics (Cohen et al. 2007; Cornwell et al. 2004). Phytoestrogens have estrogen-like properties. The bioavailability of phytoestrogens in the human body is greatly increased by gut microflora like *Lactobacillus*. Probiotics have the property to biotransform glucosides to aglycones, which changes their physiological behavior (Chien et al. 2006). A combination of probiotics and phytoestrogens can be a better option for postmenopausal women than hormone replacement therapy.

The Way Forward

Since the time of Metchnikoff, many clinical trials have been undertaken to study probiotics as an alternative for synthetic drugs. Collective data suggest the importance of probiotics in supporting human health. It is important to improve the quality of probiotics as far as their infectivity is concerned before using live bacteria as therapeutics. Along with this, quality control measures during manufacture are an issue. Above all, the health benefits of probiotics and functional foods should be critically reviewed by the food safety authorities. Research should focus on prebiotics and soluble probiotic factors along with probiotics. Probiotics have been proven to reduce infections by increasing useful flora and restoring the natural balance of the body. They have shown their efficacy in the gut and other parts of the body. It would be helpful to be able to select probiotics species or strain which work specifically in restricted groups such as women, infants, or older people. Molecular biology-based studies are required to understand the interface of probiotic micro-organisms and the host.

Conclusions

Effective probiotics as a supplement to day-to-day life is and essential part of every woman's health strategy. Probiotics maintain hormone homeostasis, promote safer pregnancy and labor, maintain vaginal ecology, ensure a strong immune system, proper digestion, mood equilibrium, and strong bones. Recent data suggest the use of probiotics for a limited number of female problems but more research is needed.

References

Abou-Saleh, M.T. and A. Coppen. (1986) The biology of folate in depression: implications for nutritional hypotheses of the psychoses. *J. Psychiat. Res.*, 20: 91–101.

Abrahamsson, T.R., G. Sinkiewicz, T. Jakobsson, M. Fredrikson and B. Björkstén (2009) Probiotic lactobacilli in breast milk and infant stool in relation to oral intake during the first year of life. *J. Pediatr. Gastroenterol. Nutr.*, 49(3): 349–354

Alander, M.R., R. Satokari, M. Korpela et al. (1999) Persistence of colonization of human colonic mucosa c fdv by a probiotic strain, *Lactobacillus rhamnosus* GG, after oral consumption. *Appl. Environ. Microbiol.*, 65: 351–354.

Allen, S.J., S. Jordan, M. Storey et al. (2010) Dietary supplementation with lactobacilli and bifidobacteria is well tolerated and not associated with adverse events during late pregnancy and early infancy. *J. Nutr.*, 140(3): 483–488.

Amdekar, S., D. Dwivedi. P. Roy, S. Kushwah and V. Singh (2010) Probiotics: multifarious oral vaccine against infectious traumas. *FEMS Immunol. Med. Microbiol.*, 58(3): 299–307.

Amdekar, S., V. Singh and D.D. Singh (2011a) Probiotic therapy: immunomodulating approach towards urinary tract infection. *Curr. Microbiol.*, 63(5): 484–490.

Amdekar, S., V. Singh, R. Singh, P. Sharma, P. Keshav and A. Kumar (2011b) *Lactobacillus casei* reduces the inflammatory joint damage associated with collagen induced arthritis by reducing the pro-inflammatory cytokines. *J. Clin. Immunol.*, 31(2): 147–154

Amdekar, S., P. Roy, V. Singh, A. Kumar, R. Singh and P. Sharma (2012) Anti-inflammatory activity of *Lactobacillus* on carrageenan induced paw edema in male Wistar rats. *Int. J. Inflamm.*, Article ID 752015.

Amdekar, S., P. Roy, V. Singh, A. Kumar, R. Singh and P. Sharma (2013) *Lactobacillus casei* and *Lactobacillus acidophilus* attenuates the severity of experimental arthritis by regulating biochemical parameters. *Biomed. Prev. Nutr.*, 3(4): 351–356.

American Psychiatric Association Work Group on Eating Disorders (2000) Practice guideline for the treatment of patients with eating disorders (revision). *Am. J. Psychol.*, 157(1 Suppl): 1–39.

Anukam, K.C., K. Hayes, K. Summers and G. Reid. (2009) Probiotic *Lactobacillus rhamnosus* GR-1 and *Lactobacillus reuteri* RC-14 may help downregulating TNF-α, IL-6, IL-8, IL-10 and IL-12 (p70) in the neurogenic bladder of spinal cord injured patient with urinary tract infection: a two case study. *Adv. Urol.*, Article ID 680363.

Asemi. Z and A. Esmaillzadeh (2013) Effect of daily consumption of probiotic yoghurt on serum levels of calcium, iron and liver enzymes in pregnant women. *Int. J. Prev. Med.*, 4(8): 949–955.

Barnhart, K.T., E.W. Freeman and S.J. Sondheimer (1995) A clinician's guide to the premenstrual syndrome. *Med. Clin. North Am.*, 79: 1457–1472.

Beaumont, P.J., T.L. Chambers, L. Rouse and S.F. Abraham (1981) The diet composition and nutritional knowledge of patients with anorexia nervosa. *J. Hum. Nutr.*, 35: 265–273.

Bhatia, V. (2008) Dietary calcium intake: a critical reappraisal. *Ind. J. Med. Res.*, 127: 269–273.

Boehm, G., J. Jelinek, B. Stahl et al. (2004) Prebiotics in infant formulas. *J. Clin. Gastroenterol.*, 38: S76–S79.

Borchert, D., L. Sheridan, A. Papatsoris et al. (2008) Prevention and treatment of urinary tract infection with probiotics: review and research perspective. *Ind. J. Urol.*, 24:139–144.

Boris, S., J.E. Suarez, F. Vazquez and C. Barbes (1998) Adherence of human vaginal *lactobacilli* to vaginal epithelial cells and interaction with uropathogens. *Infect. Immunol.*, 66(5): 1985–1989.

Botina, S.G., N.V. Koroban, K.M. Klimina et al. (2010) Genetic diversity of the genus *Lactobacillus* bacteria from the human gastrointestinal microbiome. *Genetika*, 46: 1399–1406.

Bruce, A.W. and G. Reid (1988) Intravaginal instillation of lactobacilli for prevention of recurrent urinary tract infections. *Can. J. Microbiol.*, 34: 339–343.

Bruce, A.W. and G. Reid (2003) Probiotics and the urologist. *Can. J. Urol.*, 270: 16–18.

Camadoo, L., R. Tibbott and F. Isaza (2007) Maternal vitamin D deficiency associated with neonatal hypocalcaemic convulsions. *Nutr. J.*, 6: 23.

Capcarova, M., J. Weiss, C. Hrncar, A. Kolesarova and G. Pal (2010) Effect of *Lactobacillus fermentum* and *Enterococcus faecium* strains on internal milieu, antioxidant status and body weight of broiler chickens. *J. Anim. Physiol. Anim. Nutr. (Berlin)*, 4: e215–224.

Capcarova, M., P. Hascik, A. Kolesarova, M. Kacaniova, M. Mihok and G. Pal (2011) The effect of selected microbial strains on internal milieu of broiler chickens after peroral administration. *Res. Vet. Sci.*, 91: 132–137.

Chiang, S.S. and T.M. Pan (2011) Antiosteoporotic effects of *Lactobacillus*-fermented soy skim milk on bone mineral density and the microstructure of femoral bone in ovariectomized mice. *J. Agric. Food Chem.*, 59(14): 7734–7742.

Chien, H.L., H.Y. Huang and C.C. Chou (2006) Transformation of isoflavone phytoestrogens during the fermentation of soymilk with lactic acid bacteria and bifidobacteria. *Food Microbiol.*, 23: 772–778.

Cianferoni, A. and J.M. Spergel (2009) Food allergy: review, classification and diagnosis. *Allergol. Int.*, 58: 457–466.

Cohen, L.A., J.S. Crespin, C. Wolper et al. (2007) Soy isoflavone intake and estrogen excretion patterns in young women: effect of probiotics administration. *In Vivo*, 21: 507–512.

Commane, D., R. Hughes, C. Shortt and I. Rowland (2005) The potential mechanisms involved in the anti–carcinogenic action of probiotics. *Mutat. Res. Fund. Mol. Mech. Mutagen.*, 591: 276–289.

Cornwell, T., W. Cohick and I. Raskin (2004) Dietary phytoestrogens and health. *Phytochemistry*, 65: 995–1016.

Czaja, C.A., A.E. Stapleton, Y. Yarova-Yarovaya and W.E. Stamm (2000) Phase I trail of a *Lactobacillus crispatus* vaginal suppository for prevention of recurrent urinary tract infection in women. *Infect. Dis. Obst. Gynecol.*, Article ID 35387.

de Moreno de LeBlanc, A. and G. Perdigon (2004) Yoghurt feeding inhibits promotion and progression of experimental colorectal cancer. *Med. Sci. Monitor.*, 10: 96–104.

de Moreno de LeBlanc, A. and G. Perdigon (2010) The application of probiotics fermented milks in cancer and intestinal inflammation. *Proc. Nutr. Soc.*, 69: 421–428.

de Vrese, M. (2009) Health benefits of probiotics and prebiotics in women. *Menopause Int.*. 15: 35–40.

de Vrese, M., A. Stegelmann, B. Ritcher, S. Fenselau, C. Laue and J. Schrezenmeir (2001) Probiotics: compensation for lactase insufficiency. *Am. J. Clin. Nutr.*, 73: S421–S429.

Dotterud, C.K., O. Storro, R. Johnsen and T. Oien (2010) Probiotics in pregnant women to prevent allergic disease: a randomized, double-blind trial. *Br. J. Dermatol.*, 163: 616–623.

Dubey, R.K., S. Oparil, B. Imthurn and E.K. Jackson (2002) Sex hormones and hypertension. *Cardiol. Res.*, 53: 688–708.

Elias, J., P. Bozzo and A. Einarson (2011) Are probiotics safe for use during pregnancy and lactation? *Can. Fam. Phys.*, 59: 299–301.

Elmer, G.W. (2001) Probiotics: living drugs. *Am. J. Health Syst. Pharm.*, 58: 1101–1109.

Famularo, G., M. Pieluigi, R. Coccia, P. Mastroiacovo and C. de Simone (2001) Microecology, bacterial vaginosis and probiotics: perspectives for bacteriotherapy. *Med. Hypoth.*, 56(4): 421–430.

Fong, I.W. (1994) The rectal carriage of yeast in patients with vaginal candidiasis. *Clin. Invest. Med.*, 17(5): 426–431.

Frick, J.S., K. Schenk, M. Quitadamo, F. Kah, M. Koberle and E. Bohn (2007) Apefelbacher and Autenrieth IB.*Lactobacillus fermentum* attenuates the pro-inflammatory effect of *Yersinia enterocolitica* on human epithelial cells. *Inflamm. Bowel Dis.*, 13: 83–90.

Fukushima, M. and M. Nakano (1996) Effects of a mixture of organisms *Lactobacillus acidophilus* or *Streptococcus faecalis* on cholesterol metabolism in rats fed on fat- and cholesterol-enriched diet. *Br. J. Nutr.*, 76: 857–867.

Fuller, R. (1989) Probiotics in man and animals. *J. Appl. Bacteriol.*, 66: 365–378.

Giacinto, C.M., M. Marinaro, M. Sanchez, W. Strober and M. Boirivant (2005) Probiotics ameliorate recurrent Th1-mediated murine colitis by inducing IL-10 and IL-10-dependent TGF-beta-bearing regulatory cells. *J. Immunol.*, 174: 3237–3246.

Gill, H.S., K.J. Rutherfurd, J. Prasad and P.K. Gopal (2000) Enhancement of natural and acquired immunity by *Lactobacillus rhamnosus* (HN001), *Lactobacillus acidophilus* (HN017) and Bifidobacterium lactis (HN019). *Br. J. Nutr.*, 83: 167–176.

Gill, H.S., K.J. Rutherford, M.L. Cross and P. Gopal (2001) Enhancement of immunity in the elderly by dietary supplementation with the probiotic Bifidobacterium lactis HNO19. *Am. J. Clin. Nutr.*, 74: 833–839.

Goldenberg, R.L., J.C. Hauth and W.W. Andrews (2000) Intrauterine infection and preterm delivery. *N. Engl. J. Med.*, 342: 1500–1507.

Goldenberg, R.L., W. W. Andrews and J.C. Hauth (2002) Choriodecidual infection and preterm birth. *Nutr. Rev.*, 60: S19–S25.

Goldenberg, R.L., J.D. Iams, B.M. Mercer et al. for the National Institute of Child Health and Human Development Network (2003) What we have learned about the predictors of preterm birth. *Semin. Perinatol.*, 27: 185–193.

Goldin, B.R. and S.L. Gorbach (1984) The effect of milk and *Lactobacillus* feeding on human intestinal bacterial enzyme activity. *Am. J. Clin. Nutr.*, 39: 756–761.

Gourbeyre, P., S. Denery and M. Bodinier (2011) Probiotics, prebiotics, and synbiotics: impact on the gut immune system and allergic reactions. *J. Leukoccyte Biol.*, 89: 685–695.

Grin, P.M., P.M. Kowalewska, W. Alhazzan and A.E. Fox-Robichaud (2013) *Lactobacillus* for preventing recurrent urinary tract infections in women: meta-analysis. *Can. J. Urol.*, 20: 6607–6614.

Guarner, F. and G.J. Schaafsma (1998) Probiotics. *Int. J. Food Microbiol.*, 39: 237–238.

Gupta, K., A.E. Stapleton, T.M. Hooton, P.L. Roberts, C.L. Fennell and W.E. Stamm (1998) Inverse association of H2O2-producing lactobacilli and vaginal *Escherichia coli* colonization in women with recurrent urinary tract infections. *J. Infect. Dis.*, 178: 446–450.

Havenaar, R. and J.H.J. Huisin't Veld (1992) Probiotics: a general view. In: Wood, B.J.B. (ed.) *The Lactic Acid Bacteria, Vol. 1: The Lactic Acid Bacteria in Health and Disease*. New York:Chapman and Hall, pp. 209–224.

Hemmerling, A., W. Harrison, A. Schroeder et al. (2009) Phase 1 dose-ranging safety trial of Lactobacillus crispatus CTV-05 for the prevention of bacterial vaginosis. *Sex. Trans. Dis.*, 36: 564–569.

Hernandez-Martinez, C., J. Canals, N. Aranda, B. Ribot, J. Escribano and J. Arija (2011) Effects of iron deficiency on neonatal behavior at different stages of pregnancy. *Early Hum. Dev.*, 87: 165–169.

Hilton, E., H.D. Isenberg, P. Alperstein, K. France and M.T. Borenstein (1992) Ingestion of yogurt containing *Lactobacillus acidophilus* as prophylaxis for candidal vaginitis. *Ann. Intern. Med.*, 116: 353–357.

Hirayama, K. and J. Rafter (1999) The role of lactic acid bacteria in colon cancer prevention: mechanistic considerations. *Antonie van Leeuwenhoek*, 76: 391–394.

Homayouni, A., P. Bastani, S. Ziyadi et al. (2014) Effects of probiotics on the recurrence of bacterial vaginosis: a review. *J. Low. Genital Tract Dis.*, 18: 79–86.

Hoogkamp-Korstanje, J.A.A., J.G.E.M. Lindner, J.H. Marcelis, H. den Daas-Slagt and N.M. de Vos (1979) Composition and ecology of the human intestinal flora. *Antonie van Leeuwenhoek*, 45: 35–40.

Humphries, L., B. Vivian, M. Stuart and C.J. McClain (1989) Zinc deficiency and eating disorders. *J. Clin Psychol.*, 50: 456–459.

Ishikawa, H.I., T. Akedo, T. Otani et al. (2005) Randomized trial of dietary fiber and *Lactobacillus casei* administration for prevention of colorectal tumors. *Int. J. Cancer*, 116: 762–767.

Isolauri, E., S. Salminen and A.C. Ouwehand (2004) Microbial–gut interactions in health and disease. Probiotics. *Best Pract. Res. Clin. Gastroenterol.*, 18: 299–313.

Jemal, A., R. Siegel, J. Xu and E. Ward (2010) Cancer statistics. *CA: Can. J. Clin.*, 60: 277–300.

Kajander, K., K. Hatakka and T. Poussa (2005) A probiotics mixture alleviates symptoms in irritable bowel syndrome patients: a controlled 6-month intervention. *Aliment. Pharmacol. Ther.*, 22: 387–394.

Kaminogawa, S. and M. Nanno (2004) Modulation of immune action by food. *Evid. Based Comp. Alt. Med.*, 1: 241–250.

Katz, R.L., C.L. Keen, I.F. Litt, L.S. Hurley, K.M. Kellams-Harrison and L.J. Glader (1987) Zinc deficiency in anorexia nervosa. *J. Adolesc. Health Care.*, 8: 400–406.

Kaushik, J.K., A. Kumar, R.K. Duary, A.K. Mohanty, S. Grover and K. Batish (2009) Functional and probiotic attributes of an indigenous isolate of *Lactobacillus plantarum*. *PLoS One*, 4: e8099.

Keel, P.K., D.J. Dorer, K.T. Eddy, D. Franko, D.L. Charatan and D.B. Herzog (2003) Predictors of mortality in eating disorders. *Arch. Gen. Psychiat.*, 60: 179–183.

Krauss-Silva, L., M.E. Moreira. M.B. Alves et al. (2010) Randomized controlled trial of probiotics for the prevention of spontaneous preterm delivery associated with intrauterine infection: study protocol. *Reprod. Health*, 7: 14.

Kumar, M., A. Kumar, R. Nagpal et al. (2010) Cancer-preventing attributes of probiotics: an update. *Int. J. Food Sci. Nutr.*, 61: 473–496.

Kumar, M., V. Verma, R. Nagpal et al. (2011) Effect of probiotic fermented milk and chlorophyllin on gene expressions and genotoxicity during AFB-induced hepatocellular carcinoma. *Gene*, 490: 54–59.

Langan, S.M. and P.M. Farrell (1985) Vitamin E, vitamin A and essential fatty acid status of patients hospitalized for anorexia nervosa. *Am. J. Clin. Nutr.*, 41: 1054–1060.

Lilly, D.M. and R.H. Stillwell (1965) Probiotics: growth promoting factors produced by microorganisms. *Science*, 147: 747–748.

Lojo, J., C.J. Vukasovic, I. Strahinic and L. Topisirovic (2004) Characterization and antimicrobial activity of bacteriocin 217 produced by natural isolate *Lactobacillus paracasei* subsp. *paracasei* BGBUK2-s16. *J. Food Protect.*, 67: 2727–2734.

Lykova, E.A., A.A. Vorobev, A.G. Bokovoi et al. (2000) Disruption of microbiocenosis of the large intestine and the immune and interferon status in children with bacterial complications of acute viral infections of the respiratory tract and their correction by high doses of bifidum bacterin forte. *Antibiot. Khimioter.*, 45: 22–27.

Macpherson, A.J. and N.L. Harris (2004) Interactions between commensal intestinal bacteria and the immune system. *Nat. Rev. Immunol.*, 4: 478–485.

Majamaa, H. and E. Isolauri (1997) Probiotics: a novel approach in the management of food allergy. *J. Allergy Clin. Immunol.*, 99: 179–185.

McCabe, L.R., R. Irwin, L. Schaefer and R.A. Britton (2013) Probiotic use decreases intestinal inflammation and increases bone density in healthy male but not female mice. *J. Cell. Physiol.*, 228: 1793–1798.

McFarland, L.V. (2000) Normal flora: diversity and functions. *Microb. Ecol. Health Dis.*, 12: 193–207.

Menon, R. and S.J. Fortunato (2007) Infection and the role of inflammation in preterm premature rupture of the membranes. *Best Pract. Res. Clin. Obstet. Gynaecol.*, 21: 467–478.

Mercer, B.M., R.L. Goldenberg and A. Das (1996) The preterm prediction study: a clinical risk assesment system. *Am. J. Obstet. Gynecol.*, 174: 1885–1895.

Miles, M.R., L. Olsen and A. Rogers (1977) Recurrent vaginal candidiasis. Importance of an intestinal reservoir. *JAMA*, 238: 1836–1837.

Minelli, E.B., A.M. Beghini, S. Vessentini et al. (1990) Intestinal microflora as an alternative metabolic source of estrogens in women with uterine leiomyoma and breast cancer. *Ann. NY Acad. Sci.*, 595: 473–479.

Mohammadi, R. and A. Mortazavian (2011) Review article: technological aspects of prebiotics in probiotic fermented milks. *Food Rev. Int.*, 27: 192–212.

Myhre, R., A.L. Brantsæter, S. Myking et al. (2011) Intake of probiotic food and risk of spontaneous preterm delivery. *Am. J. Clin. Nutr.*, 93: 151–157.

Naidu, N. (2009) *Bio-replenishment for Bone Health*. Pomoma: Biorep Network Media.

Narva, M., M. Collin, C. Lamberg-Allardt et al. (2004) Effects of long-term intervention with *Lactobacillus helveticus*-fermented milk on bone mineral density and bone mineral content in growing rats. *Ann. Nutr. Metab.*, 48: 228–234.

Neutra, M.R., N.J. Mantis and J.P. Kraenhenbuhl (2001) Collaboration of epithelial cells with organized mucosal lymphoid tissues. *Nat. Immunol.*, 2: 1004–1009.

Nowak-Wegrzyn, A. and H.A. Sampson (2006) Adverse reactions to foods. *Med. Clin. North Am.*, 90: 97–127.

Osset, J., R.M. Bartolome, E. Gardia and A. Andreu (2001) Assessment of the capacity of *Lactobacillus* to inhibit the growth of uropathogen and block their adhesion to vaginal epithelia cells. *J. Infect Dis.*, 183: 485–491.

Ott, S.J., T. Kuhbacher, M. Musfeldt et al. (2008) Fungi and inflammatory bowel diseases: alterations of composition and diversity. *Scand. J. Gastroenterol.*, 43: 831–841.

Parma, M., V. Stella Vanni, M. Bertini and M. Candiani (2014) Probiotics in the prevention of recurrences of bacterial vaginosis. *Altern. Ther. Health Med.*, 20: 52–57.

Parvaneh, K., R. Jamaluddin, G. Karimi and R. Erfani (2014) Effect of probiotics supplementation on bone mineral content and bone mass density. *Scient. World J.*, Article ID 595962.

Payne, S., G. Gibson, A. Wynne, B. Hudspith, J. Brostoff and K. Tuohy (2003) In vitro studies on colonization resistance of the human gut microbiota to *Candida albicans* and the effects of tetracycline and *Lactobacillus plantarum* LPK. *Curr. Iss. Intest. Microbiol.*, 4:1–8.

Peate, J.L. (2010) A pilot study to determine the effectiveness of probiotic use in elderly patients with antibiotic-associated diarrhea. PhD thesis, Department of Nutrition and Health Sciences. University of Nebraska, Lincoln.

Pelletier, X., S. Laure-Boussuge and Y. Donazzolo (2001) Hydrogen excretion upon ingestion of dairy products in lactose intolerant male subjects: importance of the live flora. *Eur. J. Clin. Nutr.*, 55: 509–512.

Rahn, D.D. (2008) Urinary tract infections: contemporary management. *Urol. Nurs. J.*, 28: 333–341.

Rao, R. and M.K. Georgieff (2007) Iron in fetal and neonatal nutrition. *Semin. Fetal Neonat. Med.*, 12: 54–63.

Rautava, S., M. Kalliomak and E. Isolauri (2002) Probiotics during pregnancy and breast-feeding might confer immunomodulatory protection against atopic disease in the infant. *J. Allergy Clin. Immunol.*, 109: 119–121.

Reckelhoff, J.F. and L.A. Fortepiani (2004) Novel mechanisms responsible for postmenopausal hypertension. *Hypertension*, 43: 918–923.

Redondo-Lopez, V., R.L. Cook and J.D. Sobel (1990) Emerging role of *Lactobacilli* in the control and maintenance of the vaginal microflora. *Rev. Infect. Dis.*, 12: 856–872.

Reid, G. and A. Seidenfeld (1997) Drug resistance amongst uropathogens isolated from women in a suburban population: laboratory findings over 7 years. *Can. J. Urol,.* 4: 432–437.

Reid, G., D.C. Beuerman, A.W. Heinemann and A.W. Bruce (2001) Probiotic *Lactobacillus* dose required to restore and maintain a normal vaginal flora. *FEMS Immunol. Med. Microbiol.*, 32: 37–41.

Ritchie, M.L. and T.N. Romanuk (2012) A meta-analysis of probiotic efficacy for gastrointestinal diseases. *PLoS ONE*, 7: e34938.

Rock, C.L. and S. Vasantharajan (1995) Vitamin status of eating disorder patients: relationship to clinical indices and effect of treatment. *Int. J. Eat. Disord.*, 18: 257–262.

Romero, R., J. Espinoza, T. Chaiworapongsa and K. Kalache (2002) Infections and prematurity and the role of preventive strategies. *Semin. Neonatol.*, 7: 259–274.

Roos, N.M. and M.B. Katan (2000) Effects of probiotic bacteria on diarrhea, lipid metabolism and carcinogenesis: a review of papers published between 1988 and 1998. *Am. J. Clin. Nutr.*, 71: 405–411.

Ross, C.C. (2007) The importance of nutrition as the best medicine for eating disorders. *Diet Nutr.*, 3: 153–157.

Russell, R.B., N.S. Green, C.A. Steiner et al. (2007) Cost of hospitalization for preterm and low birth weight infants in the United States. *Pediatrics*, 120: e1–9.

Sachan, A., R. Gupta, V. Das, A. Agarwal, P.K. Awasthi and V. Bhatia (2005) High prevalence of vitamin D deficiency among pregnant women and their newborns in northern India. *Am. J. Clin. Nutr.*, 81: 1060–1064.

Salminen, S., E. Isolauri and E. Salminen (1996) Clinical uses of probiotics for stabilizing the gut mucosal barrier: successful strains and future challenges. *Antonie van Leeuwenhoek*, 70: 347–358.

Scholl, T.O. (2005) Iron status during pregnancy: setting the stage for mother and infant. *Am. J. Clin. Nutr.*, 81: 1218S–12122S.

Schultz, M. and R.B. Sartor (2000) Probiotics and inflammatory bowel disease. *Am. J. Gastroenterol.*, 1(Suppl.): S19–S21.

Seely, E.W. (2007) Calciotropic hormones in preeclampsia: a renewal of interest. *J. Clin. Endocrinol. Metab.*, 92: 3402–3403.

Sicherer, S.H. and H.A. Sampson (2009) Food allergy: recent advances in pathophysiology and treatment. *Ann. Rev. Med.*, 60: 261–277.

Silva, A.M., E.A. Bambirra, A.L. Oliveira et al. (1999) Protective effect of bifidus milk on the experimental infection with *Salmonella enteritidis* subsp. *Typhimurium* in conventional and gnotobiotic mice. *J. Appl. Microbiol.*, 86: 331–336.

Silva, M.R., G. Dias, C.L. Ferreira, S.C. Franceschini and N.M. Costa (2008) Growth of preschool children was improved when fed an iron-fortified fermented milk beverage supplemented with *Lactobacillus acidophilus*. *Nutr. Res.*, 28: 226–232.

Singh, V., K. Singh, S. Amdekar et al. (2009) Innate and specific gut-associated immunity and microbial interference. *FEMS Immunol. Med. Microbiol.*, 55(1): 6–12.

Spiegel, C.A., R.A. Amsel, D. Eschenbach, F. Schoenknecht and K.K. Holmes (1980) Anaerobic bacteria in nonspecific vaginitis. *N. Engl. J. Med.*, 1: 601–607.

Spinillo, A., L. Carratta, G. Pizzoli et al. (1992) Recurrent vaginal candidiasis. Results of a cohort study of sexual transmission and intestinal reservoir. *J. Reprod. Med.*, 37: 343–347.

Steiner, M. (1996) Premenstrual dysphoric disorder. *An update. Gen. Hosp. Psych.*, 18: 244–250.

Strus, M., A. Kucharska, G. Kukl, M. Brzychczy-Wloch, K. Maresz and P.B. Heczko (2005) The in vitro activity of vaginal *Lactobacillus* with probiotic properties against Candida. *Infect. Dis. Obstet. Gynecol.*, 13: 69–75.

Uzar, I. (2011) Probiotics in gynecology. *Kerba Polonica*, 57: 67–71.

Vicariotto, F., M. Del Piano, L. Mogna and G. Mogna (2012) Effectiveness of the association of 2 probiotic strains formulated in a slow release vaginal product, in women affected by vulvovaginal candidiasis: a pilot study. *J. Clin. Gastroenterol.*, 46(Suppl): S73–80.

Weizman, Z., G. Asli and A. Alsheikh (2005) Effect of a probiotic infant formula on infections in child care centers: comparison of two probiotic agents. *Pediatrics*, 115: 5–9.

World Health Organization and Food and Agriculture Organization (2001) *Health and Nutritional Properties of Probiotics in Food including Powder Milk with Live Lactic Acid Bacteria*. Geneva: WHO/FAO.

World Health Organization and Food and Agriculture Organization (2002) *Guidelines for the Evaluation of Probiotics in Food. Report of a Joint FAO/WHO Working Group on Drafting Guidelines for the Evaluation of Probiotics in Food*. Available at: http://www.who.int/foodsafety/ fs_management/en/probiotic_guidelines.pdf (accessed 1 May 2017).

Xia, Y., K. Yamagata and T.L. Krukoff (2006) Differential expression of the CD14/TLR4 complex and inflammatory signaling molecules following i.c.v. administration of LPS. *Brain Res.*, 1095: 85–95.

10

Production of High-Quality Probiotics by Fermentation

Marimuthu Anandharaj[1], Rizwana Parveen Rani[2], and Manas R. Swain[3]*

[1] *Agricultural Biotechnology Research Center, Academia Sinica, Taipei, Taiwan*
[2] *Gandhigram Rural Institute, Department of Biology, Tamil Nadu, India*
[3] *Department of Biotechnology, Indian Institute of Technology Madras, Chennai, India*

*Corresponding author e-mail: manas.swain@gmail.com

Introduction

Over recent years, our knowledge of intestinal microbiota and modulating factors, and interest in supplementing various types of food products with probiotic bacteria has grown significantly. Probiotics are defined as live micro-organisms or a product containing viable micro-organisms in sufficient numbers to alter the microflora (mainly lactobacilli and bifido-bacteria) which when administered in adequate amounts confer a health benefit on the host (Lacroix and Yildirim 2007). The most extensively used probiotics are lactobacilli, bifidobacteria, bacilli, and yeasts (Hasler 2002; Kołozyn-Krajewskaa and Dolatowski 2012). These probiotic strains, which are the major microbiota of the human gut, may provide protection against gastrointestinal disorders including gastrointestinal infections and inflammatory bowel diseases (Mitsuoka 1982; Salminen et al. 1998). Probiotic lactic acid bacteria are thought to be beneficial to human health and therefore, a wide variety of lactic acid bacteria strains are available in both traditional fermented foods and in supplement form (Kourkoutas et al. 2005). A major development has been foods fortified with probiotics and prebiotics, which enhance health-promoting micro-biota in the intestine when consumed in a regular diet (Hasler 2002).

In recent years, functional food and beverage consumption has increased tremendously and the global market has grown from \$33 billion in 2000 to \$176.7 billion in 2013, which accounts for 5% of the

Microbial Functional Foods and Nutraceuticals, First Edition. Edited by Vijai Kumar Gupta, Helen Treichel, Volha (Olga) Shapaval, Luiz Antonio de Oliveira, and Maria G. Tuohy.
© 2018 John Wiley & Sons Ltd. Published 2018 by John Wiley & Sons Ltd.

overall food market (Granato et al. 2010). In the global functional food market, 60–70% are probiotic foods (Holzapfel 2006; Kołozyn-Krajewskaa and Dolatowski 2012; Stanton et al. 2001).

The viability of probiotics is a key parameter which affects therapeutic benefits and varies as a function of the strain and health effect desired. In general, a minimum level of more than 10^6 viable probiotic bacteria per milliliter or gram of food product is accepted. Another issue to be addressed is the viability of probiotic micro-organisms in the product during its manufacture and shelf-life and its resistance in upper gastrointestinal tract conditions (Roy 2011). Strains which are being isolated from human or animal intestine are able to proliferate outside their natural environment and therefore exhibit reduced technological properties (Ross et al. 2005). Furthermore, probiotic micro-organisms must exhibit their resistance against the acidic environment of the stomach and bile secreted in the duodenum in order to achieve high survival rates in the gastrointestinal tract (Chassard et al. 2011).

Currently, commercial strains are largely selected for their technological properties, ruling out some strains with promising health properties. Many strains of intestinal origin are difficult to propagate and high survival is important for both economic reasons and health effects. In addition, more efficient technologies could lead to greater product efficacy and strain diversification with the development of technologically unsuitable strains. Some fermentation technologies, which include fed-batch, continuous fermentation, membrane bioreactors, and immobilized cell technology, are considered as suitable for proliferation of probiotic bacteria. They are designed to produce increased cell yield and productivity and to decrease the downstream processing capacity required to harvest the biomass. The limitations of these technologies include operational difficulties in industrial conditions due to the susceptibility of continuous cultures to contamination; subjecting sensitive cells to low nutrient concentration, oxygen, and osmotic and mechanical stresses in membrane bioreactors; and changes in the growth, morphology, and physiology of bacterial cells in immobilized reactors (Lacroix and Yildirim 2007).

This chapter discusses the selection criteria for an assortment of probiotic strains and the latest developments in fermentation technologies for producing probiotic bacteria as well as potential new approaches for enhancing the performance of these fastidious organisms during fermentation, downstream processing, and utilization in commercial products, and for improving functionality in the gut. Processes that include sublethal stress applications during cell production and new fermentation technologies, such as immobilized cell biofilm-type fermentations, are promising in this respect and could be used to profile cell physiology to optimize survival and functionality in the gut.

Assessment of Probiotic Functionality

Although there is evidence that probiotics can play a role in disease prevention and health promotion, safety considerations should be paramount. In addition, technological performance of probiotics has to be tested and must be suitable for large-scale applications. Safety of a probiotic strain can be assessed by three approaches: studies of the strain's intrinsic properties, studies of the pharmacokinetics of the strain (survival, activity in the intestine, dose–response relationships, fecal and mucosal recovery) and studies looking for interactions between the strain and the host. The selection criteria for potential probiotic strains are illustrated in Figure 10.1.

A probiotic micro-organism should be able to withstand a range of harsh physicochemical factors during transit through the stomach and small intestine to exhibit its beneficial effects and it should be able to persist in the gut, as outlined in a joint report by the FAO and WHO in 2002 (FAO/WHO 2002).

Great viability and activity of probiotics are considered mandatory for optimal functionality. However, quite a few studies have shown

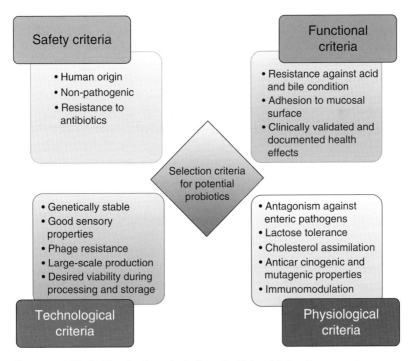

Figure 10.1 Desirable selection criteria for potential probiotic micro-organisms.

that non-viable probiotics can have positive effects such as immune modulation and carcinogen binding. Therefore, for certain probiotic strains it might be enough that they grow well during initial production processes (to obtain sufficient cell numbers in the product) but they do not necessarily need to maintain good viability during storage (Mattila-Sandholm et al. 2002).

The food industry has to fulfill consumer demand to succeed in increasing the consumption of functional probiotic products. All probiotic foods should be safe and have good sensory properties. They should also include specific probiotic strains at a suitable level during storage. They must first be able to be manufactured under industrial conditions, then survive and retain their functionality during storage as frozen or freeze-dried cultures, and also in the food products into which they are finally formulated. Some commercially available probiotic products are listed in Table 10.1. They must be able to be incorporated into foods without producing off-flavors or textures. The packaging materials used and the storage conditions are vital for the quality of products containing probiotic bacteria (Saarela et al. 2000).

Assessment of probiotic strains must focus on microbial activity following passage through the stomach as probiotics need to be functionally active in the gut. This could include physiological profiling of probiotics, i.e., bar-coding, by "omics" technologies. Bar-coding of cells involves characterization of the viable and active physiological state of a putative probiotic bacterium. During the production process, this physiological state is compared to the physiological state of the same bacterium at the final stage of fermentation. Transcriptomics, proteomics, or metabolomics can be used to compare physiological states. It is suggested that by bar-coding, new fermentation conditions can be developed that better mimic the active physiological state of a probiotic bacterium at its place of action (e.g., in various parts of the gastrointestinal tract) (Ledeboer et al. 1999).

Stress Factors Encountered by Probiotics

In order to achieve health benefits in the living system, some specific requirements regarding food products need be fulfilled. First, probiotics must have the ability to resist various stresses during the manufacturing process; next, they should retain their viability under very low storage temperatures. Several features influence the viability of probiotic micro-organisms in food products during production and storage periods, including pH, acidity, dissolved oxygen, redox potential, H_2O_2, bacteriocins, short chain fatty acids, flavoring agents, competitive

Table 10.1 Some important commercially used probiotic strains, manufacturers, and products.

Species	Strain	Manufacturer	Product
L. acidophilus	LA-1/LA-5 NCFM La1 DDS-1 SBT-2062 LA5	Chr. Hansen Rhodia Nestle Nebraska Cultures Snow Brand Milk Products Muller	TopOntbijt, Nu Trish, Biospega, Snow yoghurt + 2, Vitality, YoMi
L. delbrueckii subsp. bulgaricus	2038	Meiji	Yoghurt
L. casei immunitas		Danone	
L. casei	DN-114 001 Shirota	Danone Yakult	Yakult, DanActive, Actimel, Joie
L. plantarum	299v (DSM 9843) Lp01	Probi AB	
L. rhamnosus	GG GR-1 LB21 LC-705	Campina Urex Biotech Essum AB Valio	Vifit, Femdophilus, Bio Profit, Culturelle, Gefilac, RepHresh Pro-B
L. reuteri	SD2112/MM2 RC-14 ATCC 55730	Biogaia Chr. Hansen Biogai	Femdophilus, Stonyfield Farm yoghurts
L. paracasei	CRL 431	Chr. Hansen Nestlé, Nutricia	Chamyto, Fyos
L. fermentum	RC-14	Urex Biotech	Femdophilus, RepHresh Pro-B
L. helveticus	B02		
B. lactis	DR10™ Bb-12, Bb-02, LaftiTM, B94 DR10/ HOWARU	Parmalat Chr. Hansen DSM Danisco	Active-piu, Nu Trish, YoMi, Yo-plus yoghurt, TopOntbijt, Teddy
B. animalis	DN173 010	Danone	Activia yoghurt

(Continued)

Table 10.1 (Continued)

Species	Strain	Manufacturer	Product
B. bifidum	BF2	Cell Biotech Europe	
B. longum	BB536 SBT-2928 UCC 35624	Morinaga Snow Brand Milk Products UCCork	Bifidus yoghurt
B. infantis	35264 744 Immunitas	Procter and Gamble	Align
B. adolescentis	ATCC 15703, 94-BIM		

Source: Adapted from Muller et al. (2009), Krishnakumar and Gordon (2001), Playne et al. (2003), Holm (2003).

microbes, packaging materials and packaging conditions, rate and concentration of inoculation, fermentation conditions, microencapsulation, milk solid non-fat content, supplementation of milk with nutrients, incubation temperature and timing, storage temperature, carbonation, salt addition, sugar and sweeteners, cooling rate of the product and scale of production (Mortazavian et al. 2012).

There are two possible outcomes when probiotic micro-organisms respond to stress (Yousef and Juneja 2003):

- initiation of a sequence of proteins that repair damage, maintain the cell, or exclude the stress agent
- transient elevation in resistance or tolerance to harmful factors.

Probiotic bacteria may encounter various stress condition during transition to the gastrointestinal tract following ingestion. However, probiotics, like other bacteria, have the ability to develop defense mechanisms to withstand these stressful environmental conditions. The following section deals with the various stress factors encountered by probiotics and the cellular mechanisms used to overcome them.

Oxidative Stress (OxS)

Probiotic micro-organisms can be exposed to different levels of oxygen. Strict anaerobic conditions cannot be easily maintained during production, downstream processing, and storage of probiotic bacteria.

Probiotic cultures, generally isolated from the human gut, can be classified according to their oxygen tolerance as microaerophilic (e.g., *Lactobacillus acidophilus*) and strictly anaerobic (most of the *Bifidobacterium* spp.). However, oxygen sensitivity is species and strain dependent (Talwalkar and Kailasapathy 2004).

Oxidative stress may be defined as the disproportion between cellular production of oxidants (reactive oxygen species (ROS)) and antioxidative capacity and is known to cause acinar injury in the early course of acute pancreatitis (Rau et al. 2000). In living micro-organisms the major ROS are superoxide radicals, lipid peroxyl radicals, hydroxyl radicals and non-radical H_2O_2, and reactive nitrogen species (RNS) are nitric oxide and non-radical peroxynitrite. These hydroxyl radicals are rapidly generated through the Fenton cycle, in which free iron reacts with H_2O_2. Most of the mentioned reactive species (RS) come from endogenous sources as by-products of normal essential metabolic processes, while exogenous sources involve exposure to cigarette smoke, environmental pollutants, radiation, drugs, bacterial infections, excess of food iron, unbalanced intestinal microflora, etc. Several diseases are associated with the toxic effects of the transition metals (iron, copper, cadmium) (Halliwell and Gutteridge 1999).

A special probiotic, LfME-3, has good potential in cardiovascular health management. LfME-3, an antioxidative, antiatherogenic and antimicrobial probiotic, decreases OxS levels in the human body. Foodstuffs enriched with this probiotic decrease oxidized LDL, increase HDL, modulate postprandial lipid profile and OxS, and decrease the level of 8-isoprostanes in urine (markers of systemic OxS) and the overall OxS load, indicating an atherogenic potential (Kullisaar et al. 2003, 2011; Mikelsaar 2009; Songisepp et al. 2005).

Several studies have shown that LfME-3 alleviates inflammation and OxS-associated shifts in gut, skin, and blood (Kaur et al. 2008; Kullisaar et al. 2003). This occurs via complicated cross-talk between probiotic and host body cells via the integrated influence of several factors of LfME-3, including having complete glutathione system, the expression of antioxidative enzymes, the production of CLA and NO, etc. (Kullisaar et al. 2010, 2011; Mikelsaar and Zilmer 2009).

Some mechanisms have been proposed to explain the tolerance of lactic acid bacteria to oxygen toxicity, including reduction of the intracellular redox potential, repair of oxidative damage by overexpression of chaperones, and protection of potential targets (e.g., proteins) from oxidation (van de Guchte et al. 2002). The development of molecular tools and the increasing number of available genome sequences have permitted identification of proteins or genes involved in these mechanisms but the role of some of them remains unclear (Dubbs and Mongkolsuk 2007; Klijn et al. 2005; O'Connell-Motherway et al. 2000).

Acid Stress

Acid stress is the primary challenge faced by probiotics due to the acidic pH (>2) in the stomach. When cells are exposed to the low pH, proton accumulation inside the cell may alter the proton motive force (PMF) across the cell membrane. The changes in PMF cause cell membrane and structural damage, and acid stress also causes damage to nucleic acids and proteins (Collado et al. 2005). *Bifidobacterium* strains are less acid tolerant than *Lactobacillus* strains, and this is reflected by their poor survival in human gastric juice. Bifidobacteria are more resistant to acetate (acetic acid) rather than Hcl in milk-based media, since acetate is the major by-product after *Bifidobacterium* fermentation (Borgstrom et al. 1957). *Bacillus clausii* spore supplementation of antibiotic therapies has demonstrated efficacy, owing to the fact that the spores are resistant to acid stress and drugs (Nista et al. 2004).

The lack of standard procedures for evaluating tolerance to gastrointestinal conditions makes comparison difficult. In general, tolerance to gastrointestinal conditions is low; as a consequence, several methodologies, including those based on stress adaptation mechanisms of probiotic bacteria, are being considered as possible mechanisms to improve their acid and bile tolerance (Collado 2005; Noriega et al. 2004).

Bile Stress

Bile plays a basic role in the defense mechanisms of the intestinal tract and its inhibitory effect is determined primarily by bile salt concentration. Therefore, bile tolerance is considered as an important characteristic of *Lactobacillus* strains, which enables them to survive, grow, and exert their action in gastrointestinal transit (Argyri et al. 2013). According to Sanders et al. (1996), *Lactobacillus* strains which can to grow and metabolize in normal physical bile concentration also survive in the human gastrointestinal tract. In addition, Gänzle et al. (1999) and du Toit et al. (1998) proposed that some components of food could protect and promote the resistance of strains to bile salt. Gilliland et al. (1984) reported that a culture of *L. acidophilus* with high bile tolerance was significantly better than a strain with low bile tolerance for increasing the number of facultative lactobacilli in the upper part of the small intestines of calves.

Tanaka et al. (1999) suggested that the presence of bile salt hydrolase (BSH) had a selective advantage for the probiotic micro-organism in bile salt-rich environments. BSH activity benefits them by increasing resistance to conjugated bile salts and survival in the gastrointestinal tract for colonization (Jones et al. 2008; Ridlon et al. 2006).

Heat Stress

During the production process, the probiotic micro-organism may be exposed to heat. Several researchers have studied the effect of heat on physiological functions and the induction of a stress response by probiotics. The heat shock response is a complex process, involving various heat shock proteins (HSPs) with different roles in physiology, including chaperone activity, ribosomal stability, temperature sensing, and regulation of ribosome functions (de Angelis and Gobbetti 2004). The heat shock response by probiotics differs from strain to strain. When exposed to higher temperatures, the proteins inside the cell are denatured and start aggregating. Several studies have revealed that the heat stress response includes the destabilization of macromolecules as ribosomes and RNA as well as alterations of membrane fluidity (Earnshaw et al. 1995; Teixeira et al. 1997).

The transient induction of several heat shock proteins (such as DnaK, GroEL and GroES) in various probiotics in response to heat stress has been demonstrated (Earnshaw et al. 1995; Serrazanetti et al. 2013; Teixeira et al. 1997). The HSPs were dramatically upregulated when the cell was exposed to heat stress, induced by heat shock factor (HSF). HSPs are present in almost all living organisms, from bacteria to humans. Their major role is to bind proteins in a transient non-covalent manner during heat shock, prevent premature folding, and stabilize the proteins. The increased heat resistance was achieved by pre-exposure of cells to elevated temperatures, allowing them to adapt to these temperatures. Heat acclimation triggers the expression of HSP genes mediated by HSF and enhances tolerance to noxious high temperatures, a phenomenon known as acquired thermotolerance (Gardiner et al. 2000; Prasad et al. 2003).

Cold Stress

Probiotics bacteria can be exposed to extremely cold conditions during storage. Understanding the behavioral changes in extreme cold environments is important to increase cell viability. Probiotic micro-organisms have been subjected to various temperatures, sometimes below the optimal growth temperature. When bacteria are exposed to extreme cold conditions, several physiological changes may occur, including decreases in membrane fluidity and stabilization of secondary structures of RNA and DNA (leading to decreased transcription and translation), and decreased or no DNA replication. Similar to the heat shock response, several proteins, called cold-induced proteins (CIPs), were expressed. These play a critical role in adjusting membrane fluidity, DNA supercoiling and transcription and translation. Several researchers have demonstrated the functional contribution of CIPs in *Lactococcus lactis* and *L. plantarum* (Derzelle et al. 2000; Kim et al. 1998). Kim et al. (1998) studied various lactic acid bacteria (LAB) strains to determine the cold shock response and found that frozen or freeze-dried bacterial cells have improved cryotolerance, due to the overexpression of several CIPs. Furthermore, the response of bacterial strains to freezing may depend on exposure time and the induction of cryotolerance appears to be dependent on the growth phase in which the cold shock took place (Serrazanetti et al. 2013).

Osmotic Stress

From the production to storage process, probiotic strains encounter osmotic stresses. During storage processes, like freezing, freeze-drying, and spray drying, the osmolality of the environment increases as a result of cryoconcentration of solutes or dehydration, leading to excessive passage of water from the cell to the extracellular environment (Poolman et al. 2002). Addition of various amounts of sugars and salts also alters the osmolality of probiotic cells. Researchers found that hyperosmotic conditions occurring due to high sugar levels do very little harm to the cell and are only transient; the major reason for this is the cell's ability to balance the extra- and intracellular lactose and sucrose concentration. Probiotic strains have to adapt to these environmental changes to survive and they tackle this by taking up or synthesizing compatible solutions (organic compounds) when they sense osmotic stress (van de Guchte et al. 2002). These compatible solutes are defined as osmoprotectants, which act as a stabilizing agent for various enzymes, thereby offering protection against other stress conditions (high temperature, freezing, and drying). Water loss from the cell due to the high external osmolality was prevented by the intracellular accumulation of compatible solutes, which maintains turgor (Piuri et al. 2003). The compatible solutes synthesized by various LAB strains include carnitine, betaine, proline, and 3-methylbutanoic acid (Serrazanetti et al. 2013). LAB can synthesize these compatible solutes, at very low levels, which are then taken up from the medium by specific transporters (Poolman et al. 2002).

Competitive Stress

Almost all food fermentation is carried out by various groups of micro-organisms including LAB, yeasts, and several non-pathogenic filamentous fungi. This mixed culture fermentation has several advantages when compared with single strain fermentation, including reduced production cost and time, exchange of metabolites and growth factors or inhibiting compounds (Sieuwerts et al. 2008). The co-culture of several closely related bacterial species prevents the growth of other unwanted micro-organisms. However, this may create several types of stress for the micro-organisms including:

- nutrient competition
- generation of unfavorable environments
- competition for attachment/adhesion sites (Serrazanetti et al. 2013).

Due to the heterogeneous physicochemical composition in food fermentation, micro-organisms start to compete for available nutrients, each strain utilizing the nutrients via multiple mechanisms (Sieuwerts et al. 2008). In most cases, the concentration of available carbon sources is very high at the initial stage of fermentation, which leads to competition and rapid utilization of the carbon source which produces a higher biomass. Similarly, in dairy fermentation the available nitrogen source is very limited, so the fermentative micro-organisms compete for the utilization of free amino acids and small peptides. During fermentation, the microbes produce diffusible chemical compounds to communicate with other bacteria. Apart from the competition for available nutrients, micro-organisms also compete for adhesion sites.

Overall, in recent years many researchers have tried to determine the mechanisms behind the stress response by micro-organisms. This basic understanding about the bacterial stress response is crucial for the identification of the most efficient probiotic strains to improve food fermentation. The overall stress conditions faced by probiotics from production to consumption are illustrated in Figure 10.2.

Current Technologies Used to Increase Cell Viability

Industrial demand for new technologies that enable high cell yields at large scale and ensure probiotic stability in food remains strong, because many strains of intestinal origin are difficult to propagate and high survival is important for both economic reasons and health effects. In addition, more efficient technologies could lead to greater product efficacy and strain diversification with the development of technologically unsuitable strains into products.

Industrial processes for food culture production, including probiotics, almost exclusively use conventional batch fermentation with suspended cells. Several approaches have been investigated to enhance cell viability during downstream processing, storage, and eventually digestion. These include the application of sublethal stresses during fermentation, the addition of protectants, including compatible solutes (e.g., betaine which has been extensively studied), and cell protection by microencapsulation.

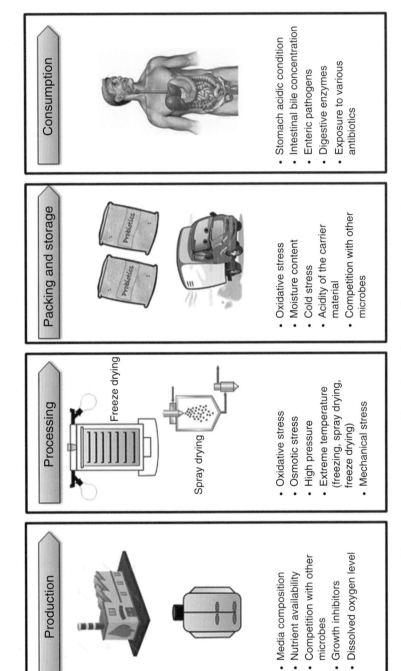

Figure 10.2 Various stressful environments encountered by probiotics from production to consumption.

The ability of micro-organisms to grow and survive depends largely on their capacity to adapt to changing environments. Adaptation to adverse environments is usually associated with the induction of a large number of genes, the synthesis of stress response proteins, and the development of cross-resistance to various stresses.

Genetic Modification of the Strain

Advancements in genomics and proteomics have made both lactobacilli and bifidobacteria more amenable to genetic manipulation by increasing their viability when exposed to stressful environments. Genetically modified (GM) probiotics are mainly concerned with the improved survival and persistence of probiotics within the human and animal gut, tolerance of packing and storage conditions of food, and the delivery of therapeutics by live probiotics. With the identification of genetic elements that confer tolerance to increased osmolarity, bile salt and reduced water activity, the improvement of probiotics for survival has become a reality. With the expression of various molecules such as antigens, enzymes, and molecules of immunological importance within probiotic microbes, the use of GM probiotics in the field of therapeutics looks promising.

Basically, there are two different approaches that can be followed:

- modify the expression/production of genes already present in the micro-organism (homologous expression)
- introduce genes from other microbial species (heterologous expression) (Desmond et al. 2004).

Trials have also been undertaken to produce mutants which overexpress protective stress metabolites, i.e., heat shock proteins GroEL and GroES, thereby improving tolerance to abusive heat treatments. Increasing the available GroES and GroEL concentration prior to the stresses associated with freezing, lyophilization, or spray-drying may offer additional protection against protein denaturation and produce a more viable and physiologically active product (Walker et al. 1999). For example, overexpression of the HSP chaperones GroES and GroEL in *Lactobacillus paracasei* NFBC 338 improved its ability to withstand thermal stress, but it did not perform as well as the heat-adapted parent culture. The heat shock response involves a consort of proteins including DnaK, DnaJ, and GrpE, tagatose-6-phosphate-aldolase; glyceraldehyde-3-phosphate dehydrogenase; and triose phosphate isomerase (Desmond et al. 2004).

The effect of homologous excess production of each of the three HSPs of *Lactobacillus plantarum* WCFS1 has been studied to determine the stress tolerance of transformed cells. A change from the optimum temperature (i.e., 28 °C) to 12 °C, 37 °C, and 40 °C lead to a decrease in cell viability ranging between 2.0 and 4.0 log CFU/mL for the untransformed cells while the three HSP-overproducing strains were not affected (Fiocco et al. 2007). The auxotrophic mutant strains of *Saccharomyces boulardii* showed functional recombinant protein even after being recovered from gastrointestinal immune tissues in mice. A UV mutagenesis approach was applied to generate three uracil auxotrophic *S. boulardii* mutants that show a low rate of reversion to wild-type growth. These mutants can express recombinant protein and are resistant *in vitro* to low pH, bile acid salts, and anaerobic conditions (Hudson et al. 2014).

However, within the latest legislative situation, i.e., Novel Food Regulations, which sets a high standard of safety, as well as the general consumer rejection of the presence of genetically modified micro-organisms in food, this approach seems unlikely to be applicable in future (Rastall and Maitin 2002).

Enhancing Stress Tolerance of Probiotic Strains

Intrinsic stress tolerance seems to be a critical factor in the overall resistance to manufacture and storage of probiotic products. Different approaches can be used to this end, which can be divided into three main categories: selection of naturally occurring strains, stress adaptation of naturally occurring strains, and genetic modification of the strains. The first two approaches use already existing diversity and genetic potential, whilst the third would imply genetic manipulation leading to a genetically modified organism (GMO) (Sanchez et al. 2012).

Physiological Adaptation and General Stress Response

A brief pretreatment of bacteria with stress can lead to physiological adaptation to forthcoming more severe stress caused by the same stressor. The storage stability of *Lactobacillus rhamnosus* HN001 was substantially increased after a sublethal stress such as heat or osmotic stress. The largest increase in the storage stability of *L. rhamnosus* HN001 was observed after sublethal heat stress during the stationary phase of growth (Prasad et al. 2003). Saarela et al. (2004) reported that sublethal treatments of acid and heat treatments of stationary phase probiotic cells (*Lactobacilli* strains) increased their viability and survival during lethal treatments and their ability to adapt at fermentor-scale production of probiotic cultures. Hence, strain-specific and sublethal treatments of specific growth phase probiotic cultures can be utilized in the production of probiotic cultures with improved viability. The methods of pretreatment, heat adaptation and other factors that affect such investigation are very important for further studies on the viability of probiotic bacteria during heat processing.

Cell responses to sublethal stresses are directly proportional to the strain selection and applied stresses. Therefore, in the absence of general mechanisms for sublethal stress adaptation, fermentation conditions must be determined for each strain. This process is labor intensive and costly. In addition, sublethal stresses can lead to decreases in cell activity and yield and eventually to changes in the functionality of probiotic cells in the intestinal tract. Clearly, there is a need to develop a more in-depth understanding of stress response mechanisms of probiotics (Lacroix and Yildirim 2007).

Methods to Increase Survival and Viability of Probiotics

Researchers have long been encouraged to find novel, proficient methods of improving the viability of probiotics in food products (especially fermented types). The most recent developments focus on fermentation technologies for producing probiotic bacteria; new approaches for enhancing the performance of these fastidious organisms during fermentation, downstream processing, and utilization in commercial products; and improving functionality in the gut (Mortazavian et al. 2007).

Fermentation Technology

The large-scale preparation of bacterial strains is difficult, time-consuming, and expensive. In addition, most of the probiotic strains show poor growth rates in milk-based media. Therefore, it can be difficult to achieve high numbers of viable cells after fermentation (Ross et al. 2005). The media composition and type of substrate used for fermentation can have a great impact on probiotic strain viability during production and downstream processing (Lacroix and Yildirim 2007). The overall fermentation process of probiotic starter culture is depicted in Figure 10.3.

Optimization of Growth Medium

The foremost demerit of probiotic strains is the rapid growth and acidification rate of typical starter cultures. Therefore, they are added to dairy products after fermentation (Champagne et al. 2005). The cultivation conditions are directly proportional to growth, culture stability, and activity as well as drying and subsequent storage. The stability and function of probiotics can be enhanced by changing the culture conditions. Usually probiotic cells are produced by large-scale fermentation. It is important to design growth media and change the fermentation technology to improve biomass yield and enhance cell stability. MRS broth (de Man et al. 1960) is the most widely used medium for cultivation of LAB or bifidobacteria. Higher cell counts of

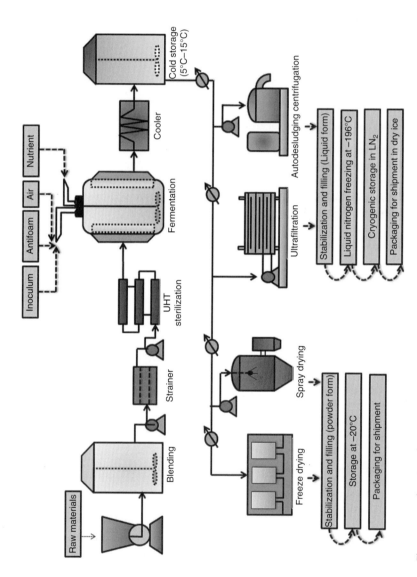

Figure 10.3 Flowchart for the production process of probiotic starter cultures.

bifidobacterial strains were produced in ultra-filtered skim milk with various protein concentrations or inclusion of milk with nitrogenous substrates such as whey and casein fractions from human or cow milk when compared to growth in skim milk only (Ventling and Mistry 1993). Soy milk or animal product-free vegetable medium (based on soy peptone, glucose, and yeast extract) resulted in lower yields or lower viability of bifidobacteria during storage because of the low buffer capacity of the vegetable medium (Heenan et al. 2002). Redox-reducing compounds like cysteine were used in growth media of bifidobacteria to enhance growth (Doleyres and Lacroix 2005). In some cases, the disulfide bonds were reduced and hence these supplements may lose their growth-enhancing properties (Ibrahim et al. 1994).

Types of Fermentation Involved in Probiotic Production

Membrane Bioreactor

Some researchers have reported on probiotic cell production in membrane bioreactors. They have supported high cell yields and volumetric productivity. Taniguchi et al. (1987) reported seven-fold higher *Bifidobacterium longum* biomass levels in a membrane bioreactor than that attained in free cell batch fermentors. Similarly, Corre et al. (1992) reported that there was a high cell yield and a 15-fold enhancement in volumetric productivity compared to free cell batch cultures of *Bifidobacterium bifidum*.

Immobilized Cell (IC) Technology in Continuous Cultures

Although the fermentation technologies, such as continuous culture and immobilized cell systems, are not common in food industries, they have the potential to improve the role of fastidious probiotic cells. Hence, the physiology and functionality of ingested probiotics can be improved with the implementation of these technologies. It helps to extend the range of commercially available probiotics (Lacroix and Yildirim 2007).

In continuous cultures, bacterial cells grow in a fermentor with continuous supplementation of fresh medium and removal of fermented broth at a given dilution rate. Metabolism, growth rate, and gene expression of bacteria of continuous culture can be monitored under constant conditions during long time periods (Hoskisson and Hobbs 2005). In continuous culture of *B. longum* SH2, there was increased volumetric biomass productivity when compared to traditional batch cultures (Kim et al. 2003). In one study, a *B. longum* NCC2705 strain grown in prolonged chemostat culture for 200 hours resulted in higher metabolic and transcriptomic stability throughout the whole culture time which suggests applying these cells as a novel method to screen for sublethal treatments (Doleyres et al. 2002). Beyond this application, contamination is considered to be the main drawback. The constant feeding of new medium must be supplemented at dilution rates lower than the maximum specific growth rate to avoid wash-out of active biomass. There is a chance of competitive contaminant strain due to low dilution rate.

Continuous Fermentation

In addition, cells produced with continuous IC cultures are in the exponential or early stationary growth phase, and exhibit both high viability and metabolic activity compared with starving cells produced with conventional batch cultures. Immobilized cell technology combined with long-term continuous culture can be used to efficiently produce, in a one-step process, cells with improved tolerance to environmental stresses, without the need for preconditioning treatments which are sometimes used for better survival of probiotics during production and use in functional foods, but eventually result in reduced cell activity and yield (Desmond et al. 2004). A change in the metabolic pathway from homofermentative to heterofermentative has also been observed during continuous cultures with immobilized cells of *Lactobacillus*. Probiotic strains produced with continuous IC fermentations could better as they pass through the gastrointestinal tract and could be used to manufacture products with new or enhanced health benefits, as well as new food or pharmaceutical bio-ingredients (Benech et al. 2002; Bouksaim et al. 2000).

Stabilization of Probiotics

Microencapsulation

Microencapsulation is a fascinating field of biopharmacy that has originated and developed rapidly in the past decade. Several micro-organisms have been immobilized within semi-permeable and biocompatible materials that modulate the delivery of cells using this technique (Vidhyalakshmi et al. 2009). Microencapsulation is a technology for packaging solids, liquids or gaseous materials in tiny, sealed capsules that can release their contents at controlled rates under specific conditions (Shahidi and Han 1993). Microencapsulation of probiotics in hydrocolloid beads has been found to enhance their viability and activity in food products and the intestinal tract by entrapping the cells inside a bead matrix, thus separating them from harsh environmental conditions, as well as protecting them against bacteriophages (Krasaekoopt et al. 2003).

A microcapsule consists of a semi-permeable, spherical, thin, strong membrane surrounding a solid or liquid core, with a diameter varying from a few microns to 1 mm (Anal and Singh 2007). Hence, the bacterial cells are retained within the microcapsule (Jankowski et al. 1997). The most widely used materials in microencapsulation of probiotic bacteria include polysaccharides originated from seaweed (alginate, κ-carrageenan), other plants (starch and its derivatives, gum arabic), bacteria (gellan, xanthan), and animal proteins (milk, gelatin). The chief techniques for microencapsulation of probiotics are extrusion techniques, spray drying, and spray cooling (Chavarri et al. 2012).

The microencapsulation techniques applied to probiotic cells for use in fermented milk products or biomass production can be classified into two groups, based on the method used to form the beads: extrusion (droplet method) and emulsion or two-phase system. Both extrusion and emulsion techniques show high survival of probiotic bacteria by up to 80–95% (Kebary et al. 1998). Lack of oxygen in the interior of the capsule may result in cell death, but conversely it may help anaerobic bacteria such as bifidobacteria (Kailasapathy 2002).

Using an electric potential, the electrostatic bead generator pulls the droplets from a needle tip. The electrostatic potential is placed between the needle feeding the mixture of alginate and bacterial cell solution and the gelling bath. Although viability of LAB could be enhanced by microencapsulation, the process may change the heat susceptibility and health benefits of the cell since the cells undergo heat treatment and dehydration during microencapsulation (Ananta et al. 2005).

Merits of Microencapsulation

Entrapment of probiotic micro-organisms in a biodegradable polymer matrix has various advantages. Once entrapped in matrix beads, the cells are easier to handle than a suspension or slurry. The number of cells in each bead can be quantified allowing dosages to be readily controlled. Cryo- and osmoprotective components can be incorporated in the polymer matrix, increasing survival of cells during processing and storage. Finally, once the matrix beads have been dried, a surface coating can be applied. This outer layer can be used to alter the esthetic and sensory properties of the product and may also be functional, providing an extra level of protection for the cells. In addition, the coating layer can have preferable dissolution properties which permit overdue release of the cells or release upon, for example, change in pH. Cells produced during continuous immobilized cell cultures show changes in cell membranes which may induce formation of cell aggregates and tolerance to bile salts and aminoglycosidic antibiotics. Cell immobilization in combination with continuous culture can be used to produce probiotic micro-organisms with improved stress tolerance.

Voluminous formation of cell aggregates has already been reported during continuous culture with immobilized lactobacilli (Bergmaier et al.

2005), whereas no cell aggregates were reported for the *B. longum* NCC2705 grown for more than 8 days in continuous mode with free cells in MRS.

A parallel decrease of the ratio of unsaturated to saturated fatty acids of cell membrane from IC cultures was reported for a bile-adapted mutant strain of *B. animalis* compared to its mother strain (Ruiz et al. 2007). These changes might be involved in restricted diffusion of bile salts into the cytoplasm (Begley et al. 2005). Doleyres et al. (2004) reported a progressive increase of survival *of B. longum* ATCC 15707 in simulated intestinal conditions and tolerance to nisin Z with fermentation time induced by cell immobilization in continuous co-culture with *Lactobacillus lactis* MD. The elevation of bile tolerance with time correlated with increased tolerance to nisin Z, which targets bacterial cell membranes.

Spray Drying

In general, dairy starter and probiotic cultures are preserved and distributed in frozen or dried form. High costs of storage, shipping, and energy are the primary reason to preserve probiotic cultures in dried forms (Johnson and Etzel 1995). Bacterial cells require water activity (aw) of about 0.98 in the resultant matrix for their growth and survival. Hence, the physical removal of water to convert the bacterial cells into a dried form is a risky process. It is necessary to maintain either high aw for metabolic activity or low aw to preserve the bacteria in live state, so that they can survive in a dormant state in powders (Paul et al. 1993).

Freeze or spray drying is the predominant method used to dry the bacterial suspension. The common method used to formulate starter and probiotic cultures is freeze drying. Its major demerit is that it is an expensive process with low yields. But in the case of inexpensive spray drying, higher production rates result (Zamora et al. 2006). Spray drying is a well-organized tool in food industries for the production of milk powders and instant coffee. Although there are several restrictive process conditions for micro-organisms (inlet reaching $\geq 180\,°C$), the rapidity of drying combined with the ability to dry large amounts of bacterial cultures has caught the attention of research and industry (Meng et al. 2008; Zamora et al. 2006).

Freeze Drying

Freeze drying is a commonly used method, which provides higher survival rates compared with spray drying. Generally, freeze drying is achieved by three important steps: freezing, primary and secondary drying. To increase the survival ratio of bacterial cultures, they are typically frozen at −196 °C in liquid nitrogen. After freezing, the frozen samples are sublimated with ice under high vacuum conditions to complete the primary freezing. In this step, high temperature under pressure causes the phase transition from solid to gas. After the primary drying step, almost 95% of the water content in the sample is removed. However, secondary drying is also important to remove the remaining hydrogen-bound water molecules to achieve a final water content below 4%, thus improving survival rates and long-term storage efficiency and preventing spoilage (Santivarangkna et al. 2007).

Freeze drying is an expensive method with a low yield. When bacterial cells are exposed to extremely low temperatures (i.e., freezing), osmotic pressure across the membrane is increased due to formation of extracellular ice and the cells dehydrate until an eutectic point is reached (Fowler and Toner 2005). During the freeze-drying process, the bacteria face various stresses which damage the cell; generally, longer rod-shaped lactobacilli are more susceptible to damage than small round enterococci, because of their larger surface area (Fonseca et al. 2000). The cell membrane lipids are more sensitive and easily damaged during freezing, and destabilization of nucleic acids also limits several important growth functions, including replication of DNA, transcription, and translation (van de Guchte et al. 2002).

To reduce these problems during freeze drying, several approaches have been developed, including the addition of protectants, such as trehalose, betaine, adonitol, sucrose, skim milk powder, lactose, commercial cryoprotectants (e.g., Unipectine, Satialgine), and several commercially available polymers also increase the stability of bacteria after freezing (Burns et al. 2008).

Fluidized Bed and Vacuum Drying

Fluidized bed dryers use an upward-moving flow of heated air and mechanical shaking to create a fluidized effect in a solid product. Particles are freely suspended in air and are dried by rapid heat exchange (Santivarangkna et al. 2007). This method is more economical than others, including spray drying. In this method, the length of bacterial exposure to heat is easily controlled, which reduces the risk of heat inactivation. Using the fluidized bed drying method, several yeast strains have been successfully dried and this method is also employed for LAB (Bayrock and Ingledew 1997). It is suitable for granular particles so bacterial solutions are encapsulated with alginate, potato starch, skim milk or casein before being introduced into the drier. The desired moisture content is achieved by drying the granulated particles under gentle temperatures.

The vacuum drying method is used to dry heat-sensitive compounds; here the water molecules in the samples are removed at low temperature under vacuum conditions. This method is similar to freeze drying but the temperature is maintained as low as $-2\,°C$. Due to the high vacuum conditions, oxidation reactions are reduced in the vacuum drying method, which is more suitable for oxygen-sensitive bacteria, but this method is not extensively used for drying LAB strains. It has several limitations including long drying time (10–100 hours) compared with the spray drying or fluidized bed method (Santivarangkna et al. 2007). However, these problems can be overcome by modifications such as using continuous vacuum drying. Continuous vacuum drying is used in large-scale industries for drying enzymes, food additives, and other pharmaceutical products (Hayashi et al. 1983).

Enumeration of Viable Probiotic Cells

Minimum levels of viable cells in probiotic products are essential so quantification methods are required to detect viable probiotic strains in functional food and dairy products. The quality of probiotic products can be evaluated using fast and reliable methods to enumerate probiotic cells in mono- and mixed cultures. Plate counting on selective media is the most commonly used method but cell clumping, inhibition by adjacent cells, and time consumption are limitations (Bergmaier et al. 2005). The lack of a standard medium to discriminate bifidobacteria from other LAB has prompted the search for alternative, non-culture-based technologies for cell enumeration (Masco et al. 2005). Fluorescent stains alone or combined with enzymatic assays help to distinguish between live, metabolically active, injured, dormant, viable but not cultivable, and dead bacterial cells and also enable quantification of viable bacteria (Alakomi et al. 2005). Molecular tools such as fluorescence *in situ* hybridization (FISH) and flow cytometry are applied to estimate viable cells in probiotic products (Maukonen et al. 2006). Flow cytometry can distinguish between subpopulations of stressed and unstressed bifidobacteria (Ben Amor et al. 2002). However, single cell suspension is required for culture-based and fluorescent methods because cells in clusters or chains may cause distortion (Bibiloni et al. 2001).

Several other methods exist to detect or quantify bifidobacteria in fecal samples and pharmaceutical or probiotic products. DNA fingerprinting approaches such as denaturing gradient gel electrophoresis (DGGE) and pulsed-field gel electrophoresis (PFGE) are used to detect bifidobacteria. Quantitative approaches include the use of quantitative real-time PCR (qRT-PCR) (Masco et al. 2007; Matsuda et al. 2007). Since DNA and rRNA are very stable and can be used as viability markers, these methods cannot provide information regarding metabolic activity or viability of microbial cells. mRNA with short half-life times is a useful viable cell marker (Hellyer et al. 1999). qRT-PCR has been used in medical applications to enumerate viable pathogenic micro-organisms by quantifying mRNA (Birmingham et al. 2008) but no such method with mRNA has been applied to probiotic bacteria. Increasing counts of accessible genome sequences of food-related micro-organisms and upcoming microarray technology will extend our knowledge regarding gene expression profiles of micro-organisms in various environments which will increase the number of target mRNAs to accurately quantify viable micro-organisms.

Conclusion

The ability of probiotic micro-organisms to survive the severe environments existing during processing and gastrointestinal transit has been a primary factor in their use. Undoubtedly, initiation of the probiotic stress response through preadaptation approaches may not always ensure the improved performance of a culture in compromising environments. Further research will be needed to discriminate the influence of specific stressors and their combined effect on cell growth and survival, particularly distinguishing between changes due to growth stage and the impact of specific stressors, given that most studies to date have not controlled variables.Therefore, a complete approach is required for incorporating the emerging food processing technologies, which may improve and maintain survival of probiotics during processing and storage, with recent knowledge on genotypes and expressed characteristics of probiotics. The benefits of the technology should benefit the consumer rather than supporting corporate profit. Genetically modified probiotics should directly benefit the consumer but studies which investigate the safety of engineered probiotics are vital. Hence, it is very important to understand the bacterial stress response which helps scientists to modify probiotic strains to obtain their maximum potential. To enhance the viability and health functionality of probiotics, novel cultivation technologies need to be devised with the ability to produce functional probiotics, which is not achieved by traditional fermentation methods. Microencapsulation of probiotics is a promising technology which can protect bacterial cells from various environmental stresses and is also useful for the targeted delivery of probiotics to desired regions.

References

Alakomi, H.L., Mättö, J., Virkajärvi, I. and Saarela, M. (2005) Application of a micro plate scale fluorochrome staining assay for the assessment of viability of probiotic preparations. *J. Microbiol. Methods*, 62: 25–35.

Anal, A.K. and Singh, H. (2007) Recent advances in microencapsulation of probiotics for industrial applications and targeted delivery. *Trends Food Sci. Technol.*, 18: 240–251.

Ananta, E., Volkert, M. and Knorr, D. (2005) Cellular injuries and storage stability of spray-dried *Lactobacillus rhamnosus* GG. *Int. Dairy J.*, 15(4): 399–409.

Argyri, A., Zoumpopoulou, G., Karatzas, K.A., Tsakalidou, E., Nychas, G.J.E. and Panagou, E.Z. (2013) Selection of potential probiotic lactic acid bacteria from fermented olives by in vitro tests. *Food Microbiol.*, 33: 282–291.

Bayrock, D. and Ingledew, W.M. (1997) Mechanism of viability loss during fluidized bed drying of baker's yeast. *Food Res. Int.*, 30: 417–425.

Begley, M., Gahan, C.G.M. and Hill, C. (2005) The interaction between bacteria and bile. *FEMS Microbiol. Rev.*, 29: 625–651.

Ben Amor, K., Breeuwer, P., Verbaarschot, P. et al. (2002) Multi parametric flow cytometry and cell sorting for the assessment of viable, injured, and dead *Bifidobacterium* cells during bile salt stress. *Appl. Environ. Microbiol.*, 68: 5209–5216.

Benech, R.O., Kheadr, E.E., Lacroix, C. and Fliss, I. (2002) Antibacterial activities of nisin Z encapsulated in liposomes or produced in situ by mixed culture during Cheddar cheese ripening. *Appl. Environ. Microbiol.*, 68: 5607–5619.

Bergmaier, D., Champagne, C.P. and Lacroix, C. (2005) Growth and exopolysaccharide production during free and immobilized cell chemostat culture of *Lactobacillus rhamnosus* RW-9595M. *J. Appl. Microbiol.*, 98: 272–284.

Bibiloni, R., Zavaglia, A.G. and de Antoni, G. (2001) Enzyme-based most probable number method for the enumeration of *Bifidobacterium*in dairy products. *J. Food Protect.*, 64: 2001–2006.

Birmingham, P., Helm, J.M., Manner, P.A. and Tuan, R.S. (2008) Simulated joint infection assessment by rapid detection of live bacteria with real-time reverse transcription polymerase chain reaction. *J. Bone Joint Surg.*, 90A: 602–608.

Borgstrom, B., Dahlqvist, A., Lundh, G. and Sjovall, J. (1957) Studies of intestinal digestion and absorption in the human. *J. Clin. Invest.*, 36: 1521–1536.

Bouksaim, M., Lacroix, C., Audet, P. and Simard, R.E. (2000) Effects of mixed starter composition on nisin Z production by *Lactococcus lactis* subsp. lactisbiovar. diacetylactis UL 719 during production and ripening of Gouda cheese. *Int. J. Food Microbiol.*, 59: 141–156.

Burns, P., Vinderola, G., Molinari, F. and Reinheimer, J. (2008) Suitability of whey and buttermilk for the growth and frozen storage of probiotic lactobacilli. *Int. J. Dairy Technol.*, 61: 156–164.

Champagne, C.P., Gardner, N.J. and Roy, D. (2005) Challenges in the addition of probiotic cultures to foods. *Crit. Rev. Food Sci. Nutr.*, 45: 61–84.

Chassard, C., Grattepanche, F. and Lacroix, C. (2011) Probiotics and health claims: challenges for tailoring their efficacy. In: Kneifel, W. and Salminen, S. (eds) *Probiotics and Health Claims*. Oxford: Wiley-Blackwell.

Chavarri, M., Maranon, I. and Villara, M.C. (2012) Encapsulation technology to protect probiotic bacteria. In: Rigobelo, E.C. (ed.) *Probiotics*. InTech, 501–540.

Collado, M.C., Hernandez, M. and Sanz, Y. (2005) Production of bacteriocin-like inhibitory compounds by human fecal *Bifidobacterium* strains. *J. Food Protect.*, 68: 1034–1040.

Corre, C., Modec, M.N. and Boyaval, P. (1992) Production of concentrated *Bifidobacterium bifidum*. *J. Chem. Technol. Biotechnol.*, 53: 189–194.

De Angelis, M. and Gobbetti, M. (2004) Environmental stress responses in *Lactobacillus*: a review. *Proteomics*, 4: 106–122.

De Man, J.D., Rogosa, M. and Sharpe, M.E. (1960) A medium for the cultivation of Lactobacilli. *J. Appl. Bacteriol.*, 23: 130–135.

Derzelle, S., Hallet, B., Francis, K.P., Ferain, T., Delcour, J. and Hols, P. (2000) Changes in cspL, cspP, and cspC mRNA abundance as a function of cold shock and growth phase in *Lactobacillus plantarum*. *J. Bacteriol.*, 182: 5105–5113.

Desmond, C., Fitzgerald, G.F., Stanton, C. and Ross, R.P. (2004) Improved stress tolerance of GroESL-overproducing *Lactococcus lactis* and probiotic *Lactobacillus paracasei* NFBC 338. *Appl. Environ. Microbiol.*, 70: 5929–5936.

Doleyres, Y. and Lacroix, C. (2005) Technologies with free and immobilised cells for probiotic bifidobacteria production and protection. *Int. Dairy J.*, 15: 973–988.

Doleyres, Y., Paquin, C., LeRoy, M. and Lacroix, C. (2002) *Bifidobacterium longum* ATCC 15707 cell production during free- and immobilized-cell cultures in MRS-whey permeate medium. *Appl. Microbiol. Biotechnol.*, 60: 168–173.

Doleyres, Y., Fliss, I. and Lacroix, C. (2004) Increased stress tolerance of *Bifidobacterium longum* and *Lactococcus lactis* produced during continuous mixed-strain immobilized-cell fermentation. *J. Appl. Microbiol.*, 97: 527–539.

Dubbs, J.M. and Mongkolsuk, S., (2007) Peroxiredoxins in bacterial antioxidant defense. *Subcell. Biochem.*, 44: 143–193.

Du Toit, M., Franz, C.M., Dicks, L.M. et al. (1998) Characterization and selection of probiotic lactobacilli for a preliminary minipig feeding trial and their effect on serum cholestrerol levels, faeces pH and faeces moisture content. *Int. J. Food Microbiol.*, 40: 93–104.

Earnshaw, R.G., Appleyard, J. and Hurst, R.M. (1995) Understanding physical inactivation processes: combined preservation opportunities using heat, ultrasound and pressure. *Int. J. Food Microbiol.*, 28: 197–219.

FAO/WHO (2002) *Guidelines for the Evaluation of Probiotics in Food.* Available at: www.scribd.com/doc/110123033/Guidelines-for-the-Evaluation-of-Probiotics-in-Food (accessed 2 May 2017).

Figueroa-Gonzalez, I., Quijano, G., Ramirez, G. and Cruz-Guerrero, A. (2011) Probiotics and prebiotics – perspectives and challenges. *J. Sci. Food Agric.*, 91: 1341–1348.

Fiocco, D., Capozzi, V., Goffin, P., Hols, P. and Spano, G. (2007) Improved adaptation to heat, cold, and solvent tolerance in Lactobacillus plantarum. *Appl. Microbiol. Biotechnol.*, 77: 909–915.

Fonseca, F., Beal, C. and Corrieu, G. (2000) Method of quantifying the loss of acidification activity of lactic acid starters during freezing and frozen storage. *J. Dairy Res.*, 67: 83–90.

Fowler, A. and Toner, M. (2005) Cryo-injury and bio-preservation. *Ann. NY Acad. Sci.*, 1066: 119–135.

Gänzle, M.G., Hertel, C., van der Vossen, J.M. and Hammes, W.P. (1999) Effect of bacteriocin-producing lactobacilli on the survival of *Escherichia coli* and *Listeria* in a dynamic model of the stomach and the small intestine. *Int. J. Food Microbiol.*, 48(1): 21–35.

Gardiner, G.E., O'Sullivan, E., Kelly, J. et al. (2000) Comparative survival rates of human-derived probiotic *Lactobacillus paracasei* and *L. salivarius* strains during heat treatment and spray drying. *Appl. Environ. Microbiol.*, 66: 2605–2612.

Gilliland, S.E., Staley, T.E. and Bush, L.J. (1984) Importance of bile tolerance of Lactobacillus acidophilus used as a dietary adjunct. *J. Dairy Sci.*, 67: 3045–3051.

Granato, D., Branco, G.F., Cruz, A.G., Faria, J.A.F. and Nazzaro, F. (2010) Functional foods and nondairy probiotic food development: trends, concepts and products. *Compr. Rev. Food Sci. F.*, 9: 292–302.

Halliwell, B. and Gutteridge, J.M.C. (1999) *Free Radicals in Biology and Medicine*, 3rd edn. New York: Oxford University Press, pp. 617–783.

Harzallah, D. and Belhadj, H. (2013) Lactic acid bacteria as probiotics: characteristics, selection criteria and role in immunomodulation of human gi muccosal barrier. Available at: http://cdn.intechopen.com/pdfs/42329/intech-lactic_acid_bacteria_as_probiotics_characteristics_selection_criteria_and_role_in_immunomodulation_of_human_gi_muccosal_barrier.pdf (accessed 2 May 2017).

Hasler, C.M. (2002) Functional foods: benefits, concerns and challenges – a position paper from the American Council on Science and Health. *J. Nutr.*, 132(12): 3772–3781.

Hayashi, H., Kumazawa, E., Saeki, Y. and Lahicka, Y. (1983) Continuous vacuum dryer for energy saving. *Drying Technol.*, 1: 275–284.

Heenan, C.N., Adams, M.C., Hosken, R.W. and Fleet, G.H. (2002) Growth medium for culturing probiotic bacteria for applications in vegetarian food products. *Food Sci. Technol.*, 35: 171–176.

Hellyer, T.J., DesJardin, L.E., Hehman, G.L., Cave, M.D. and Eisenach, K.D. (1999) Quantitative analysis of mRNA as a marker for viability of *Mycobacterium tuberculosis*. *J. Clin. Microbiol.*, 37: 290–295.

Holm, F. (2003) Gut health and diet: the benefits of probiotic and prebiotics on human health. *World Ingredients*, 2: 52–55.

Holzapfel, W.H. (2006) Introduction to prebiotics and probiotics. In: Goktepe, I., Juneja, V.K. and Ahmedna, M. (eds) *Probiotics in Food Safety and Human Health*. New York: CRC Press, pp. 1–35.

Hoskisson, P.A. and Hobbs, G. (2005) Continuous culture – making a comeback. *Microbiology*, 151: 3153–3159.

Hudson, L.E., Fasken, M.B., McDermott, C.D. et al. (2014) Functional heterologous protein expression by genetically engineered probiotic yeast *Saccharomyces boulardii*. *PLoS One*, 9(11): e112660.

Ibrahim, S.A. and Bezkorovainy, A. (1994) Growth promoting factors for *Bifidobacterium longum*. *J. Food Sci.*, 59: 189–191.

Jankowski, T., Zielinska, M. and Wysakowska, A. (1997) Encapsulation of lactic acid bacteria with alginate/starch capsules. *Biotechnol. Technol.*, 11: 31–34.

Johnson, J.A.C. and Etzel, M.R. (1995) Properties of *Lactobacillus helveticus* CNRZ-32 attenuated by spray drying, freeze drying, or freezing. *J. Dairy Sci.*, 78: 761–768.

Jones, B.V., Begley, M., Hill, C., Gahan, C.G.M. and Marchesi, J.R. (2008) Functional and comparative metagenomic analysis of bile salt hydrolase activity in the human gut microbiome. *Proc. Natl Acad. Sci. USA*, 105: 13580–13585.

Kailasapathy, K. (2002) Microencapsulation of probiotic bacteria: technology and potential applications. *Curr. Iss. Intest. Microbiol.*, 3: 39–48.

Kaur, S., Kullisaar, T., Mikelsaar, M. et al. (2008) Successful management of mild atopic dermatitis in adults with probiotics and emollients. *Cent. Eur. J. Med.*, 3: 215–220.

Kebary, K.M.K., Hussein, S.A. and Badawi, R.M. (1998) Improving viability of Bifidobacteria and their effect on frozen ice milk. *Egypt. J. Dairy Sci.*, 26(2): 319–337.

Kim, T.B., Song, S.H., Kang, S.C. and Oh, D.K. (2003) Quantitative comparison of lactose and glucose utilization in *Bifidobacterium longum* cultures. *Biotechnol. Progr.*, 19: 672–675.

Kim, W.S., Khunajakr, N. and Dunn. N.W. (1998) Effect of cold shock on protein synthesis and on cryotolerance of cells frozen for long periods in *Lactococcus lactis*. *Cryobiology*, 37: 86–91.

Klijn, A., Mercenier, A. and Arigoni, F. (2005) Lessons from the genomes of bifidobacteria. *FEMS Microbiol. Rev.*, 29: 491–509.

Kołozyn-Krajewskaa, D. and Dolatowski, Z.J. (2012) Probiotic meat products and human nutrition. *Process. Biochem.*, 47: 1761–1772.

Kourkoutas Y., Xolias V., Kallis M., Bezirtzoglou E. and Kanellaki M.(2005) *Lactobacillus casei* cell immobilization on fruit pieces for probiotic additive, fermented milk and lactic acid production. *Proc. Biochem.*, 40(1): 411–416.

Krasaekoopt, W., Bhandari, B. and Deeth, H.F. (2003) Evaluation of encapsulation techniques of probiotics for yoghurt. *Int. Dairy J.*, 13: 3–13.

Krishnakumar, V. and Gordon, I. R. (2001) Probiotics: challenges and opportunities. *Dairy Ind. Int.*, 66: 38–40.

Kullisaar, T., Songisepp, E., Mikelsaar, M., Zilmer, K., Vihalemm, T. and Zilmer, M. (2003) Antioxidative probiotic fermented goats' milk decreases oxidative stress-mediated atherogenicity in human subjects, *Br. J. Nutr.*, 90: 449–456.

Kullisaar, T., Songisepp, E., Aunapuu, M. et al. (2010) Complete glutathione system in probiotic *L. fermentum* ME-3. *Appl. Biochem. Microbiol.*, 46: 527–531.

Kullisaar, T., Shepetova, J., Zilmer, K. et al. (2011) An antioxidant probiotic reduces postprandial lipemia and oxidative stress. *Cent. Eur. J. Biol.*, 6: 32–40.

Lacroix, C. and Yildirim, S. (2007) Fermentation technologies for the production of probiotics with high viability and functionality. *Curr. Opin. Biotechnol.*, 18: 176–183.

Ledeboer, A.M., Nauta, A., Sikkema, J. and Laulund, E. (1999) Technological aspects of making live, probiotic-containing gut health foods. In: Halliwell, B. and Gutteridge, J.M.C. (eds) *Free Radicals in Biology and Medicine*. New York: Oxford University Press, pp. 1–7.

Masco, L., Huys, G., de Brandt, E., Temmerman, R. and Swings, J. (2005) Culture-dependent and culture-independent qualitative analysis of probiotic products claimed to contain bifidobacteria. *Int. J. Food Microbiol.*, 102: 221–230.

Masco, L., Vanhoutte, T., Temmerman, R., Swings, J. and Huys, G. (2007) Evaluation of realtime PCR targeting the 16S rRNA and recA genes for the enumeration of bifidobacteria in probiotic products. *Int. J. Food Microbiol.*, 113: 351–357.

Matsuda, K., Tsuji, H., Asahara, T., Kado, Y. and Nomoto, K. (2007) Sensitive quantitative detection of commensal bacteria by rRNA-targeted reverse transcription-PCR. *Appl. Environ. Microbiol.*, 73: 32–39.

Mattila-Sandholm, T., Myllarinen, P., Crittenden, R., Mogensen, G., Fonden, R. and Saarela. M. (2002) Technological challenges for future probiotic foods. *Int. Dairy J.*, 12: 173–182.

Maukonen, J., Alakomi, H.L., Nohynek, L. et al. (2006) Suitability of the fluorescent techniques for the enumeration of probiotic bacteria in commercial non-dairy drinks and in pharmaceutical products. *Food Res. Int.*, 39: 22–32.

Meng, X.C., Stanton, C., Fitzgerald, G.F., Daly, C. and Ross, R.P. (2008) Anhydrobiotics: the challenges of drying probiotic cultures. *Food Chem.*, 106: 1406–1416.

Mikelsaar, M. and Zilmer, M. (2009) *Lactobacillus fermentum* ME-3 an antimicrobial and antioxidative probiotic. *Microb. Ecol. Health Dis.*, 21: 1–27.

Mitsuoka, T. (1982) Recent trends in research on intestinal flora. *Bifidobacteria Microflora* 1: 3–24.

Mortazavian, A., Razavi, S.H., Ehsani, M.R. and Sohrabvandi, S. (2007) Principles and methods of microencapsulation of probiotic microorganisms, *Iran. J. Biotechnol.*, 5(1): 1–18.

Mortazavian, A.M., Mohammadi, R. and Sohrabvandi, S. (2012) Delivery of probiotic microorganisms into gastrointestinal tract by food products. In: Brzozowski, T. (ed.) *New Advances in Basic and Clinical Gastroenterology*. InTech, pp. 122–146.

Muller, J.A., Ross, R.P., Fitzgerald, G.F. and Stanton, C. (2009) Manufacture of probiotic bacteria. In: Charalampopoulos, D. and Rastall, R.A. (eds) *Prebiotics and Probiotics: Science and Technology*. Berlin: Springer, pp. 725–759.

Nista, E.C., Candelli, M., Cremonini, F. et al. (2004) *Bacillus clausii* therapy to reduce side-effects of anti-*Helicobacter pylori* treatment: randomized, double-blind, placebo controlled trial. *Aliment. Pharmacol. Ther.*, 20: 1181–1188.

Noriega, L., Gueimonde, M., Sanchez, B., Margolles, A. and de los Reyes-Gavilan, C.G. (2004) Effect of the adaptation to high bile salts concentrations on glycosidic activity, survival at low pH and cross-resistance to bile salts in *Bifidobacterium*. *Int. J. Food Microbiol.*, 94: 79–86.

O'Connell-Motherway, M., van Sinderen, D., Morel-Deville, F., Fitzgerald, G.F., Ehrlich, S.D. and Morel, P. (2000) Six putative two-component regulatory systems isolated from *Lactococcus lactis* subsp. *cremoris* MG1363. *Microbiology*, 146: 935–947.

Paul, E., Fages, J., Blanc, P., Goma, G. and Pareilleux, A. (1993) Survival of alginate-entrapped cells of *Azospirillum lipoferum* during dehydration and storage in relation to water properties. *Appl. Microbiol. Biotechnol.*, 40: 34–39.

Piuri, M., Sanchez-Rivas, C. and Ruzal, S.M. (2003) Adaptation to high salt in *Lactobacillus*: role of peptides and proteolytic enzymes. *J. Appl. Microbiol.*, 95: 372–379.

Playne, M.J., Bennet, L.E. and Smithers, G.W. (2003) Functional dairy foods and ingredients. *Aust. J. Dairy Technol.*, 58: 242–264.

Poolman, B., Blount, P., Folgering, J.H.A., Friesen, R.H.E., Moe, P. and van der Heide, T. (2002) How do membrane proteins sense water stress? *Mol. Microbiol.*, 44: 889–902.

Prasad, J., McJarrow, P. and Gopal, P. (2003) Heat and osmotic stress responses of probiotic *Lactobacillus rhamnosus* HN001 (DR20) in relation to viability after drying. *Appl. Environ. Microbiol.*, 69: 917–925.

Rastall, R.A. and Maitin, V. (2002) Prebiotics and synbiotics: towards the next generation. *Curr. Opin. Biotechnol.*, 13: 490–498.

Rau, B., Poch, B., Gansauge, F. et al. (2000) Pathophysiologic role of oxygen free radicals in acute pancreatitis: initiating event or mediator of tissue damage. *Ann. Surg.*, 231: 352–360.

Ridlon, J.M., Kang, D.J. and Hylemon, P.B. (2006) Bile salt bio transformations by human intestinal bacteria. *J. Lipid Res.*, 47: 241–259.

Ross, R.P., Desmond, C., Fitzgerald, G.F. and Stanton, C. (2005) Overcoming the technological hurdles in the development of probiotic foods. *J. Appl. Microbiol.*, 98: 1410–1417.

Roy, D. (2011) Probiotics. In: Butler, M., Webb, C., Moreira, A. et al. (eds) *Comprehensive Biotechnology*, vol. 4, 2nd edn. Amsterdam: Elsevier, pp. 591–600.

Ruiz, L., Sanchez, B., Ruas-Madiedo, P., de los Reyes-Gavilan, C.G. and Margolles, A. (2007) Cell envelope changes in *Bifidobacterium animalis* ssp. lactis as a response to bile. *FEMS Microbiol. Lett.*, 274: 316–322.

Saarela, M., Rantala, M., Hallamaa, Nohynek, L., Virkajärvi, I. and Mättö, J. (2004) Stationary-phase acid and heat treatments for improvement of the viability of probiotic lactobacilli and bifidobacteria. *J. Appl. Microbiol.*, 96: 1205–1214.

Salminen, S., Bouley, C., Boutron-Ruault, M.C. et al. (1998) Functional food science and gastrointestinal physiology and function. *Br. J. Nutr.*, 80(1): S147–S171.

Sanchez, B., Ruiz, L., Gueimonde, M., Ruas-Madiedo, P. and Margolles, A. (2012) Toward improving technological and functional properties of probiotics in foods. *Trends Food Sci. Technol.*, 26: 56–63.

Sanders, M.E., Walker, D.C., Walker, K.M., Aoyama, K. and Klaenhammer, T.R. (1996) Performance of commercial cultures in fluid milk applications. *J. Dairy Sci.*, 79(6): 943–955.

Santivarangkna, C., Kulozik, U. and Foerst, P. (2007) Alternative drying processes for the industrial preservation of lactic acid starter cultures. *Biotechnol. Prog.*, 23(2): 302–315.

Serrazanetti, D.I., Gottardi, D., Montanari C. and Gianotti, A. (2013) Dynamic stresses of lactic acid bacteria associated to fermentation processes. In: Kongo, M. (ed.) *Lactic Acid Bacteria – R & D for Food, Health and Livestock Purposes.* InTech, pp. 539–570.

Shahidi, F. and Han, X.Q. (1993) Encapsulation of food ingredients. *Crit. Rev. Food Sci. Nutr.*, 33: 501–547.

Sieuwerts, S., de Bok, F.A.M., Hugenholtz, J. and van Hylckama Vlieg, J.E.T. (2008) Unraveling microbial interactions in food fermentations: from classical to genomics approaches. *Appl. Environ. Microbiol.*, 74: 4997–5007.

Songisepp, E., Kals, J., Kullisaar, T. et al. (2005) Evaluation of the functional efficacy of an antioxidative probiotic in healthy volunteers. *Nutr. J.*, 4: 22.

Stanton, C., Gardiner, G., Meehan, H. et al. (2001) Market potential for probiotics. *Am. J. Clin. Nutr.*, 73: 476S–483S.

Talwalkar, A. and Kailasapathy, K. (2004) The role of oxygen in the viability of probiotic bacteria with reference to *L. acidophilus* and *Bifidobacterium* spp. *Curr. Issues Intest. Microbiol.*, 5: 1–8.

Tanaka, H., Doesburg, K., Iwasaki, T. and Mierau, I. (1999) Screening of lactic acid bacteria for bile salt hydrolase activity. *J. Dairy Sci.*, 82: 2530–2535.

Taniguchi, K., Urasawa, T., Morita, Y., Greenberg, H.B. and Urasawa, S. (1987) Direct serotyping of human rotavirus in stools by an enzyme-linked immunosorbent assay using serotype 1-, 2-, 3-, and 4-specific monoclonal antibodies to VP7. *J. Infect. Dis.*, 155: 1159–1166.

Teixeira, P., Castro, H., Mohacsi-Farkas, C. and Kirby, R.J. (1997) Identification of sites of injury in *Lactobacillus bulgaricus* during heat stress. *Appl. Microbiol.*, 83: 219–226.

Van de Guchte, M., Serror, P., Chervaux, C., Smokvina, T., Erhlich, S.D. and Maguin, E. (2002) Stress responses in lactic acid bacteria. *Antonie van Leeuwenhoek*, 82: 187–216.

Ventling, B.L. and Mistry, V.V. (1993) Growth characteristics of bifidobacteria in ultrafiltered milk. *J. Dairy Sci.*, 76: 962–971.

Vidhyalakshmi, R., Bhakyaraj, R. and Subhasree, R.S. (2009) Encapsulation "The future of probiotics": a review. *Adv. Biol. Res.*, 39: 96–103.

Walker, D.C., Girgis, H.S. and Klaenhammer, T.R. (1999) The groESL chaperone operon of *Lactobacillus johnsonii. Appl. Environ. Microbiol.*, 65(7): 3033–3041.

Yousef, A.E. and Juneja, V.K. (2003) *Microbial Stress Adaptation and Food Safety*. Boca Raton: CRC Press.

Zamora, L.M., Carretero, C. and Pares, D. (2006) Comparative survival rates of lactic acid bacteria isolated from blood, following spray-drying and freeze-drying. *Food Sci. Technol. Int.*, 12: 77–84.

11

Probiotics and Their Health Benefits

*Michelle Fleet and Pattanathu K.S.M. Rahman**

School of Science and Engineering, Teesside University, Middlesbrough, UK

*Corresponding author e-mail: p.rahman@tees.ac.uk

Introduction

With growing interest in the use of probiotics for maintaining good health, we have examined the far-reaching scope of their potential use. Research surrounding probiotics has developed at a rapid pace and has continued to be of great interest to consumers and scientists over recent years. While some reservations regarding their efficacy are consistently raised with emerging research, we examine these issues and discuss their implications. Recent advances in the area and how such developments may be utilized to facilitate further advancement are described. Potential barriers to interpreting current clinical data are discussed, as is the possible future direction of suitable experimental design. For the purposes of this review, the legislative issues surrounding the manufacturing of probiotics have not been covered. While there is ongoing interesting research in the field of probiotics, examining strain-specific activity, enumeration, robust human trials and the mechanisms of action of probiotics will provide us with a great deal of information. Traditional analytical chemistry methods, advances in metagenomics and culturing techniques for gut bacteria should serve as key tools in the research and development of probiotics for the food and pharmaceutical industries. In the interests of public health, a co-ordinated, multifaceted, and multidisciplinary approach is required to give full assurance about the health benefits to human subjects of probiotics, administered at specific doses and defined to strain level.

Over recent years, we have heard more and more about probiotics and their health benefits. Our knowledge of both microbiology and nutrition has developed dramatically since their health benefits were initially

Microbial Functional Foods and Nutraceuticals, First Edition. Edited by Vijai Kumar Gupta, Helen Treichel, Volha (Olga) Shapaval, Luiz Antonio de Oliveira, and Maria G. Tuohy.
© 2018 John Wiley & Sons Ltd. Published 2018 by John Wiley & Sons Ltd.

recognized. For thousands of years before we were aware of their benefits, people were consuming fermented foods containing micro-organisms.

In recent times, there has been ever-increasing interest in the so-called "friendly bacteria." To quote some figures, a Science Direct database search of the word "probiotic" has shown an increase to 2187 publications, just under 400 more than in 2013. This has come a long way since 1999, when only 192 papers were added to the database. In 2013, 115 newspaper articles contained the word "probiotic" in comparison with just 10 in a 1999 News Bank search. Looking at consumer demand, Mintel research conducted in July 2014 discovered that two in five of those who drink probiotic yoghurt do so because of digestive health.

Probiotics are non-pathogenic live active micro-organisms which improve microbial balance, particularly in the gastrointestinal (GI) system. Health benefits are provided to the host when consumed in specific quantities and are believed to be largely strain specific (Butel 2014). Typical probiotics in use are lactobacilli (*L. acidophilus*, *L. casei*, *L. fermentum*, *L. gasseri*, *L. johnsonii*, *L. paracasei*, *L. plantarum*, *L. rhamnosus*, and *L. salivarius*) and bifidobacteria (*B. adolescentis*, *B. animalis*, *B. bifidum*, *B. breve*, and *B. longum*). The aforementioned species, when presented in food at a level of 1×10^9 colony forming units (CFU) per serving, have been accepted by Health Canada (2009) to be given the term "probiotic" for which non-strain-specific claims might be made. The general benefit of supporting a healthy gut microbiota is a core benefit of probiotics as considered by Canadian regulatory authorities; these represent well-researched species, likely to impart some health benefit (Hill et al. 2014). Other genera for which health benefits have been documented (Fijan et al. 2014) are strains of *Saccharomyces*, *Enterococcus*, *Streptococcus*, *Pediococcus*, *Leuconostoc*, *Bacillus*, and *Escherichia coli*.

The proposed health benefits of probiotics are wide-ranging and research is continuing to establish a comprehensive picture of the microbiota, host cells, and nutrients which interact with one another. The mechanisms by which probiotics operate are only just beginning to be understood. Butel (2014) reports that research in relation to mechanisms has been conducted largely *in vitro* or on animal models. As subsequent interpretation of their extensive value to human subjects may be met with a degree of caution, we anticipate there will be further research in this area following careful examination of the applicable ethical issues. In addition to attempting to modulate the gut microbiota for preventive purposes, research has focused on the feasibility of treating diseases with probiotics. The Royal Pharmaceutical Society (2013a) applauds the recognition of the need for research into use of non-antibiotic infection management techniques, which include probiotics. In terms of prescription as a pharmaceutical agent, a combination of various probiotics proved to be more efficacious than single strains for prophylactic activities (Sarker 2013).

This review will focus on the present view of scientific research in the area of probiotics and reflect upon current issues which need further examination. Magrone and Jirillo (2013) identified the three main activities of probiotics: immune regulation through microbial activity; alteration of the microbiota; and prevention of bacterial infection.

The Gastrointestinal System

A "healthy" intestine has been described by Amara and Shilb (2015) as one which contains a balance of lactobacilli, streptococci, *Clostridia*, coliform bacteria, and *Bacterioides*. It is an imbalance or dysbiosis which is often implicated in various health problems. A range of factors have been identified which may influence this balance, from modern diet, lifestyle, and environment (Sun and Chang 2014) to therapeutic disruption factors (antibiotics, probiotics, prebiotics/dietary intervention, and fecal transplantation) identified by Walker and Lawley (2013). The influence of the mother in the case of infants has also been recognized by Chassard et al. (2014), who state that the infant microbiome is partly inherited from the mother and the environment during the first 2 years of life. The majority of research has focused on the large intestine and fecal microbiota, but the significance of the small intestine is becoming apparent, with recent work completed by El Aidy et al. (2014).

Link Between the Gastrointestinal System and Probiotics

Probiotics have been indicated in a number of GI conditions including irritable bowel disease (IBD) and, more recently, irritable bowel syndrome (IBS) by Shanahan and Quigley (2014). A 2014 mini-review by Sherban suggests that although there is an ample body of evidence that gut microbiota contributes to GI tumorigenesis, there is not enough evidence from human studies to strongly support the use of these biotherapeutics in prevention/therapy of GI cancers.

In a clinical setting, we have seen the use of probiotics in the treatment and prevention of diarrhea, in children (Wanke and Szajewska 2014) and patients undergoing radiotherapy (Demersa et al. 2014). *Clostridium difficile* infection is more likely to occur in the elderly and sick and can occur following antibiotic treatment. Furthermore, it can lead to more serious infections of the intestines with severe inflammation of the bowel (pseudomembranous colitis). Tojo et al. (2014) suggest that treatment with probiotics for *C. difficile* infections has an "undoubtedly proven role." Lenoir-Wijnkoop et al. (2014) concluded that use of a fermented milk dairy product containing *L. paracasei* CNCM I-1518 to prevent antibiotic-associated diarrhea (ADD) in older hospitalized patients could lead to substantial cost savings. Public Health England (PHE) (2013) states that it cannot at present recommend the use of probiotics for the prevention of AAD or *C. difficile* infection. The Royal Pharmaceutical Society (2013b) concludes that evidence is of moderate quality but consistently shows that probiotics prevent *C. difficile*-associated diarrhea and are safe. In its Clinical Knowledge Summary, NICE (2013a) reiterates PHE's position due to insufficient evidence about the treatment and prevention of *C. difficile* with probiotics.

In terms of preventing or treating AAD, NICE (2013b) offers no formal guidance on the use of probiotics and suggests that future studies should address factors such as strains or blends of probiotics, patient characteristics, and type of antibiotic. In contrast, the Royal College of Nursing (RCN 2013), in its guidance to nursing staff, advises on the consumption of yoghurt drinks containing probiotics to reduce the length of a diarrhea episode.

Functional Foods

Although there is no universally accepted term for a functional food (McKevith 2013), there is a loose definition of a food that has disease-preventing properties over and above its usual nutritional value and/or which has health-promoting benefits.

The British Nutrition Foundation (BNF 2014) states that although functional foods and drinks may provide benefits in health terms, they should not be seen as an alternative to a varied and balanced diet and a healthy lifestyle. This echoes the position of Adam et al. (2012), who state that probiotics should not be treated as an "elixir of life" but as part of a balanced diet. As shown in Table 11.1, the products available are milk and dairy products and, with the exception of a yoghurt produced by a family dairy, are made by large multinational companies.

The use of comparative genomics and genome sequencing of probiotic bacteria is expected to become a "gold standard" method for characterization and typing of isolates (Bull 2013). DairyCo (2013) informs us that food is developing towards nutritional or functional needs in developed countries, so we expect to see some interesting changes to the current UK probiotic market in coming years. As an example of recent innovation, Ganeden Biotech US (2015) introduced single serving disposable

Table 11.1 Probiotic dairy products available in supermarkets in the UK. Supermarket own brands are not included.

Brand	Product	Presentation	Volume	Probiotic	Amount
Danone	Actimel®	Yoghurt drink	100 g	*Lactobacillus casei* DN-114001 (Brand name: Danone)	Minimum 10 billion
	Activia®	Yoghurt	125 g	*B. lactis* DN 173010 (Brand name: ActiRegularis)	4 billion
Lancashire Farm	Bio-yoghurt	Yoghurt	500 g/1 Kg	*Bifidobacterium* BB12®, *Lactobacillus acidophilus*	Not known
Yakult	Yakult®	Fermented milk drink	65 mL	*Lactobacillus casei* Shirota	Minimum 6.5 billion
Müller	Vitality®	Yoghurt drink	100 g	*Bifidobacterium* BB12®	Not known
		Yoghurt	120 g		Not known

coffee K-cups containing probiotics; this is patented and defined to strain level with Food and Drug Administration (FDA) Generally Recognized as Safe (GRAS) status. As more and more consumers ask about the benefits of such products, it is the role of health professionals, especially state registered dieticians, to provide appropriate advice and guidance within the remit of their professional practice.

Challenges and Developments

Strain Selection

Since the effects of probiotics are determined largely by their specific type, sometimes down to strain level, their value is often considered in terms of use by practitioners. For example, in an expert review, Boyonova and Mitov (2012) concluded that further studies evaluating strain selection by molecular methods will reveal the full potential of probiotics as an adjunct to antibiotic therapy for co-administration.

A recent review by Bron et al. (2013) identified that in order to optimize strain selection for specific interventions and maximize the efficacy of probiotics, a more precise understanding of the molecular mechanisms by which probiotic effector molecules elicit host immune responses is essential. Focusing on the genus *Lactobacillus*, the advantages of determining this information were cited by these authors as follows.

- It would allow prediction of which part of the population would respond most effectively.
- Knowledge of these mechanisms of action could support the identification of adequate efficacy markers that can be measured during trials.
- Potential to provide another model of quality control, rather than that currently of viable cells.
- A tool to consider other ingredients contained in products, since the signaling pathways induced by probiotic strains and other nutritional ingredients are key for the design of efficient mixtures that deliver an enhanced health benefit.

Most interesting was the possibility that the probiotic-derived effector molecules may have an application as bioactives administered in a non-viable form. In a later review by Ruiz et al. (2014), this was explored further, and it was concluded that such isolated metabolites or molecules may offer an alternative treatment in inflammatory disorders.

In a systematic review by Bermudez-Brito et al. (2012), to explore probiotic modes of action focusing on how gut microbes influence the host, the authors concluded that experiments on different intestinal epithelial cells require careful interpretation due to the fact that different cell lines vary in their characteristics. Bron et al. (2013) look towards future human studies in order to fill the gap between preclinical research and large efficacy clinical trials. If we take a proactive and pre-emptive approach to possible challenges in terms of their subsequent future application in the food and pharmaceutical industries, this will be most beneficial to public health and assist in achieving the goals identified by Bermudez-Brito et al. (2012) of fostering the development of novel strategies for the treatment or prevention of gastrointestinal and autoimmune diseases, in addition to improving the credibility of the probiotic concept.

Metagenomics

As knowledge is rapidly advancing in the field of metagenomics, we can expect to see interesting developments in terms of understanding the gut microbiota and probiotic use. In a recent review by Walker et al. (2014), the continual need for microbial cultivation in addition to metagenomics was highlighted. It was concluded that metagenomics and culturing can provide further insight and essentially serve to complement each other.

An outstanding question the authors posed was whether many important new microbial isolates could be obtained simply by varying substrates and nutrients in anaerobic growth media. This would lead to an interesting avenue of research, since it would allow us to predict changes to gut bacteria diversity if conditions were manipulated. Moreover, it has been recommended to employ a combination of phenotypic and genetic techniques to accomplish the identification, classification, and typing of micro-organisms (Adam et al. 2012). If this is achieved, we can expand our knowledge of the behavior of probiotic strains and applications in both maintaining optimum health condition and the potential for treating specific disease.

Regarding the above discussion about a complementary approach to metagenomics and culturing, we must also consider the point that it is culture-independent techniques which can enumerate probiotic cells, since cell culture estimates only the replicating strains and not those in the viable but non-culturable (VBNC state. By a culture-independent technique, we mean fluorescent *in situ* hybridization, nucleic acid amplification techniques such as real-time quantitative PCR, reverse transcriptase PCR, propidium monoazide PCR, and cell sorting techniques such as flow cytometry/fluorescent activated cell sorting. The dose of required probiotic is an important factor when critically evaluating any study or considering its potential use in a clinical setting. Davis (2014) recommends these methods to enumerate VBNC bacteria and culturable bacteria and also calls for a more holistic definition of viable probiotic bacteria. Future research will hopefully be able to more carefully examine this issue and fine-tune this interpretation of viable cells in accordance with evolving techniques.

In a comprehensive literature review (1985–2013), McFarland (2014) echoed previous work in concluding that future trials should include a description of the probiotic down to strain level. Different molecular microbiological typing techniques can be applied such as pulsed-field gel electrophoresis, amplified fragment length polymorphism or multilocus sequence typing (Rijkers 2011). Linking this with industrial developments, Nestlé have sequenced several probiotic genomes (2014) and ongoing partnerships between researchers and manufacturers will undoubtedly give rise to diversification of products.

Additional Biotechnological Methods

Novel and interesting developments have been seen within analytical chemistry, particularlythe use of high performance liquid chromatography (HPLC) when researching probiotics. For example, Mogna et al. (2014) stated the possible hypothesis that the administration of selected oxalate-degrading probiotics could be an alternative approach to reducing the intestinal absorption of oxalate and the resulting urinary excretion. The methodology included testing oxalate-degrading activity of 13 lactobacilli and five bifidobacteria using a novel reverse-phase (rp) HPLC method.

Similarly, in a study by Dubey et al. (2012), the biohydrogenation of unsaturated linoleic acid to its conjugated form by *Pediococcus* spp. GS4 was measured using UV scanning rp-HPLC and gas chromatography-mass spectrometry. HPLC has also been used to analyze folate compounds produced by strains of bifidobacteria, concluding that some may contribute to folate intake (d'Aimmo et al. 2012). The investigation of oligosaccharide utilization of probiotic strains was recently examined by Sims et al. (2014) using high performance anion exchange chromatography and HPLC and it was found that strains utilized them differently.

We can therefore foresee the increasing use of these techniques in probiotic research to determine the specific phenotypical traits of strains of bacteria. This in turn can lead to further studies which could determine their possible use and required dosage.

Conclusion

If Italy is successful in obtaining permission to use "probiotic" as a generic term on its food labeling, the UK could also do this, getting an extension as a "concerned Member State" (Pendrous 2014). This could have interesting consequences for the food industry since it would pave the way for simple labeling and subsequent ease of consumer recognition.

The need for more comprehensive research in this area cannot be stressed enough. There are currently 54 clinical trials being conducted within the NHS (2014), showing the potential of probiotics in a wide range of treatment areas. However, unlike several other nutritional issues (Public Health England 2014), there is no government Scientific Advisory Committee for probiotics. Additionally, NICE guidelines in which probiotics are mentioned, as shown in Table 11.2, are meager in relation to the range of conditions they cover.

With little comprehensive guidance from NICE and no government scientific advisory committee for probiotics, there is a potential vacuum in terms of professional knowledge dissemination within the UK. Furthermore, the use and promotion of probiotics by health professionals are likely to remain limited until these important steps are taken.

There are several issues which we have identified in this review which, if adequately addressed, could assist our advancement in this field tremendously. Since *in vitro* culture conditions may influence the

Table 11.2 NICE guidelines in which probiotics are mentioned.

Condition	Name and date of NICE Clinical Guideline (CG)	Guidance on probiotic use	Status
Otitis media with effusion (OME) in children	CG60 February 2008	Dietary modification, including probiotics not recommended	Last reviewed August 2011 Next review date to be confirmed Placed on static list February 2014
Diarrhoea and vomiting in children	CG84 April 2009	Highlights methodological limitations of studies, treatment regimens used and variation in the specific probiotics evaluated. Also states many studies completed in developing countries, where response to probiotics may differ. Good-quality randomized controlled trials should be conducted in the UK to evaluate the effectiveness and safety of specific probiotics, using clearly defined treatment regimens and outcome measures	Placed on static list 2014 Review date to be confirmed
Irritable bowel syndrome (IBS) in adults	CG61 February 2008	States "People with IBS who choose to try probiotics should be advised to take the product for at least 4 weeks while monitoring the effect. Probiotics should be taken at the dose recommended by the manufacturer"	This evidence has not been reviewed since 2008. Placed on static list December 2013 Clinical Guidance updated on April 2017 Next review: 2019

expression of certain molecular characteristics, and results obtained in animal models cannot be directly transferred to humans due to physiological differences (Bermudez-Brito et al. 2012), robust, well-designed and reproducible clinical trials on human subjects, with a tightly defined description of both strain and viable cells, would secure confidence and assist in making the vital transition into clinical practice. With the rapidly emerging techniques and knowledge to allow us to understand the bacteria within our bodies, the effectiveness of specific probiotic strains at accurate doses is a key consideration for industry, regulators, and health professionals alike.

References

Adam, J.K., B. Odhav and K.S.B. Naidu (2012) Probiotics: recent understandings and biomedical applications. *Curr. Trends Biotechnol. Pharm.*, 6(1): 1–14.

Amara, A. and A.A. Shilb (2015) Role of probiotics in health improvement, infection control and disease treatment and management. *Saudi Pharm. J.*, 23(2): 107–114.

Bermudez-Brito, M., J. Plaza-Díaz, C. Muñoz-Quezada, C. Gómez-Llorente and A. Gil (2012) Probiotic mechanisms of action. *Ann. Nutr. Metab.*, 61(2): 160–174.

Boyonova, L. and I. Mitov (2012) Co-administration of probiotics with antibiotics: why, when and for how long? *Expert Rev. Antiinfect. Ther.*, 10(4): 407–409.

British Nutrition Foundation (2014) *Functional Foods*. Available at: www.nutrition.org.uk/nutritionscience/foodfacts/functional-foods.html (accessed 3 May 2017).

Bron, P.A., S. Tomita, A. Mercenier and M. Kleerebezem (2013) Cell surface-associated compounds of probiotic lactobacilli sustain the strain-specificity dogma. *Curr. Opin. Microbiol.*, 16: 262–269.

Bull, M.J. (2013) Molecular genetic characterisation of probiotic bacteria: Lactobacillus acidophilus and Bifidobacterium species. PhD thesis, Cardiff University. Available at: orca.cf.ac.uk/53985/ (accessed 3 May 2017).

Butel, M.J. (2014) Probiotics, gut microbiota and health. *Méd. Malad. Infect.*, 44(1): 1–8.

Chassard, C., T. de Wouters and C. Lacroix (2014) Probiotics tailored to the infant: a window of opportunity. *Curr. Opin. Biotechnol.*, 26: 141–147.

D'Aimmo, M.R., P. Mattarelli, B. Biavati, N.G. Carlsson and T. Andlid (2012) The potential of bifidobacteria as a source of natural folate. *J. Appl. Microbiol.*, 112(5): 975–984.

DairyCo (2013) Dairy Market Update: Can the UK benefit from the growing appetite for dairy? *DairyCo Datum*, 28 November: 1–2.

Danone (2014a) Actimel. Available at: www.actimel.co.uk/all-about-actimel/ (accessed 3 May 2017).

Danone (2014b) Activia. Available at: www.danoneactivia.co.uk/ (accessed 3 May 2017).

Davis, C. (2014) Enumeration of probiotic strains: review of culture-dependent and alternative techniques to quantify viable bacteria. *J. Microbiol. Methods*, 103: 9–17.

Demersa, M., A. Dagnaulta and J. Desjardins (2014) A randomized double-blind controlled trial: impact of probiotics on diarrhea in patients treated with pelvic radiation. *Clin. Nutr.*, 33(5): 761–767.

Dubey, V., A.R. Ghosh and B.K. Mandal (2012) Appraisal of conjugated linoleic acid production by probiotic potential of Pediococcus spp. *GS4*, 168(5): 1265–1276.

El Aidy, S., B. van den Bogert and M. Kleerebezem (2014) The small intestine microbiota, nutritional modulation and relevance for health. *Curr. Opin. Biotechnol.*, 32: 14–20.

Fijan, S. (2014) Microorganisms with claimed probiotic properties: an overview of recent literature. *Int. J. Environ. Res. Public Health*, 11: 4745–4767.

Ganeden Biotech (2015) Ganeden BC30. Available at: www.ganedenbc30.com/site (accessed 3 May 2017).

Health Canada (2009) *Accepted Claims about the Nature of Probiotic Microorganisms in Food*. Available at: www.hc-sc.gc.ca/fn-an/label-etiquet/claims-reclam/probiotics_claims-allegations_probiotiques-eng.php (accessed 3 May 2017).

Hill, C., F. Guarner, G. Reid et al. (2014) Expert consensus document: the International Scientific Association for Probiotics and Prebiotics consensus statement on the scope and appropriate use of the term probiotic. *Nat. Rev. Gastroenterol. Hepatol.*, 11: 506–514.

Lancashire Farm (2014) Bio yoghurt. Available at: http://lancashirefarm.com/ (accessed 3 May 2017).

Lenoir-Wijnkoop, I., M.J.C. Nuijten, J. Craig and C.C. Butler (2014) Nutrition economic evaluation of a probiotic in the prevention of antibiotic-associated diarrhea. *Front. Pharmacol.*, 5: 13.

Magrone, T. and E. Jirillo (2013) The interaction between gut microbiota and age-related changes in immune function and inflammation. *Immun. Ageing*, 10: 31.

McFarland, L.V. (2014) Use of probiotics to correct dysbiosis of normal microbiota following disease or disruptive events: a systematic review. *BMJ*, 4: 8.

McKevith, B. (2013) *An Introduction to Functional Foods and Drinks.* Available at: www.fdin.org.uk/output/Brigid%20McKevith's%20 Definitions.pdf (accessed 3 May 2017).

Mintel (2014) *Yogurt and Yogurt Drinks: The Consumer – Reasons to Eat Yogurt/Drink Yogurt Drinks.* Available at: academic.mintel.com/ display/709912/?highlight#hit1 (accessed 3 May 2017).

Mogna, L., M. Pane, S. Nicola and E. Raiteri (2014) Screening of different probiotic strains for their in vitro ability to metabolise oxalates: any prospective use in humans? *J. Clin. Gastroenterol.*, 48: 1.

Nestlé (2014) *Food Science and Technology.* Available at: www.nestle.com/ randd/technologies (accessed 3 May 2017).

NHS (2014) Clinical Trials and Medical Research. Available at: www.nhs.uk/ Conditions/Clinical-trials/Pages/clinical-trial.aspx?Condition=probiotics &CT=0&Rec=0&Countries=United+Kingdom (accessed 3 May 2017).

NICE (2008a) *Surgical Management of Otitis Media with Effusion in Children.* Clinical Guideline 60. Available at: www.nice.org.uk/guidance/ cg60/resources/guidance-surgical-management-of-otitis-media-with-effusion-in-children-pdf (accessed 3 May 2017).

NICE (2008b) *Irritable Bowel Syndrome in Adults: Diagnosis and Management of Irritable Bowel Syndrome in Primary Care.* Clinical Guideline 61. Available at: www.nice.org.uk/guidance/cg61 (accessed 3 May 2017).

NICE (2009 *Diarrhoea and Vomiting in Children: Diarrhoea and Vomiting Caused By Gastroenteritis: Diagnosis, Assessment and Management in Children Younger Than 5 Years.* Clinical Guideline 84. Available at: www. nice.org.uk/guidance/cg84/chapter/4-research-recommendations#other-therapies-probiotics (accessed 3 May 2017).

NICE (2013a) *Diarrhoea – Antibiotic Associated.* Clinical Knowledge Summary. Available at: cks.nice.org.uk/diarrhoea-antibiotic-associated#!scenario (accessed 3 May 2017).

NICE (2013b) *Probiotics in Antibiotic-Associated Diarrhoea.* Eyes on Evidence: Expert Commentary on Important New Evidence 49.

Pendrous, R. (2014) Probiotic term could reappear on yogurt in UK. Available at: www.foodmanufacture.co.uk/Regulation/Probiotic-generic-descriptor-application-moves-on (accessed 3 May 2017).

Public Health England (2013) *Updated Guidance on the Management and Treatment of Clostridium difficile Infection.* Available at: www.gov.uk/ government/uploads/system/uploads/attachment_data/file/321891/ Clostridium_difficile_management_and_treatment.pdf (accessed 3 May 2017).

Public Health England (2014) Reports, position statements and annual reports of the Scientific Advisory Committee on Nutrition. Available at: www.gov.uk/government/collections/sacn-reports-and-position-statements#position-statements (accessed 3 May 2017).

RCN (2013) *The Management of Diarrhoea in Adults.* Available at: www. rcn.org.uk/__data/assets/pdf_file/0016/510721/004371.pdf (accessed 3 May 2017).

Rijkers, G. T., W.M. de Vos, R.J. Brummer, L. Morelli, G. Corthier and P. Marteau (2011) Health benefits and health claims of probiotics: bridging science and marketing. *Br. J. Nutr.*, 106: 1291–1296.

Royal Pharmaceutical Society (2013a) RPS responds to UK Antimicrobial Resistance Strategy: Royal Pharmaceutical Society What's Happening? News. Available at: www.rpharms.com/news/details//RPS-responds-to-UK-Antimicrobial-Resistance-Strategy (accessed 8 May 2017).

Royal Pharmceutical Society (2013b) *Are Probiotics Useful for Preventing and Treating C difficile Diarrhoea?*Available at: www.pharmaceutical-journal.com/learning/learning-article/are-probiotics-useful-for-preventing-and-treating-c-difficile-diarrhoea/11121705.article (accessed 3 May 2017).

Ruiz, L., A. Hevia, D. Bernardo, A. Margolles and B. Sánchez (2014) Extracellular molecular effectors mediating probiotic attributes. *FEMS Microbiol. Lett.*, 359(1): 1–11.

Sarker, S. (2013) Potential of probiotics as pharmaceutical agent: a review. *Br. Food J.*, 115: 11.

Shanahan, F. and E.E.M. Quigley (2014) *Manipulation of the Microbiota for Treatment of IBS and IBD – Challenges and Controversies.* Available at: www.sciencedirect.com.ezproxy.tees.ac.uk/science/article/pii/S0016508514001413 (accessed 3 May 2017).

Sherban, D.A. (2014) Gastrointestinal cancers: influence of gut microbiota, probiotics and prebiotics. *Cancer Lett.*, 345: 258–270.

Sims, I.M., J.L.J. Ryan and S.H. Kim (2014) In vitro fermentation of prebiotic oligosaccharides by Bifidobacterium lactis HN019 and Lactobacillus spp. *Anaerobe*, 25: 11–17.

Sun, J. and E.B. Chang (2014) Exploring gut microbes in human health and disease: pushing the envelope. *Genes Dis.*, 1: 2.

Tojo, R., A. Suárez, M.G. Clemente et al. (2014) Intestinal microbiota in health and disease: role of bifidobacteria in gut homeostasis. *World J. Gastroenterol.*, 20(41): 15163–15176.

Walker, A.W. and T.D. Lawley (2013) Therapeutic modulation of intestinal dysbiosis. *Pharmacol. Res.*, 69: 75–86.

Walker, A.W., S. H. Duncan, P. Louis and H. J. Flint (2014) Phylogeny, culturing, and metagenomics of the human gut microbiota. *Trends Microbiol.*, 22: 267–274.

Wanke, M. and H. Szajewska (2014) Probiotics for preventing healthcare-associated diarrhoea in children: a meta-analysis of randomized controlled trials. *Pediatr. Polska*, 89(1): 8–16.

Yakult (2014) Yakult. Available at: www.yakult.co.uk/ (accessed 3 May 2017).

12

Nutritional Potential of *Auricularia auricula-judae* and *Termitomyces umkowaan* – The Wild Edible Mushrooms of South-Western India

Namera C. Karun, Kandikere R. Sridhar*, and Cheviri N. Ambarish

Department of Biosciences, Mangalore University, Mangalagangotri, Mangalore, Karnataka, India

*Corresponding author e-mail: kandikere@gmail.com

Introduction

Protein-energy malnutrition (PEM) is one of the major challenges in developing countries. It has been intensified due to increased population, insufficient food production, more dependence on starch- or monocarbohydrate-based diets, transformation of agricultural lands into industrial yards and loss of phytodiversity (FAO/WHO 1991). Besides PEM, mineral deficiencies in food (e.g., Fe, K, Zn, Se, Ca, Mg, and Cu) also affect human nutrition (Frossard et al. 2000; Grusak and Cakmak 2005; Thacher et al. 2006; Welch and Graham 2005). There has therefore been an upsurge in focused research efforts exploring underutilized, indigenous, and non-conventional food sources as alternatives to expensive animal protein sources. Fermented foods (cereals and legumes) with non-toxic fungi (e.g., *Rhizopus oligosporus*) provide several nutrients and health benefits (e.g., proteins, essential amino acids, fiber, vitamin B, low fat, and zero cholesterol) (Ghorai et al. 2009; Niveditha and Sridhar 2015). Among the alternative food sources used to overcome PEM, indigenous legumes and macrofungi are at the forefront of research.

Macrofungi (e.g., mushrooms, puffballs, and morels) are attractive sources of nutrition and pharmaceuticals worldwide (Boa 2004; Halpern and Miller 2002). About 25 species of mushrooms are accepted as a food source and some are commercially cultivated to harness their benefits for human health (Lindequist et al. 2005), which include proteins, essential amino acids, fibers, minerals, fatty acids, vitamins, and flavors (Oboh and

Microbial Functional Foods and Nutraceuticals, First Edition. Edited by Vijai Kumar Gupta, Helen Treichel, Volha (Olga) Shapaval, Luiz Antonio de Oliveira, and Maria G. Tuohy.
© 2018 John Wiley & Sons Ltd. Published 2018 by John Wiley & Sons Ltd.

Shodehinde 2009; Sadler 2003). Although the Western Ghats is one of the hotspots of diversity of several wild macrofungi (Bhosle et al. 2010; Farook et al. 2013; Manoharachary et al. 2005; Mohanan 2011; Senthilarasu, 2014), very little is known about their nutritional attributes (Johnsy et al. 2011; Meghalatha et al. 2014; Sudheep and Sridhar 2014). This chapter focuses on the assessment of the nutritional qualities of two wild mushrooms, *Auricularia auricula-judae* and *Termitomyces umkowaan*, collected from the Western Ghats and the west coast of southern India, respectively.

Wild Mushrooms

Auricularia auricula-judae (Bull.) Quél. (Auriculariaceae-Basidiomycotina) and *Termitomyces umkowaan* (Cooke and Massee) D.A. Reid (Lyophyllaceae-Basidiomycotina) (Figure 12.1) were collected from Kadnur, in the Western Ghats (12°13′N, 75°46′E; 891 m asl) and from mixed forest at the Mangalore

(a)

(b)

Figure 12.1 *Auricularia auricula-judae* grown on decomposing logs in the Western Ghats (a) and *Termitomyces umkowaan* grown on termite mounds (b). (*See color plate section for the color representation of this figure.*)

University Campus, Mangalore, west coast (12°48'N, 74°55'E; 112.4 m asl) during July–August 2012, respectively. *A. auricula-judae* was plentiful on decomposing standing dead and fallen logs, bark, and twigs (e.g., *Artocarpus heterophyllus*), while *T. umkowaan* was common in and around the termite mounds in mixed forests. They were identified based on macro- and micro-morphological features (Buczacki 2012; Jordan 2004; Mohanan 2011; Pegler and Vanhaecke 1994; Phillips 2006; Tibuhwa et al. 2010) and comparison with herbarium specimens at the Department of Botany, Goa University, Goa, India.

For assessment of proximal features, minerals, protein qualities and fatty acids, freshly collected mushroom samples (n=5) were blotted and grouped into two portions in each replicate. One portion of each replicate was oven dried (50–55 °C) and another portion was cooked using a household pressure-cooker with limited water (6.5 L, Deluxe stainless steel, TTK Prestige™, Prestige Ltd, Hyderabad, India) and the oven-dried on aluminum foil (50–55 °C). Dried mushroom samples were milled (Wiley Mill, mesh #30) and refrigerated (4 °C) in airtight containers.

Nutritional Assessment

Proximal Properties

Proximal qualities including crude protein, total lipids, crude fiber, ash, carbohydrates,and energy of the cooked and uncooked mushroom samples were determined employing standard methods. The micro-Kjeldahl method was used to determine the crude protein (N × 6.25) of uncooked and cooked mushrooms (Humphries 1956). The Soxhlet method was employed for total lipid extraction of mushroom samples using petroleum ether (60–80 °C), while the quantities of crude fiber and ash were determined gravimetrically (AOAC 1990). Total carbohydrate (%) content was calculated according to Müller and Tobin (1980):

$$\left[\text{Total carbohydrates}\,(\%)=100-\left(\begin{array}{l}\%\,\text{crude protein}+\\ \%\,\text{total lipids}+\%\,\text{crude fiber}\,\%\,\text{ash}\end{array}\right)\right]$$

Calorific value (kJ/100 g) of mushroom samples was calculated based on Ekanayake et al. (1999):

$$\left[\begin{array}{l}\text{Calorific value}\,\bigl(\text{kJ}\,/\,100\,\text{g}\bigr)=\bigl(\text{crude protein}\times16.7\bigr)+\\ \bigl(\text{total lipids}\times37.7\bigr)+\bigl(\text{carbohydrate}\times16.7\bigr)\end{array}\right].$$

Minerals

Mushroom flours were digested in a mixture of concentrated HNO_3, H_2SO_4, and $HClO_4$ (10:0.5:2 v/v) (AOAC 1990). Quantities of minerals (Na, K, Ca, Mg, Fe, Cu, Zn, Se) in digested samples were assessed by AAS (GBC Scientific Equipment Pty Ltd, Melbourne, Australia). To determine the total phosphorus in the samples, the vanadomolybdophosphoric acid method was employed by measuring the absorbance at 420 nm using KH_2PO_4 as standard (AOAC 1990). The ratios of Na/K and Ca/P were calculated.

Amino Acids

The method proposed by Hofmann et al. (1997, 2003) was followed to determine the amino acid content of the samples. A defined quantity of flour was hydrolyzed using HCl (6N, 15 ml) up to 4 hours (145 °C). Alkaline extracted and oxidized samples were used to determine tryptophan and sulfur amino acids, respectively. On cooling, HCl was removed by rotoevaporator (Büchi Laboratoriumstechnik AG RE121, Switzerland) using a diaphragm vacuum pump (MC2C, Vacuubrand GmbH, Germany). Internal standard trans-4-(aminomethyl)-cyclohexanecarboxylic acid (Aldrich, purity 97%) was added to each sample. The derivatization step includes esterification by trifluoroacetylation (Brand et al. 1994). To eliminate traces of water, standards were weighed in reaction vials and dried in CH_2Cl_2 with a gentle stream of helium by slow heating in an oil bath (40–60 °C). Fresh acidified isopropanol (acetyl chloride 3 mL + 2-propanol 12 mL) (12 mL) was added and heated (100 °C, 1 hour). Reagent was removed from cooled samples by a gentle helium stream (60 °C). Evaporation with three successive aliquots of CH_2Cl_2 followed to remove propanol and water. The dry residue was trifluoroacetylated overnight at laboratory temperature using trifluoroacetic anhydride (200 mL). An aliquot of this solution without treatment was fed to a gas chromatography combustion-isotope ratio mass spectrometer (GC-C-IRMS/MS). Measurements of GC-C-IRMS/MS were undertaken using a gas chromatograph (Hewlett-Packard 58590 II, Germany) connected via a split with a combustion interface to the IRMS system (GC-C-II to MAT 252, Finnigan MAT, Germany) to determine nitrogen and via a transfer line to a mass spectrometer (GCQ, Finnigan MAT Germany) for analysis of the amino acids. A capillary column of GC (50 m × 0.32 mm i.d. × 0.5 μm BPX5, SGE) was operating with carrier gas flow (1.5 mL/min): initial 50 °C (1 min), raised to 100 °C by 10 °C/min (10 min), raised to 175 °C by 3 °C/min (10 min) and to 250 °C/min (10 min) (head pressure, 13 psi, 90 kPa). The ratio of total essential amino acids (TEAA) to total amino acids (TAA) was calculated: [TEAA/TAA ratio = (Total EAA ÷ TAA)].

Protein Digestibility and Qualities

Estimation of *in vitro* protein digestibility (IVPD) was performed based on Akeson and Stahmann (1964). Defatted mushroom flours (100 mg) were treated at 37 °C for 3 hours with pepsin (Sigma, 3165 units/mg protein) (1.5 mg/2.5 mL 0.1N HCl) and immediately inactivated (0.25 mL 1N NaOH). Incubation was continued for 24 hours at 37 °C with trypsin (Sigma, 16 100 units/mg protein) and α-chymotrypsin (Sigma, 76 units/mg protein) (2 mg each/2.5 mL potassium phosphate buffer, pH 8.0, 0.1M) and inactivated immediately (0.7 mL, 100% TCA). Zero-time control was maintained by inactivation of enzyme prior to addition of mushroom flour.

Supernatant was collected after centrifugation of the inactivated mixture, and the remaining residue was washed (10% TCA, 2 mL) followed by centrifugation. The supernatant was pooled twice with diethyl ether (10 mL) and the ether layer was removed by aspiration. The aqueous layer was kept in a boiling water bath for 15 min to eliminate traces of ether. On attaining room temperature, the solution was made up to 25 mL in distilled water. Nitrogen content in 5 mL aliquots was determined using micro-Kjeldahl apparatus (Humphries 1956) and protein in digest was estimated:

$$\left[\text{IVPD}\,(\%) = \left(\text{protein in digest} \div \text{protein in defatted flour} \right) \times 100 \right].$$

The essential amino acid score (EAAS) was determined according to FAO/WHO EAA requirements for adults (FAO/WHO 1991):

$$\left[\begin{array}{l} \text{EAAS}\,(\%) = \left(\text{amino acid content in the test protein in mg/g} \right) \div \\ \left(\text{FAO/WHO EAA reference pattern in mg/g} \right) \times 100 \end{array} \right].$$

The protein digestibility corrected amino acid score (PDCAAS) for adults was calculated:

$$\left[\begin{array}{l} \text{PDCAAS} = \left(\text{EAA in test protein in mg / g} \right) \div \\ \left(\text{FAO / WHO EAA reference pattern in mg / g} \right) \times \text{IVPD}\,(\%) \end{array} \right].$$

The protein efficiency ratio (PER) was calculated based on the amino acid composition of mushroom flours according to Alsmeyer et al. (1974):

$$\begin{array}{l} \text{PER}_1 = -0.684 + 0.456 \times \text{Leu} - 0.047 \times \text{Pro}; \; \text{PER}_2 = -0.468 + \\ 0.454' \, \text{Leu} - 0.105 \times \text{Tyr}; \; \text{PER}_3 = -1.816 + 0.435 \times \text{Met} + \\ 0.78 \times \text{Leu} + 0.211 \times \text{His} - 0.944 \times \text{Tyr}. \end{array}$$

Fatty Acids

Total lipids of mushroom flours extracted by the Soxhlet method were subjected to analysis of fatty acid methyl esters (FAMEs) based on the method outlined by Padua-Resurreccion and Benzon (1979). In brief, HCl (5%, 0.2 mL) and acetyl chloride (8.3 mL) were added drop by drop to absolute methanol (100 mL) kept in an ice jacket. This mixture (0.2 mL) was added to extracted lipids of mushroom flour (200 mg) in screw-cap glass vials (15 mL capacity), mixed, vortexed, and incubated at 70 °C for up to 10 hours followed by cooling to room temperature. Later, it was suspended in distilled water (500 μL), n-hexane (HPLC grade, l00 μL) was added, vortexed and kept for separation. The top hexane layer was transferred into air-tight microcentrifuge tubes and preserved in the refrigerator (4 °C) for FAME assay.

Esterified samples (100 μL) in vials were diluted with n-hexane (HPLC grade, 900 μL). One μL aliquot was injected into the gas chromatograph (GC-2010, Shimadzu, Japan) equipped with autoinjector (AOI) and capillary column (BPX-70). The capillary column was conditioned for 10 hours before use. Elutants were detected by flame ionization detector (FID), amplified signals were transferred and followed by monitoring GC-Solutions software (http://files.instrument.com.cn/FilesCenter/20110318/Gc_reference.pdf). Analytical conditions of the autosampler, injection port settings, column oven settings, and column information of the gas chromatograph were set up according to Nareshkumar (2007). The FAMEs quantification was based on the standard mixture (C_4–C_{24}) (Sigma, USA) run in similar conditions as the sample. The concentration and area of each peak of FAME were computed using GC Post-run analysis software (www.umich.edu/~mssgroup/docs/GCinstructions.pdf). The ratio of polyunsaturated to saturated fatty acids was calculated.

Data Analysis

The difference in proximal characteristics, minerals, protein fractions, amino acids, *in vitro* protein digestibility, and fatty acids between uncooked and cooked mushroom flours was assessed by *t*-test using Statistica version # 8.0 (StatSoft Inc. 2008).

Nutritional Comparison

Even though *A. auricula-judae* and *T. umkowaan* are utilized as food sources, their nutritional features have not been systematically studied. *A. auricula-judae* occurs in large quantities especially on standing wood or fallen logs on the forest floor. *T. umkowaan* erupts in large numbers in termite mounds. Mature *A. auricula* occurs under south-west monsoon conditions and is available for up to 7–10 days for collection, while *T. umkowaan* shows up all of a sudden and remains up to 3–5 hours for sampling. However, both mushrooms serve as traditional food sources in the Western Ghats and west coast of India.

Proximate Qualities

Crude protein content in uncooked and cooked *A. auricula-judae* was low (6.1–6.4%) and did not change significantly on cooking, matching the protein content of cereals (Table 12.1). Protein content reported by Crisan and Sands (1978) was low (4.2%), but was higher in *A. auricula* from the Western Ghats and Nagaland (36.3%) (Johnsy et al. 2011; Kumar et al. 2013). It is likely the protein content of *Auricularia* is dependent on the type of woody substrate on which it grows and this needs further insight. As for crude proteins, crude lipids were also low (1.6–1.7%) and not significantly altered on cooking, which agrees with a report from the Western Ghats (Johnsy et al. 2011) (1.6%). However, crude lipid content as high as 8.3% in *A. auricula* has been reported by Crisan and Sands (1978). Crude fiber significantly decreased on cooking (24% versus 18.5%), which corresponds with a report by Crisan and Sands (1978) (19.8%), but was lower in *A. auricula* from the Western Ghats (8.4%) and Nagaland (2.8%) (Johnsy et al. 2011; Kumar et al. 2013). Ash content ranged between 2.7% and 3.9% without significant change on cooking, however, its quantity was lower compared to *A. auricula* (4.7–7.1%) (Crisan and Sands 1978; Johnsy et al. 2011; Kumar et al. 2013), which was reflected in low mineral composition in *A. auricula*. Total carbohydrate content of *A. auricula* was significantly increased on cooking, which is higher compared to reports by Johnsy et al. (2011) and Kumar et al. (2013) (33.2%), but lower than the report by Crisan and Sands (1978) (82.8%).

In *T. umkowaan*, all proximal properties were significantly higher in uncooked than in cooked samples except for crude lipid which was not altered significantly (see Table 12.1). Protein content is comparable with several edible legumes (*Cajanus*, *Cicer*, *Phaseolus*, and *Vigna*) and *T. mammiformis* (Jambunathan and Singh 1980; Khan et al. 1979; Kumari 2012; Nwokolo 1987; Nwokolo and Oji 1985; Singdevsachan et al. 2014). It is higher than *T. striatus* (18.9–21.5% versus 13%) and comparable with *T. eurrhizus and T. globulus* (19.5–23.8%), but lower than many termitomycetes (*T. badius, T. heimii, T. medius, T. microcarpus, T. radicatus*, and *T. robustus*) (34.2–46.5%) (Aletor and Aladetimi 1995; Johnsy et al. 2011; Kumari 2012; Singdevsachan et al. 2014; Sudheep and Sridhar 2014). Crude lipid content in *T. umkowaan* is higher than other termitomycetes (*T. badius, T. eurrhizus, T. globulus, T. heimii, T. medius, T. mammiformis*,

Table 12.1 Proximate composition of uncooked and cooked mushrooms on dry weight basis (n = 5; mean±SD).

	Auricularia auricula-judae		*Termitomyces umkowaan*	
	Uncooked	**Cooked**	**Uncooked**	**Cooked**
Crude protein (g/100 g)	6.44±0.16[a]	6.06±0.32[a]	21.49±1.16[a]	18.89±0.14[b*]
Crude lipid (g/100 g)	1.63±0.15[a]	1.67±0.02[a]	4.62±0.58[a]	4.50±0.29[a]
Crude fiber (g/100 g)	23.96±4.62[a]	18.51±2.70[b*]	28.76±2.08[a]	15.73±0.41[b**]
Ash (g/100 g)	3.85±0.10[a]	2.66±0.84[a]	7.28±0.83[a]	2.58±0.01[b**]
Total carbohydrates (g/100 g)	64.07±4.71[a]	71.14±1.74[b*]	50.89±1.04[a]	45.26±1.67[b*]
Calorific value (kJ/100 g)	1241±76.6[a]	1351±28.19[b*]	1383±17.50[a]	1241±40.87[b**]

Figures across the uncooked and cooked mushrooms with different letters are significantly different (*t*-test: *p<0.05; **p<0.01).

T. microcarpus, T. radicatus. T. robustus, T. shimperi, and *T. tylerianus*) (4.5–4.6% versus 0.1–4.3%) (Aletor and Aladetimi 1995; Johnsy et al. 2011; Kavishree et al. 2008; Kumari 2012; Singdevsachan et al. 2014; Sudheep and Sridhar 2014). The crude fiber in *T. umkowaan* is also substantially higher than other termitomycetes (15.7–28.8% versus 2.5–11.5%). Ash content in *T. umkowaan* was lower compared to other termitomycetes (2.6–7.3% versus 5–36%), which is reflected in the low minerals profile. Total carbohydrate content was higher than *T. heimii, T. globulus*, and *T. microcapus* (45.3–50.9% versus 39–46.5%), but lower than *T. eurrhizus* (63%) (Johnsy et al. 2011; Singdevsachan et al. 2014; Sudheep and Sridhar 2014).

The present study revealed that protein content is moderate in *T. umkowaan*, and crude lipid content in *A. auricula* is lower than *T. umkowaan*. However, uncooked and cooked samples of both mushrooms possess high quantities of crude fiber, which has several health benefits (Cheung 1997). A high proportion of insoluble fiber is expected in mushrooms due to chitin and other polysaccharides and this fiber is nutritionally beneficial (Kalač 2009). A high-fiber diet helps to lower blood cholesterol and decreases the risk of large bowel cancer (Anderson et al. 1995; Slavin et al. 1997). The moderate quantity of carbohydrates in these mushrooms helps to control intestinal cancer and has a low glycemic index which is advantageous in type 2 diabetes (Venn and Mann 2004).

Minerals

Seven minerals (sodium, potassium, calcium, magnesium, phosphorus, iron, and copper) were significantly decreased on cooking *A. auricula*, with the exception of zinc (Table 12.2). Sodium, potassium, calcium, magnesium, iron, and zinc were lower compared to *A. polytricha* and only the copper content is comparable (Manjunathan et al. 2011). As seen in this study, the minerals profile was also low as in *A. polytricha* from Nigeria (Afiukwa et al. 2013).

Mineral levels in uncooked and cooked *T. umkowaan* were similar to *A. auricula* (see Table 12.2). Calcium, magnesium, iron, and copper were lower in *T. umkowaan* compared to other termitomycetes (*T. badius, T. heimii, T. medius, T. mammiformis, T. microcarpus, T. radicatus,* and *T. striatus*) with the exception of calcium in *T. striatus* (15 versus 15.6–20.1 mg/100 g) and magnesium in *T. microcarpus* (6 versus 25.1–28.5 mg/100 g) (Kumari 2012). However, phosphorus, iron, and zinc contents are higher and copper is comparable with uncooked and cooked *T. globulus* from the Western Ghats (Sudheep and Sridhar 2014).

Mushrooms are generally known for high levels of potassium (Dursun et al. 2006; Sudheep and Sridhar 2014). Although potassium was high in both mushrooms studied, it was not up to the expected level and most of the mineral was considerably drained on cooking. None of the minerals in these mushrooms are comparable with the NRC-NAS (1989) recommended pattern. However, the Na/K ratio (0.28–0.47) in both mushrooms and Ca/P ratio (1.9–2) in *A. auricula* are favorable for human health. The Na/K ratio <1 helps to control high blood pressure, while the Ca/P ratio >1 helps to prevent loss of calcium in urine (Shills and Young 1988; Yusuf et al. 2007).

Amino Acids Profile and Protein Bioavailability

Alanine was highest in both mushrooms followed by glycine and glutamic acid. Cooking significantly decreased most of the amino acids, but tryptophan was below detectable levels (Table 12.3). Many non-essential amino acids in uncooked and cooked mushrooms are higher or comparable with soybean and wheat. Among EAA, isoleucine, leucine, lysine, tyrosine, threonine, and valine in uncooked as well as cooked *A. auricula* were higher compared to soybean, wheat, and the FAO/WHO (1991) recommended standard; the same was true of histidine, lysine, tyrosine, threonine, and valine in uncooked and cooked samples of *T. umkowaan*. Similarly, leucine, lysine, tyrosine, tryptophan, and valine contents were higher than *T. globulus*, while phenylalanine is comparable with *T. globulus* (Sudeep and Sridhar 2014). Both mushrooms fulfill the requirement for several EAA, and the ratio of total EAA and total amino

Table 12.2 Mineral composition of uncooked and cooked mushrooms on dry weight basis (mg/100g) (n = 5; mean±SD).

	Auricularia auricula-judae		Termitomyces umkoowan		Recommended pattern[¥]
	Uncooked	Cooked	Uncooked	Cooked	
Sodium	39.10±1.1[a]	31.80±0.3[b**]	41.50±0.6[a]	26.20±0.6[b**]	120–200
Potassium	121.50±1.4[a]	68.30±1.1[b***]	146.20±0.9[a]	75.40±1.2[b***]	500–700
Calcium	29.80±1.3[a]	18.60±0.8[b**]	20.10±0.5[a]	15.60±0.5[b*]	600
Magnesium	29.20±1.2[a]	24.10±1.1[b***]	28.50±0.7[a]	25.10±0.8[b***]	60
Phosphorus	14.86±1.8[a]	9.90±1.0[b*]	96.00±3.0[a]	63.73±3.4[b***]	500
Iron	7.80±0.6[a]	6.50±3.0[b**]	8.10±0.3[a]	6.80±0.3[b**]	10
Copper	0.31±0.02[a]	0.11±0.01[b***]	0.23±0.02[a]	0.15±0.02[b**]	0.6–0.7
Zinc	3.40±0.1[a]	3.10±0.1[a]	2.80±0.2[a]	2.20±0.2[a]	5.0
Na/K ratio	0.32	0.47	0.28	0.35	0.24–0.29
Ca/P ratio	2.01	1.88	0.21	0.25	1.2

Figures across the cooked and fermented mushrooms with different letters are significantly different (t-test: *p<0.01; **p<0.001; ***p<0.001).
[¥] NRC-NAS (1989) pattern.

Table 12.3 Amino acid composition of uncooked and cooked mushrooms in comparison with soybean, wheat, and FAO/WHO pattern (g/100 g protein; n = 5, mean±SD).

	Auricularia auricula-judae		*Termitomyces umkoowan*		Soybean[#]	Wheat[¥]	FAO/WHO[§]
	Uncooked	Cooked	Uncooked	Cooked			
Essential amino acids							
His	1.87±0.03[a]	1.40±0.08[b**]	3.21±0.09[a]	2.95±0.04[b**]	2.50	1.9–2.6	1.9
Ile	5.00±0.50[a]	4.81±0.03[a]	4.83±0.09[a]	4.41±0.03[b*]	4.62	3.4–4.1	2.8
Leu	8.58±0.09[a]	8.07±0.07[b***]	6.43±0.03[a]	6.36±0.03[a]	7.72	6.5–7.2	6.6
Lys	12.4±0.40[a]	7.19±0.01[b**]	8.22±0.04[a]	6.52±0.03[b***]	6.08	1.8–2.4	5.8
Met	0.76±0.02[a]	0.59±0.01[b**]	0.96±0.02[a]	0.67±0.02[b*]	1.22	0.9–1.5	2.5[£]
Cys	0.21±0.01[a]	0.13±0.01[b*]	0.27±0.04[a]	0.06±0.002[b**]	1.70	1.6–2.6	
Phe	4.62±0.04[a]	4.51±0.03[a]	3.97±0.08[a]	3.72±0.02[b*]	4.84	4.5–4.9	6.3[¥]
Tyr	1.82±0.03[a]	1.77±0.05[b*]	2.18±0.02[a]	2.01±0.01[b***]	1.24	1.8–3.2	
Thr	6.97±0.07[a]	4.86±0.14[b***]	5.98±0.04[a]	5.57±0.07[b*]	3.76	2.2–3.0	3.4
Trp	ND	ND	ND	ND	3.39	0.7–1.0	1.1
Val	7.22±0.08[a]	6.35±0.05[b**]	6.06±0.07[a]	5.63±0.03[b**]	4.59	3.7–4.5	3.5
Non-essential amino acids							
Ala	11.2±0.16[a]	10.65±0.03[b*]	12.92±0.04[a]	10.11±0.08[b***]	4.23	2.8–3.0	

(Continued)

Table 12.3 (Continued)

	Auricularia auricula-judae		Termitomyces umkoowan		Soybean[#]	Wheat[¥]	FAO/WHO[§]
	Uncooked	Cooked	Uncooked	Cooked			
Arg	3.99 ± 0.02^a	$2.57\pm0.03^{b***}$	3.56 ± 0.08^a	$2.85\pm0.03^{b**}$	7.13	3.1–3.8	
Asp	6.97 ± 0.04^a	$3.91\pm0.03^{b***}$	7.87 ± 0.04^a	$7.10\pm0.20^{b**}$	11.30	3.7–4.2	
Glu	7.36 ± 0.06^a	7.17 ± 0.19^a	12.59 ± 0.03^a	$8.23\pm0.24^{b***}$	16.90	35.5–36.9	
Gly	10.69 ± 0.55^a	9.99 ± 0.01^a	12.53 ± 0.13^a	$9.21\pm0.04^{b***}$	4.01	3.2–3.5	
Pro	9.25 ± 0.07^a	$7.01\pm0.02^{b***}$	5.63 ± 0.03^a	$4.15\pm0.06^{b***}$	4.86	11.4–11.7	
Ser	7.58 ± 0.04^a	$4.06\pm0.04^{b**}$	8.37 ± 0.02^a	$7.18\pm0.28^{b**}$	5.67	3.7–4.8	
TEAA/TAA ratio[Ψ]	0.35	0.67	0.47	0.49	0.56	0.65–0.69	

Figures across the uncooked and cooked mushrooms with different letters are significantly different (*t*-test: $^*p<0.05$, $^{**}p<0.00$, $^{***}p<0.001$). [#], USDA (1999); [¥], FAO/WHO (1991) pattern; [£], methionine+cystine; [ξ], phenylalanine+tyrosine; [Ψ], total essential amino acid/total amino acid; ND, not detectable.
[§] Bau et al. (1994);

acids (TAA) increased on cooking in both mushrooms, indicating their nutritional potential. The EAA score of both mushrooms is higher than the FAO/WHO (1991) requirement with a few exceptions (*A. auricula*: histidine and methionine+cystine; *T. umkowaan*: leucine, methionine+cystine and phenylalanine+tyrosine) (Table 12.4).

Table 12.4 *In vitro* protein digestibility (IVPD), essential amino acid score (EAAS), protein digestibility corrected amino acid score (PDCAAS), and protein efficiency ratio (PER) of uncooked and cooked mushrooms.

	Auricularia auricula-judae		Termitomyces umkoowan	
	Uncooked	Cooked	Uncooked	Cooked
IVPD (%)	39.10 ± 1.1^a	$31.80 \pm 0.3^{b*}$	41.50 ± 0.6^a	$26.20 \pm 0.6^{b*}$
EAAS				
His	0.98	0.73	1.68	1.55
Ile	1.78	1.71	1.73	1.58
Leu	1.30	1.22	0.97	0.96
Lys	2.13	1.23	1.41	1.12
Met + Cys	0.38	0.29	0.49	0.29
Phe + Tyr	1.02	0.99	0.97	0.90
Thr	2.05	1.42	1.75	1.63
Val	2.06	1.81	1.73	1.60
PDCAAS				
His	0.201	0.122	0.370	0.213
Ile	0.248	0.194	0.256	0.147
Leu	0.077	0.059	0.061	0.038
Lys	0.143	0.067	0.101	0.050
Met + Cys	0.059	0.036	0.081	0.030
Phe + Tyr	0.063	0.050	0.064	0.037
Thr	0.235	0.133	0.214	0.126
Val	0.230	0.164	0.205	0.120
PER				
PER_1	2.79	2.66	1.98	2.02
PER_2	3.23	3.00	2.22	2.20
PER_3	3.88	3.36	2.23	2.16
Mean of $PER_{1,2,3}$	3.30	3.01	2.14	2.13

Figures for the IVPD of uncooked and cooked mushrooms with different letters are significantly different (*t*-test: *, $p < 0.01$).

Besides the EAA and EAA score, the IVPD serves as an index of protein quality in foodstuffs. Cooking in a pressure cooker significantly decreased the IVPD in both mushrooms, suggesting that partial cooking or other methods of cooking may be recommended. The PDCAAS serves as an important parameter in the evaluation of protein quality of foodstuffs. The EAA and PDCAAS in both mushrooms were decreased on cooking in a pressure cooker, so this should be avoided to gain more benefit from these mushrooms. According to Friedman (1996), if the PER in a food is >2, it is of high quality; if between 1.5 and 2, it is of moderate quality; if <1.5, it is of poor quality. The PER_{1-3} in both mushrooms ranged between 1.98 and 3.88, indicating the benefits of these mushrooms in protein nourishment.

Fatty Acids

Uncooked and cooked *A. auricula* has the highest amount of stearic acid followed by oleic, palmitic, and linolelaidic acids (Table 12.5). Uncooked and cooked *T. umkowaan* showed highest quantity of linoleic acid followed by oleic, palmitic, and stearic acids. Oleic acid was the second most common fatty acid in both mushrooms, which was significantly increased in cooked *T. umkowaan*. The dominance of oleic and palmitic acids in both mushrooms is comparable with other wild mushrooms in India (Kavishree et al. 2008; Longvah and Deosthale 1998). Among the saturated fatty acids, palmitic and stearic acids were high in both mushrooms and they are important in mammalian nutrition, especially to prevent liver damage by alcohol (Cha and Sachan 1994; Hayes 2002). Both mushrooms contain many essential fatty acids. In cooked *A. auricula*, the presence of linolenic acid ($\omega3$) indicating its improved quality. Although *T. umkowaan* is devoid of linolenic acid, in uncooked as well as cooked samples high quantities of linoleic acid ($\omega6$) were present and significantly increased on cooking. In addition, it contains docosahexaenoic acid ($\omega3$), which also significantly increased on cooking.

None of the saturated and unsaturated fatty acids in both mushrooms were comparable with soybean and wheat. Total saturated fatty acids were higher than total unsaturated fatty acids in *A. auricula*, while the opposite was true in *T. umkowaan*, indicating its superiority. Foodstuffs with high levels of unsaturated fatty acids rather than high carbohydrate are preferable for combating the risks of coronary heart disease (Bentley 2007). Cooked *T. umkowaan* demonstrated a significant decrease in saturated fatty acids and increase in TUFA/TSFA ratio, indicating its nutritional value. Both mushrooms showed an increased ratio of TPUFA/TMUFA, which is one of the desired traits in foodstuffs. Besides, *T. umkowaan* also showed favorable nutritional quality by decreased

Table 12.5 Fatty acid methyl esters of uncooked and cooked mushrooms in comparison with soybean and wheat (mg/100 g lipid) (n = 5, mean±SD).

	Auricularia auricula-judae		*Termitomyces umkowaan*		Soybean[#]	Wheat[¥]
	Uncooked	Cooked	Uncooked	Cooked		
Saturated fatty acid						
Lauric acid (C12:0)	–	–	–	–	Trace–45	
Myristic acid (C14:0)	–	2.99±0.002	1.03±0.006[a]	0.68±0.002[b]***	Trace–45	
Pentadecanoic acid (C15:0)	1.76±0.07[a]	1.55±0.15[a]	0.29±0.014[a]	–		
Palmitic acid (C16:0)	20.71±0.19[a]	18.74±0.19[b]***	17.82±0.019[a]	14.9±0.09[b]***	110–116	110–320
Heptadecanoic acid (C17:0)	–	–	0.56±0.019[a]	0.35±0.004[b]***		
Stearic acid (C18:0)	29.86±0.04[a]	29.60±0.004[b]**	6.79±0.006[a]	7.18±0.004[b]**	25–41	0–46
Arachidic acid (C20:0)	1.85±0.045[a]	1.76±0.01[b]*	0.59±0.04[a]	0.67±0.01[a]	Trace	
Heneicosanoic acid (C21:0)	2.75±0.064[a]	2.59±0.038[a]	0.31±0.007[a]	0.31±0.008[a]		
Behenic acid (C22:0)	–	–	0.85±0.004[a]	0.89±0.014[b]*		
Lignoceric acid (C24:0)	3.73±0.007[a]	3.69±0.005[b]***	0.82±0.009[a]	0.89±0.018[b]**		
Unsaturated fatty acids						
Palmitoleic acid (C16:1)	–	–	1.17±0.009	–	Trace	
Cis-10-heptadecanoic acid (C17:1)	–	–	0.33±0.04	–		
Oleic acid (C18:1)	24.03±0.018[a]	22.76±0.106[b]***	22.23±0.029[a]	23.37±0.56[b]	211–220	110–290

(Continued)

Table 12.5 (Continued)

	Auricularia auricula-judae		Termitomyces umkowaan		Soybean[#]	Wheat[¥]
	Uncooked	Cooked	Uncooked	Cooked		
Linoleic acid (C18:2)	–	–	46.23±0.036[a]	50.35±0.052[b***]	524–540	440–740
Linolelaidic acid (C18:2)	14.13±0.011[a]	13.16±0.008[b***]	–	–		
Linolenic acid (C18:3)	–	0.95±0.007	–	–	71–75	7–44
Eicosenoic acid (C20:1)	1.18±0.011[a]	1.07±0.033[b**]	–	–		
Docosahexaenoic acid (C22:6)	–	–	0.26±0.021[a]	0.31±0.007[b*]		
Total saturated fatty acids	60.40±0.26[a]	60.64±0.29[b**]	29.06±0.11[a]	25.87±0.12[b***]		
Total unsaturated fatty acids	39.34±0.04[a]	37.94±0.12[b**]	70.22±0.13[a]	74.04±0.61[b**]		
TUFA÷TSFA	0.65	0.63	2.42	2.86		
TPUFA÷TMUFA	0.56	0.59	1.96	2.17		
$C_{14:0}+C_{15:0}+(C_{16:0}÷C_{18:0})$	2.46	5.17	3.94	2.75		
$C_{18:1}÷C_{18:2}$	1.70	1.73	0.48	0.46		

Figures across the uncooked and cooked mushrooms with different letters are significantly different (t-test: [*]p<0.05, [**]p<0.01, [***]p<0.001).
[#] Wahnon et al. (1988), Cho (1989); [¥], Pomeranz (1998); –, not detectable.

ratios of $C_{14:0}+C_{15:0}+(C_{16:0}\div C_{18:0})$, $C_{18:1}\div C_{18:2}$ and $\omega 6 \div \omega 3$ on cooking, which was not evident in *A. auricula.*

Conclusion

In contrast to the assessment of nutritional quality in cultivated mushrooms, its assessment in wild mushrooms is difficult. Studies have revealed the presence of considerable amounts of fiber, low lipid levels, moderate quantity of carbohydrates, and high levels of essential amino acids and unsaturated fatty acids in two wild mushrooms *Auricularia auricula-judae* and *Termitomyces umkowaan*. In addition, *T. umkowaan* possesses levels of total proteins comparable to edible legumes. These features qualify both mushrooms as ideal food sources for human consumption. Future studies should use suitable technology to collect, process, and preserve these mushrooms without loss of nutritional attributes for human health benefits.

Acknowledgments

NCK acknowledges Mangalore University for partial financial support through fellowship under the Promotion of University Research and Scientific Excellence (PURSE), Department of Science Technology, New Delhi. KRS acknowledges the University Grants Commission, New Delhi, and Mangalore University for the award of UGC-BSR Faculty Fellowship.

References

Afiukwa, C.A., Oko, A.O., Afiukwa, J.N., Ugwu, O.P.C., Ali, F.U. and Ossai, E.C. (2013) Proximate and mineral element compositions of five edible wild grown mushroom species in Abakaliki, Southeast Nigeria. *Res. J. Pharm. Biol. Chem. Sci.*, 4: 1056–1064.

Akeson, W.R. and Stahmann, M.A. (1964) A pepsin pancreatin digest index of protein quality. *J. Nutr.*, 83: 257–261.

Aletor, V.A. and Aladetimi, O.O. (1995) Compositional studies on edible tropical species of mushrooms. *Food Chem.*, 54: 265–268.

Alsmeyer, R.H., Cunningham, A.E. and Happich, M.L. (1974) Equations predict PER from amino acid analysis. *Food Technol.*, 28: 34–38.

Anderson, J.W., Johnstone, B.M. and Cook-Newell, M.E. (1995) Meta-analysis of the effects of soy protein intake on serum lipids. *N. Engl. J. Med.*, 333: 276–282.

AOAC (1990) *Official Methods of Analysis*, 15th edn. Washington, DC: Association of Official Analytical Chemists.

Bau, H.M., Vallaume, C.F., Evard, F., Quemener, B., Nicolas, J.P. and Mejean, L. (1994) Effect of solid state fermentation using *Rhizophus oligosporus* sp. T-3 on elimination of antinutritional substances and modification of biochemical constituents of defatted rape seed meal. *J. Sci. Food Agric.*, 65: 315–322.

Bentley G. (2007) The health effects of dietary unsaturated fatty acids. *Nutr. Bull.*, 32: 82–84.

Bhosle, S., Ranadive, K., Bapat, G., Garad, S., Deshpande, G. and Vaidya, J. (2010) Taxonomy and diversity of *Ganoderma* from the western parts of Maharashtra (India). *Mycosphere*, 1: 249–262.

Boa, E.R. (2004) *Wild Edible Fungi: A Global Overview of their Use and Importance to People.* Rome: Food and Agricultural Organization.

Brand, W.A., Tegtmeyer, A.R. and Hilkert, A. (1994) Compound-specific isotope analysis, extending towards 15N/14N and 13C/12C. *Org. Geochem.*, 21: 585–594.

Buczacki, S. (2012) *Collins Fungi Guide.* London: Harper-Collins.

Cha, Y.S. and Sachan, D.S. (1994) Opposite effects of dietary saturated and unsaturated fatty acids on ethanol pharmacokinetics, triglycerides and carnitines. *J. Am. Coll. Nutr.*, 13: 338–343.

Cheung, P.C.K. (1997) Dietary fibre content and composition of some edible fungi determined by two methods of analysis. *J. Sci. Food Agric.*, 73: 255–260.

Cho, B.H.S. (1989) *Soybean Oil: Its Nutritional Value and Physical Role Related to Polyunsaturated Fatty Acid Metabolism.* Creve Couer: American Soybean Association.

Crisan, E.V. and Sands, A. (1978) Nutritional value. In: Chang, S.T. and Hayes, W.A. (eds) *The Biology and Cultivation of Edible Mushrooms.* New York: Academic Press, pp. 137–168.

Dursun, N., Özcan, M.M., Kaşık, G. and Öztürk, C. (2006) Mineral contents of 34 species of edible mushrooms growing wild in Turkey. *J. Sci. Food Agric.*, 86: 1087–1094.

Ekanayake, S., Jansz, E.R. and Nair, B.M. (1999) Proximate composition, mineral and amino acid content of mature *Canavalia gladiata* seeds. *Food Chem.*, 66: 115–119.

FAO/WHO (1991) *Protein Quality Evaluation. Report of a Joint FAO-WHO Expert Consultation.* Food and Nutrition Paper # 51. Rome: Food and Agriculture Organization of the United Nations.

Farook, V.A., Khan, S.S. and Manimohan, P. (2013) A checklist of agarics (gilled mushrooms) of Kerala State, India. *Mycosphere*, 4: 97–131.

Friedman, M. (1996) Nutritional value of proteins from different food sources – a review. *J. Agric. Food Chem.*, 44: 6–29.

Frossard, E., Bucher, M., Mächler, F., Mozafar, A. and Hurrell, R. (2000) Potential for increasing the content and bioavailability of Fe, Zn and Ca in plants for human nutrition. *J. Sci. Food Agric.*, 80: 861–879.

Ghorai, S., Banik, S.P., Verma, D., Chowdhury, S., Mukherjee, S. and Khowala, S. (2009) Fungal biotechnology in food and feed processing. *Food Res. Int.*, 42: 577–587.

Grusak, M.A. and Cakmak, I. (2005) Methods to improve the crop-delivery of minerals to humans and livestock. In: Broadley M.R. and White, P.J. (eds) *Plant Nutritional Genomics*. Oxford: Blackwell, pp. 265–286.

Halpern, G.M. and Miller, A.H. (2002) *Medicinal Mushrooms*. New York: M. Evans.

Hayes, K.C. (2002) Dietary fat and heart health: in search of the ideal fat. *Asia Pac. J. Clin. Nutr.*, 11: 394–400.

Hofmann, D., Jung, K., Bender, J., Gehre, M. and Schüürmann, G. (1997) Using natural isotope variations of nitrogen in plants an early indicator of air pollution stress. *J. Mass Spectr.*, 32: 855–863.

Hofmann, D., Gehre, M. and Jung, K. (2003) Sample preparation techniques for the determination of natural 15N/14N variations in amino acids by gas chromatography-combustion-isotope ratio mass spectrometry (GC-C-IRMS). *Isot. Environ. Health Stud.*, 39: 233–244.

Humphries, E.C. (1956) Mineral composition and ash analysis. In: Peach, K. and Tracey, M.V. (eds) *Modern Methods of Plant Analysis*. Berlin: Springer, pp. 468–502.

Jambunathan, R. and Singh, U. (1980) Studies on Desi and Kabuli chickpea (*Cicer arietinum*) cultivars 1. Chemical composition. In: *Proceedings of the International Workshop on Chickpea Improvement*, ICRISAT, Hyderabad, India, pp. 61–66.

Johnsy, G., Sargunam, D., Dinesh, M.G. and Kaviyarasan, V. (2011) Nutritive value of edible wild mushrooms collected from the Western Ghats of Kanyakumari District. *Bot. Res. Int.*, 4: 69–74.

Jordan, M. (2004) *The Encyclopaedia of Fungi of Britain and Europe*. London: Francis Lincoln.

Kalač, P. (2009) Chemical composition and nutritional value of European species of wild growing mushrooms: a review. *Food Chem.*, 113: 9–16.

Kavishree, S., Hemavathy, J., Lokesh, B.R., Shashirekha, M.N. and Rajarathnam, S. (2008) Fat and fatty acids of Indian edible mushrooms. *Food Chem.*, 106: 597–602.

Khan, M.A., Jacobsen, I. and Eggum, B.O. (1979) Nutritive value of some improved varieties of legumes. *J. Sci. Food Agric.*, 30: 395–400.

Kumar, R., Tapwal, A., Pandey, S., Borah, R.K., Borah, D. and Borgohain, J. (2013) Macro-fungal diversity and nutrient content of some edible mushrooms of Nagaland, *India. Nusantara Biosci.*, 5:1–7.

Kumari, B. (2012) Diversity, sociobiology and conservation of lepiotoid and termitophilous mushrooms of North West India. PhD thesis, Panjabi University, Patiala, India.

Lindequist, U., Niedermeyer, T.H.J. and Jülich, W.D. (2005) The pharmacological potential of mushrooms. *Evid-Based Compl. Alt. Med.*, 2: 285–299.

Longvah, T. and Deosthale, Y.G. (1998) Compositional and nutritional studies on edible wild mushroom from Northeast India. *Food Chem.*, 63: 331–334.

Manjunathan, J., Subbulakshmi, N., Shanmugapriya, R. and Kaviyarasan, V. (2011) Proximate and mineral composition of four edible mushroom species from South India. *Int. J. Biodivers. Conserv.*, 3: 386–388.

Manoharachary, C., Sridhar, K.R., Singh, R. et al. (2005) Fungal biodiversity: distribution, conservation and prospecting of fungi from India. *Curr. Sci.*, 89: 58–71.

Meghalatha, R., Ashok, C., Nataraja, S. and Krishnappa, M. (2014) Studies on chemical composition and proximate analysis of wild mushrooms. *World J. Pharm. Sci.*, 2: 357–363.

Mohanan, C. (2011) *Macrofungi of Kerala*. Handbook # 27. Peechi: Kerala Forest Research Institute.

Müller, H.G. and Tobin, G. (1980) *Nutrition and Food Processing*. London: Croom Helm.

Nareshkumar, S. (2007) Capillary gas chromatography method for fatty acid analysis of coconut oil. *J. Pl. Crops*, 35: 23–27.

Niveditha, V.R. and Sridhar, K.R. (2015) Nutritional qualities of fermented beans of coastal sand dune wild legume *Canavalia maritima*. In: Watson, R.R., Tabor, J.A., Ehiri, J.E. and Preedy, V.R. (eds) *Handbook of Public Health in Natural Disasters: Nutrition, Food, Remediation and Preparation*. Wageningen: Academic Publishers, pp. 443–464.

NRC-NAS (1989) *Recommended Dietary Allowances*. Washington DC: National Academy Press.

Nwokolo, E. (1987) Nutritional evaluation of pigeon pea meal. *Pl. Foods Hum. Nutr.*, 37: 283–290.

Nwokolo, E. and Oji, D.I.M. (1985) Variation in metabolizable energy content of raw or autoclaved white and brown varieties of three tropical grain legumes. *Ann. Food Sci. Technol.*, 13: 141–146

Oboh, G. and Shodehinde, S.A. (2009) Distribution of nutrients, polyphenols and antioxidant activities in the pilei and stipes of some commonly consumed edible mushrooms in Nigeria. *Bull. Chem. Soc. Ethiop.*, 23: 391–398.

Padua-Resurreccion, A.B. and Banzon, J.A. (1979) Fatty acid composition of the oil from progressively maturing bunches of coconut. *Philip. J. Coconut Stud.*, 4: 1–15.

Pegler, D.N. and Vanhaecke, M. (1994) *Termitomyces* of Southeast Asia. *Kew Bull.*, 49: 717–736.

Phillips, R. (2006) *Mushrooms*. London: Pan Macmillan.

Pomeranz, Y. (1998) Chemical composition of kernel structures. In: Pomeranz, Y. (ed.) *Wheat Chemistry and Technology*. St Paul: American of Cereal Chemists, pp. 97–158.

Sadler, M. (2003) Nutritional properties of edible fungi. *Nutr. Bull.*, 28: 305–308.

Senthilarasu, G. (2014) Diversity of agarics (gilled mushrooms) of Maharashtra, *India. Curr. Res. Environ. Appl. Mycol.*, 4: 58–78.

Shills, M.E.G. and Young, V.R. (1988) Modern nutrition in health and disease. In: Neiman, D.C., Buthepodorth, D.E. and Nieman, C.N. (eds) *Nutrition*. Dubuque: Wm C. Brown, pp. 276–282.

Singdevsachan, S.K., Patra, J.K., Tayung, K., Sarangi, K. and Thatoi, K. (2014) Evaluation of nutritional and nutraceutical potentials of three wild edible mushrooms from Similipal Biosphere Reserve, Odisha, *India. J. Verbr. Lebensm.*, 9: 111–120.

Slavin, J., Jacobs, D.R. and Marquart, L. (1997) Whole grain consumption and chronic disease: protective mechanisms. *Nutr. Canc.*, 27: 14–21.

StatSoft (2008) Statistica Version # 8. Tulsa: StatSoft Inc.

Sudheep, N.M. and Sridhar, K.R. (2014) Nutritional composition of two wild mushrooms consumed by tribals of the Western Ghats of India. *Mycology*, 5: 64–72.

Thacher, T.D., Fischer, P.R., Strand, M.A. and Pettifor, J.M. (2006) Nutritional rickets around the world: causes and future directions. *Ann. Trop. Paediatr.*, 26: 1–16.

Tibuhwa, D.D., Kivaisi, A.K. and Magingo F.S.S. (2010) Utility of the macro-micromorphological characteristics used in classifying the species of *Termitomyces. Tanz. J. Sci.*, 36: 31–45.

USDA (1999) *Nutrient Data Base for Standard Reference Release 13, Food Group 20: Cereal Grains and Pasta*. Agriculture Handbook # 8–20. Washington, DC: US Department of Agriculture, Agricultural Research Service.

Venn, B.J. and Mann, J.I. (2004) Cereal grains, legumes and diabetes. *Eur. J. Clin. Nutr.*, 58: 1443–1461.

Wahnon, R., Mokady, S. and Cogan, U. (1988) Fatty acid composition in soybean oil. Proceedings of 19th World Congress of the International Society for Fat Research, Tokyo.

Welch, R.M. and Graham, R.D. (2005) Agriculture: the real nexus for enhancing bioavailable micronutrients in food crops. *J. Tr. Elem. Med. Biol.*, 18: 299–307.

Yusuf, A.A. Mofio, B.M. and Ahmed, A.B. (2007) Proximate and mineral composition of *Tamarindus indica* Linn 1753 seeds. *Sci. World J.*, 2: 1–4.

Index

Microbial Functional Foods and Nutraceuticals, First Edition. Edited by Vijai Kumar Gupta, Helen Treichel, Volha (Olga) Shapaval, Luiz Antonio de Oliveira, and Maria G. Tuohy.
© 2018 John Wiley & Sons Ltd. Published 2018 by John Wiley & Sons Ltd.